Lecture Notes in Computer Science 2078

Edited by G. Goos, J. Hartmanis and J. van Leeuwen

T0254133

Springer
Berlin
Heidelberg
New York
Barcelona
Hong Kong
London
Milan
Paris
Singapore
Tokyo

Rick Reed   Jeanne Reed (Eds.)

SDL 2001:
Meeting UML

10th International SDL Forum
Copenhagen, Denmark, June 27-29, 2001
Proceedings

Springer

Series Editors

Gerhard Goos, Karlsruhe University, Germany
Juris Hartmanis, Cornell University, NY, USA
Jan van Leeuwen, Utrecht University, The Netherlands

Volume Editors

Rick Reed
Jeanne Reed
Telecommunications Software Engineering Ltd.
13 Weston House, 18-22 Church Street
Lutterworth, Leicestershire, LE17 4AW, UK
E-mail: {rickreed/jeanne}@tseng.co.uk

Cataloging-in-Publication Data applied for

Die Deutsche Bibliothek - CIP-Einheitsaufnahme

Meeting UML ; proceedings / SDL 2001, 10th International SDL Forum,
Copenhagen, Denmark, June 27 - 29, 2001. Rick Reed ; Jeanne Reed (ed.). -
Berlin ; Heidelberg ; New York ; Barcelona ; Hong Kong ; London ; Milan ;
Paris ; Singapore ; Tokyo : Springer, 2001
    (Lecture notes in computer science ; Vol. 2078)
    ISBN 3-540-42281-1

CR Subject Classification (1998): C.2, D.2, D.3, F.3, C.3, H.4

ISSN 0302-9743
ISBN 3-540-42281-1 Springer-Verlag Berlin Heidelberg New York

Springer-Verlag Berlin Heidelberg New York
a member of BertelsmannSpringer Science+Business Media GmbH

http://www.springer.de

© Springer-Verlag Berlin Heidelberg 2001
Printed in Germany

Typesetting: Camera-ready by author, data conversion by Steingräber Satztechnik GmbH, Heidelberg
Printed on acid-free paper   SPIN: 10781828   06/3142   5 4 3 2 1 0

Preface

This volume contains the papers presented at the Tenth SDL Forum, Copenhagen.

SDL is the Specification and Description Language first standardized by the world telecommunications body, the International Telecommunications Union (ITU), more than 20 years ago in 1976. While the original language and domain of application has evolved significantly, the foundations of SDL as a graphical, state-transition and process-communication language for real-time systems have remained. Today SDL has also grown to be one notation in the set of unified modelling languages recommended by the ITU (ASN.1, MSC, SDL, ODL, and TTCN) that can be used in methodology taking engineering of systems from requirements capture through to testing and operation.

The SDL Forum is held every two years and has become the most important event in the calendar for anyone involved in SDL and related languages and technology. The SDL Forum Society that runs the Forum is a non-profit organization whose aim it is to promote and develop these languages.

At the first SDL Forum in Florence in September 1982, few imagined that SDL would be an important real-time software engineering language with applications used outside the telecommunications industry in 2000. SDL was graphical before low cost supporting graphics computers existed. SDL was defined in terms of an abstract grammar (or meta-model) and concrete syntax in 1984, long before the currently fashionable UML collection of notations with similar formulation came into existence. The basic SDL part of the Telelogic Tau tool was shown at the second SDL Forum in 1985. Objects were introduced at the third SDL Forum in 1987. The idea of Message Sequence Charts (MSC) as a separate standard was introduced at the fourth Forum 1989. The fifth Forum focused on methods, and the sixth on the use of objects and introduced integration of ASN.1 and SDL. The seventh and eighth covered the use of SDL and the new MSC standard in computer aided software engineering (CASE) and testing. The ninth Forum showed how SDL-2000 and MSC-2000 provide object modelling and harmonization with OMG UML.

The SDL Forum has always been the primary event for the discussion of the evolution of SDL and MSC, and the event that demonstrates how the notations adapt to meet market requirements. The Tenth SDL Forum is no exception, and sees further unification between SDL and other languages, and in particular the meeting of SDL with UML in methods, notations, and tools. The programme covers a broad range of languages used from requirements capture to implementation: as well as SDL and MSC these include UML, ASN.1, TTCN, and new developments associated with these languages.

The content of this volume will be of interest to engineers and researchers who are applying or are considering using the languages mentioned above. While the Forum is essentially a technical conference, the selection of items for the programme covers both theoretical topics and the practical application of the languages in real product development.

As an editor of this volume, I have read through every paper and have been pleased with the interesting and varied selection made by the Programme Committee. However, the papers show that perhaps the acronym *SDL* in the name of the Society should be re-assigned to stand for *System Design Languages*, as the real domain of concern of the Society and the participants of the conference is systems engineering.

April 2001

Rick Reed
`secretary@sdl-forum.org`

Organization

Each SDL is organized by the SDL Forum Society with the help of local organizers. The Organizing Committee consists of the Board of the SDL Forum Society plus the local organizers and others as needed depending on the actual event. For SDL 2001 the local local organizers were Cinderella and Telelogic, who need to be thanked for their effort to ensure that everything was in place for the papers in the book to presented.

Organizing Committee

Chairman, SDL Forum Society: Yair Lahav (Textology)
Treasurer, SDL Forum Society: Uwe Glässer (Paderborn University)
Secretary, SDL Forum Society: Rick Reed (TSE Ltd.)
Local Organizer, Telelogic: Anna Abramson
Local Organizer, Cinderella: Anders Olsen

Programme Committee

Rolv Bræk, Norwegian University of Science and Technology Trondheim, Norway
Rachida Dssouli, University of Montreal, Canada
Fabrice Dubois, France Telecom, France
Anders Ek, Telelogic Technologies, Sweden
Joachim Fischer, Humboldt University Berlin, Germany
Uwe Glässer, Visiting Fellow Microsoft Research, Redmond, USA
Reinhard Gotzhein, University of Kaiserslautern, Germany
Jens Grabowski, Institute for Telematics Lübeck, Germany
Øysten Haugen, Ericsson, Norway
Dieter Hogrefe, Institute of Telematics Lübeck, Germany
Eckhardt Holz, Humboldt University Berlin, Germany
Olle Hydbom, Telelogic Technologies, Sweden
Clive Jervis, Motorola, United Kingdom
Ferhat Khendek, Concordia University, Canada
Yair Lahav, Textology, Israel
Philippe Leblanc, Telelogic Toulouse, France
Sjouke Mauw, University of Eindhoven, The Netherlands
Birger Møller-Pederson, Ericsson, Norway
Ostap Monkewich, Nortel Networks, Canada
Anders Olsen, Cinderella, Denmark
Andreas Prinz, DResearch, Germany
Bob Probert, University of Ottawa, Canada
Rick Reed, TSE Ltd., United Kingdom

Ekkart Rudolph, Technical University of Munich, Germany
Amardeo Sarma, Deutsche Telekom,
Richard Sanders, SINTEF, Norway
Ina Schieferdecker, GMD Fokus, Germany
Daniel Vincent, France Telecom, France
Thomas Weigert, Motorola, USA
Milan Zoric, ETSI, France.

Thanks

A volume such as this could not, of course, exist without the contributions of the authors who are thanked for their work.

Table of Contents

Tools

Combining SDL with Synchronous Data Flow Modelling
for Distributed Control Systems 1
   J.-L. Camus, T. Le Sergent (Telelogic)

Using Message Sequence Charts to Accelerate Maintenance
of Existing Systems 19
   N. Mansurov (KLOCwork Solutions, ISPRAS),
   D. Campara (KLOCwork Solutions)

2001 and Beyond: Language Evolution

From MSC-2000 to UML 2.0 – The Future of Sequence Diagrams 38
   Ø. Haugen (Ericsson Research NorARC)

SDL and Layered Systems: Proposed Extensions to SDL
to Better Support the Design of Layered Systems 52
   R. Arthaud (Telelogic)

Combined SDL and UML

Collaboration-Based Design of SDL Systems........................... 72
   F. Roessler, B. Geppert (Avaya Inc.),
   R. Gotzhein(University of Kaiserslautern)

Using UML for Implementation Design of SDL Systems................ 90
   J. Floch, R. Sanders, U. Johansen (SINTEF),
   R. Bræk (Trondheim University)

Deployment of SDL Systems Using UML............................... 107
   N. Bauer (Telelogic)

Unified Testing

Invited Presentation: ETSI Testing Activities and the Use of TTCN-3 123
   A. Wiles (ETSI)

HyperMSCs with Connectors
for Advanced Visual System Modelling and Testing 129
   J. Grabowski (Universität zu Lübeck)
   P. Graubmann (Siemens)
   E. Rudolph (Technische Universität München)

Graphical Test Specification – The Graphical Format of TTCN-3 148
  P. Baker (Motorola),
  E. Rudolph (Technische Universität München)
  I. Schieferdecker (GMD FOKUS)

Some Implications of MSC, SDL and TTCN Time Extensions
for Computer-Aided Test Generation 168
  D. Hogrefe, B. Koch, H. Neukirchen
  (Institute for Telematics, University of Lübeck)

Timing

Verification of Quantitative Temporal Properties of SDL Specifications ... 182
  I. Ober (IRIT and Telelogic)
  A. Kerbrat (Telelogic)

A General Approach for the Specification
of Real-Time Systems with SDL 203
  R. Münzenberger, F. Slomka, M. Dörfel, R. Hofmann
  (University of Erlangen-Nuremberg)

Timed Extensions for SDL 223
  M. Bozga, S. Graf, L. Mounier (VERIMAG)
  I. Ober , J.-L. Roux (Telelogic)
  D. Vincent (France-Telecom R&D)

Unified ITU-T Languages

ASN.1 Is Reaching Out! 241
  J. Larmouth (Salford University)

Distributed Systems: From Models to Components 250
  F. Dubois (France Telecom R&D),
  M. Born (GMD FOKUS),
  H. Böhme, J. Fischer, E. Holz, O. Kath, B. Neubauer, F. Stoinski
  (Humboldt-Universität zu Berlin)

Deriving Message Sequence Charts
from Use Case Maps Scenario Specifications 268
  A. Miga, F. Bordeleau, M. Woodside (Carleton University),
  D. Amyot (Mitel Networks),
  D. Cameron (Nortel Networks)

SDL Application

An SDL Implementation Framework
for Third Generation Mobile Communications System 288
  J. Sipilä, V. Luukkala (Nokia)

OSPF Efficient LSA Refreshment Function in SDL 300
  O. Monkewich (Nortel Networks),
  I. Sales, R. Probert (University of Ottawa)

Using SDL in a Stateless Environment 316
  V. Courzakis (Siemens),
  M. von Löwis, R. Schröder (Humboldt-Universität)

MSC

An MSC Based Representation of DiCons 328
  J.C.M. Baeten, H.M.A. van Beek, S. Mauw (Eindhoven University)

Some Pathological Message Sequence Charts, and How to Detect Them ... 348
  L. Hélouët (France Télécom R&D)

An Execution Semantics for MSC-2000 365
  B. Jonsson (Uppsala University),
  G. Padilla (Telelogic)

Test and Verification

Comparing TorX, Autolink, TGV and UIO Test Algorithms 379
  N. Goga (Eindhoven University)

Verifying Large SDL-Specifications Using Model Checking 403
  N. Sidorova (Eindhoven University),
  M. Steffen (Christian-Albrechts-Universität)

Applying SDL Specifications and Tools to the Verification of Procedures .. 421
  W. Zhang (Institute for Energy Technology, Norway)

Author Index 439

Combining SDL
with Synchronous Data Flow Modelling
for Distributed Control Systems*

Jean-Louis Camus and Thierry Le Sergent

Telelogic Technologies Toulouse
Jean-Louis.Camus@telelogic.com, Thierry.LeSergent@telelogic.com

Abstract. Engineers are faced nowadays with the challenge of designing strongly distributed control systems, with complex interactions. There is little theory and tool support to address this recent challenge. Control engineering and telecom engineering have dedicated but unrelated techniques, each for their specific domain. In this paper, we explore an approach where we combine two complementary formal methods, with good tool support and industrial acceptance: SCADE/Lustre from the Control Engineering domain, and SDL, from the Telecom domain.

1 Introduction

Traditional real time computerised systems were designed and implemented as fairly independent systems, having a strong interaction with the process they controlled, but very limited interaction with other computers. Also, they used to fall into specific categories with dedicated approaches, tools and techniques, such as:

- Programmable Logic Controllers, with focus on combinatorial or sequence logic;
- Regulation Controllers with focus on regulation laws;
- Telecom Systems (switches, phones) with focus on protocols.

For each of these categories, engineers have developed specific approaches, supported by dedicated notations and tools, such as SDL (Telecom), relay logic or SCADE (control engineering).

Things are getting more complex as the functions of computerised systems are becoming more and more sophisticated, and as these systems need to communicate via more complex protocols. As an example, new aircraft have to combine:

- Flight Control, involving both regulation and complex logic;
- Communication between embedded computers;
- Automatic communication with the environment (ground, satellites).

---

* Work supported by the European project CRISYS, EP 25.514 – Critical Instrumentation and control SYStems.

R. Reed and J. Reed (Eds.): SDL 2001, LNCS 2078, pp. 1–18, 2001.

Cars constitute another example, where synchronous, clock driven systems control the engine, gears, brakes, airbags, while at the other end of the scale asynchronous systems manage routes using GPS, and all computing units communicate through buses.

In this paper we show why it is not possible to design such systems by using just one of the above mentioned techniques. It is necessary to design systems using a global, consistent approach. Rather than reinventing the wheel, we prefer to rely on existing proven techniques, and use them in a sound way.

We introduce a formal technique, SCADE, which is suitable for the computer based, control engineering domain and show some advances of that technique in distributed control, and its limitations. We reiterate why SDL alone is generally not suited to the design of control system.

The contribution of this paper is to present a global approach, where we cleanly combine SDL and SCADE, and illustrate that approach with a simple example. Note: This approach is part of a larger project, where all notations are part of a UML framework (as is SDL-2000 already), but these UML aspects are beyond the scope of the current paper.

2   The SCADE/Lustre Technology

*Telelogic Tau SCADE is the name for both the notation and the toolset maintained by Telelogic (*http://www.telelogic.com/SCADE*). It is based on the formal language Lustre. We will call it SCADE in the remainder of this paper.*

2.1   Application Domain

Currently, SCADE is mostly used for the development of safety critical control systems: Airbus and Eurocopter embedded computers [1], Nuclear Power Plant control, Railway signalling systems [2], Car engine control.

2.2   The Language

The formal language Lustre is the basis of SCADE and was created to provide a simple, formal notation for the description of digital control systems. It is a bridge between control engineering and computer science. Lustre is a synchronous, data flow programming language [3]. It belongs to the family of synchronous programming languages, like Statecharts [4], Esterel [5] and SIGNAL [6].

A synchronous system computes its output from its input and state in "zero time": this does not mean that the real program takes zero real time, but rather that computation is completed before the output is communicated outside, and that there is no communication during the computation. In practice, this is generally realised by using sample and hold interfaces. Note that this property is relative to a clock and that several (sub-)systems may run at different speeds.

Nodes. The basic structuring unit is the **Node**. A node is modular. It has an interface with formal parameters. A node can be composed with other nodes by calls and by passing data flows (which are actual parameters) without recursion. Node compositions preserve their properties; in particular their time related properties. A model is a set of nodes (transitively) called from a root node. Like SDL, SCADE has both a textual and a graphical notation. A node can be expressed as:

- A set of **equations** This matches well the Z transform of signal processing and control engineering;
- A **block diagram**;
- A graphical **finite state machine**] This is convenient for state/event dominated nodes.

Time Constructs. Time is logical, based on clocks. Every data flow and node is attached to a **clock**, a Boolean variable; the data is defined when the clock is true. All clocks are derived from the basic clock of the model.

The language supports time related operators: sample (when), hold (current), delay (pre), initialization (init), activation condition (condact: a node under activation condition C is computed when C is true). These time operators are used in the equations of nodes.

Concurrency, Simultaneity, Sequencing. Data flow dependencies are automatically analysed and taken into account by the SCADE tool set checking causality loops. The code generator ensures a correct sequential implementation.

Parts of a SCADE model are functionally concurrent as long as there is no instantaneous (that is, in the same time interval) data flow dependency between them: the case for parallel automata.

It is possible to express explicitly simultaneity of events and their processing.

2.3   An SDL View of a SCADE Model

From an SDL point of view the execution of a SCADE model can be schematically seen as follows. There is one SDL process per SCADE model, with 2 transitions. The initialization transition initializes state variables according to the "init" statements of SCADE. The remainder is a loop, triggered by the basic clock:

1. Acquisition of input and hold;
2. Compute one step, based on the SCADE model;
3. Release output (deliver result).

This transition has special characteristics: there is no interleaving with any other transition and in particular there is no state inside the procedures.

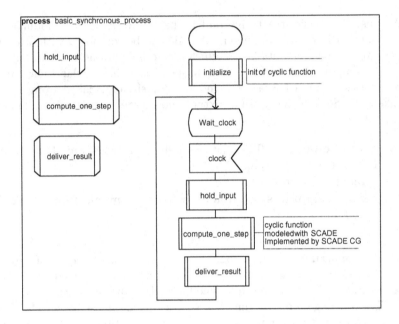

Fig. 1. Schematic SDL view for the execution of a SCADE program

2.4   A Simple Example

The synchronous system is a simple heating controller based on a digital thermostat.

The controller inputs the temperature from a sensor, and outputs a command to activate the boiler. The controller can also warn that there is a risk of freezing, when the temperature is below 4 degrees Celsius.

The *Simple_filter* node (not detailed here) improves the temperature signal by filtering noise and keeping it within bounds. As an example of dynamic function, the *regulator* node (not detailed here) contains an integrator which can simply be written (or drawn) as:

```
X=input+pre(X)
```

which means add the current input value to the value that X had at the previous sample.

2.5   Benefits of SCADE

SCADE is a **natural** notation for control engineers, since its textual form fits directly the Z transform notation for sampled control systems, and its graphical form is based on block diagrams.

It is **formal**, thus allowing formal verification of the controller under design.

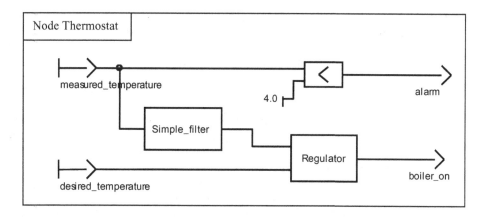

Fig. 2. SCADE Block Diagram for the heating Controller

The auto code generator saves large amounts of time, effort and money, in particular when changing the system. The generated code is fully **safe** and **deterministic** and has a **predictable** execution time.

2.6   Tool Support

The Lustre language is currently supported by the SCADE™ toolset of Telelogic (`http://www.telelogic.com/SCADE`). The current commercial toolset comprises:

- A graphical editor;
- A simulator;
- A code generator that is qualified for level A of DO 178B (Civil avionics standard) - the most demanding software quality standard.

Links to formal verification tools, such as LESAR (Verimag, as described at `http://www-verimag.imag.fr/SYNCHRONE`) and NP-Tools (Prover Technology, see `http://www.prover.com`) are also available. Links with the BOOST tool from Siemens, and the HYBRID tool from OFFIS are also under construction under the European project SafeAir.

3   SDL and Control Engineering

3.1   Origin

SDL was created by and for telecom engineers to describe event-driven, asynchronous communicating systems such as switches, mobile phones, and networks. This formal notation [7] has been widely accepted by the telecom community, and is supported by tools providing editing, simulation, formal verification and auto coding, such as Telelogic Tau SDL suite.

Is SDL Suited to Process Control? As SDL is based on extended finite state machines, one may ask whether this technique is suited to automatic process control as is. The answer is "yes" in a few cases, but "no" in many others. In most cases, SDL is **not** suited to control engineering, because:

- SDL does not match the control engineering notation and theory for control loops: these are based on sampling and transfer functions. The theory with its related tools (such as Matlab/Simulink, connected to SCADE) is necessary for essential activities such as frequency/stability analysis and parameter tuning.
- Asynchrony and queues introduce complex, non-deterministic behaviour that is needed to reflect the behaviour of telecom switches, but is not expected from a controller. Asynchronous systems are inherently difficult to analyse and non-predictive (response time cannot be guaranteed). Safety critical systems, such as the flight control computer of an Airbus, or the control of nuclear power plants are required to be deterministic and predictive.

Recent Uses of SDL at the Border of Control Engineering. A good reason for using SDL is that modern control computers need to communicate with other computers, and that the communication, which is getting more complex needs to be mastered. For a system that is mainly event driven, the SDL tools would even allow a very detailed state exploration of the global system to be performed.

Recently, SDL has been used at the edge of control engineering, not for control engineering, but for the communication part of control systems:

- For the communication protocols of the FANS (Future Air Navigation System, (see `http://www.sae.org/products/books/B-787.htm`) in the civil avionics domain; ARINC used it for the specification, and AEROSPATIALE for the design and implementation of these protocols [8];
- For the specification of the communication layer of the OSEK operating system (see `http://www-iiit.etec.uni-karlsruhe.de/~osek`) for the automotive domain.

4   Approaches for the Design of Distributed Communicating Control Systems

4.1   Reasons and Problems of Distribution

Distribution may be introduced in control systems for one or several of the following reasons:

Physical constraints: the system elements of necessity need to be distributed in different locations; for instance - aircraft or the ground and satellites, or the heating at home and the user in his office as in the example in 5.2;

Redundancy: control computers may be replicated to increase robustness with respect to failures;

Load balancing: it may be necessary to distribute the computing power over several computers to achieve performance requirements;

While bringing some benefits, distribution may also introduce serious problems, such as:

- Communication overheads;
- Unpredictable and/or erroneous behaviour;
- Failures introduced by protocol design errors or by communication medium failures, such as Byzantine faults.

A part of these errors and failures are well known to telecom engineers; they have developed notations (e.g. SDL), techniques (protocol engineering) and tools (e.g. Telelogic Tau) to carefully design (layers, error detection/correction) and validate their systems (conformance verification, deadlock detection...). Undoubtedly these techniques have to be taught to control engineers.

These techniques are necessary but not sufficient to design safe control systems, because control systems are more sensitive than telecom systems to timelines. While most telecom systems require some average response time and tolerate deviations from time to time, a flight control computer must strictly ensure response time, otherwise the control laws are changed and the plane may crash. This is not just a matter of making the computer "faster", or with more redundancy, but this requires a well-suited design method.

4.2   Distribution/Communication Architectures

In practice only a limited set of architectures is used when implementing distributed control systems, at least regarding the type of service they offer. They generally have properties that ensure bounded delay, non-blocking read/write, and can be easily used for redundant implementations.

The most typical ones are:

- "Shared Memory" service (in the logical sense) as offered by field buses: the buses of several standards such as for aircraft ARINC 429, and for industrial or automotive applications CAN/WAN. They just "pack" together several serial links and buffers, and provide a kind of "shared memory" service, on top of which synchronisation services can be implemented on need.
- Time Triggered Architecture, based on the Time Triggered Protocol [9]: this is a more recent architecture, which may develop in the future.

4.3   The Quasi-synchronous Architecture and Approach

The so-called "Quasi-Synchronous" architecture/approach concerns periodic systems, having nearly the same base clock, with limited frequency and phase drifts, and communicating through a kind of shared memory (in a functional sense). For instance "field buses" such as CAN buses in cars, or the ARINC 429 in aircraft support this type of communication, where there is one writer and N readers for a message, without other interaction.

The quasi-synchronous approach uses the characteristics of that architecture for safe implementation of control systems. The quasi-synchronous theory has been formalized and experimented in the European project CRISYS [10]. This approach should be used wherever applicable, since it provides a simple, rigorous framework for the analysis and design of such systems, and formally guarantees several key properties for safety critical systems, such as conformance to a reference synchronous model, robustness, bounded delay, ... when using protocols identified in this theory.

4.4   GALS Architectures

Another possible architecture is a network of synchronous systems, communicating asynchronously via networks involving possibly complex protocols. This is sometimes referred to as a "GALS" (Globally Asynchronous, Locally Synchronous) situation. An in-depth discussion of the relationship between synchrony and asynchrony is given in [11].

To be able to analyse such GALS systems, we first need clear semantics of this class of systems. Second, we need a way of taking into account realistic hypotheses on the environment (such as frequency bounds); indeed, without such hypotheses, hardly any model would be acceptable.

The remainder of this paper addresses a way of describing and implementing such GALS systems by combining existing notations (SDL and SCADE) and related tools.

5   Principle of SCADE/SDL Combination

5.1   Basis

This combination addresses the GALS situation in the case where only the synchronous parts are safety critical.

The major motivations for selecting and combining SCADE and SDL are the following:

- They have a formal basis;
- They are executable;
- They address complementary needs;
- They have a significant acceptance and use within the industry;
- Commercial tools support them.

The combination of two languages such as SDL and SCADE requires a kind of "glue": a specification of the combination. We chose to describe the glue in the SCADE and in the SDL frameworks, instead of inventing a new formalism.

Each synchronous model is seen as one SDL process that can be used at any place a normal SDL process is used. These processes are defined in SCADE. Each is running at its own speed, independently of the speed of the other synchronous process or the activity in the SDL model. This is defined more formally in the semantics section.

The communication is possible in both directions, with natural semantics for SDL and SCADE. Communicating by message exchanges is natural in the SDL framework. Message communication is less obvious in the SCADE framework because of the constraint imposed by the synchronous hypothesis: computation in "zero time" on a basic clock. It implies that a synchronous process cannot be blocked waiting for a message, cannot be interrupted for additional work such as treating the reception of an unknown number of messages. The solution we propose is to use SCADE imported operators to model the "glue". The purpose of the SCADE imported operators is to allow definition of operators that are external to SCADE and that can be used as normal SCADE nodes. The imported operators defined must respect the synchronous hypothesis.

Three complementary communication means are proposed: two from SCADE to SDL, and one from SDL to SCADE.

1. *SCADE to SDL triggered by SCADE:* A SCADE node should be able to generate messages for the SDL system. It is mandatory because it is the only way to wake up an SDL system that is idle. The reception by SDL of messages sent from a SCADE node can be modelled as the usual SDL reception of messages.

   The solution we propose is to define a SCADE imported operator, called the *sendToSDL* operator, whose inputs are the SCADE variables (typically outputs of the SCADE root node) to be sent. The *sendToSDL* operator has no output. It is embedded within a SCADE activation condition. So, the user can specify precisely when a SCADE node must send an SDL message.

2. *SDL to SCADE initiated by SDL:* Like SCADE to SDL communication by message, SDL to SCADE communication is based on SDL messages. They are sent in the usual way by SDL processes, and directed via the SDL communication paths to the SDL process that is a SCADE node.

   The parameters of the message carry the values. A SCADE imported operator, called *receiveFromSDL* operator, produces on each of its outputs the value of the corresponding parameter of the message received. Several policies are possible to treat the possible case of several messages received between two computations. They are discussed in the semantics section.

   It is mandatory to have initial values, so the *receiveFromSDL* operator must have a single input per output to carry an initial value, exactly like the SCADE "initialization" operator.

   The *receiveFromSDL* operator can also have a Boolean output, set to true during one cycle, when a new message has been received. This possibility is discussed in the semantics section.

3. *SCADE to SDL initiated by SDL:* In some cases, it is useful to allow the SDL world to initiate the transfer of information from the synchronous process. The model given in the next section shows an example of its usage. The requesting SDL process will wait for the reception of a message carrying the requested value from SCADE. This is exactly the SDL Remote Procedure Call mechanism, but in the SCADE framework it is sensible to not handle it with the two previous operators (*receiveFromSDL* and *sendToSDL*),

but instead to introduce a new operator for that purpose, the *exportToSDL* operator. The motivations of this choice are given in the semantics section.

Note that communication from *SDL to SCADE initiated by SCADE* is not relevant. Information going from SDL to SCADE has only one meaning in the SCADE framework, an input port has a value that may be new. For any input, the value is taken into account for the next computation, performed on the basic clock.

5.2   A Simple Example

Here is a very simple example to show how the different communication modes between an asynchronous and a synchronous system can be used. The system models a "domotic" (domestic automation) application based on the model of the digital thermostat given in the SCADE presentation section of the paper.

The asynchronous system can change the desired temperature set on the thermostat and read the actual temperature captured by the heating controller. The heating controller can also alert the asynchronous system that the temperature is below 4 degrees Celsius.

The SCADE Model. The net view shown in Fig. 3 of the heating controller (seen as the heating process in the SDL model presented afterwards) is composed of the thermostat and the glue for the combination with the asynchronous world. The three operators introduced above are used:

- *thermostat_val* is an instance of the *receiveFromSDL* operator: it models the reception of an SDL signal that has the same name. The value of its output is the value of the parameter of the SDL message received. It is initialised (before the first reception of the *thermostat_val* SDL signal) to 20.0 degrees.
- *alarm_temp* is an instance of the *sendToSDL* operator: it models the send of an SDL message with the same name. The value of parameters of the message is the value of the inputs of the operator (not useful here - shown only for demonstration). The sending is executed if and only if the condition of the activation condition that encapsulates it is true. Here the condition is realised when the alarm computed by the thermostat goes from false to true.
- *actual_temp* is an instance of the *exportToSDL* operator: it allows the SDL system to read the value of its input (*temperature*) at any time, via the SDL RPC. The name of the Remote Procedure is the name of that SCADE operator.

The SDL Model. The system modelling the application consists of two blocks: *home* manages the synchronous temperature controller, and *office* represents the remote user.

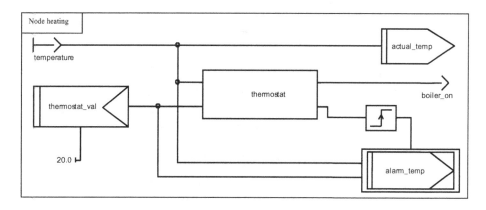

Fig. 3. Communicating Heating Controller

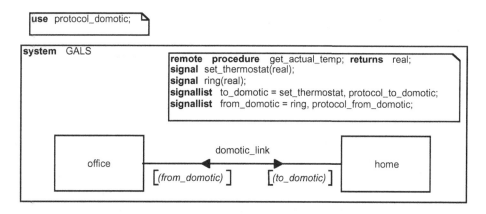

Fig. 4. SDL System "GALS"

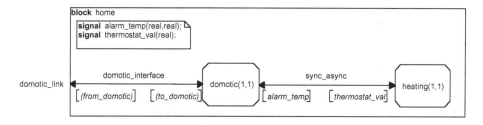

Fig. 5. SDL Block "Home"

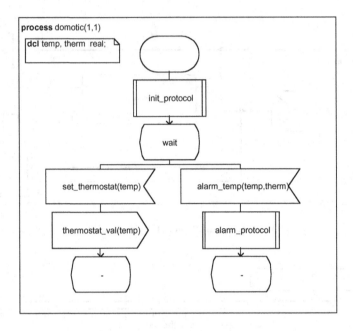

Fig. 6. SDL Process "Domotic"

Block *home* is composed of two processes, the *heating* process, whose behaviour is the SCADE node described above, and the *domotic* process in charge of the communication between the *heating* and the *office*.

The behaviour of the *domotic* process is very simple: after initialization of a protocol to communicate with its environment (the block *office* here), it waits for commands from its single state.

The main role of the *domotic* process is to set the desired value of the heating thermostat. To that purpose *domotic* sends the *thermostat_val* message to the *heating* process each time a *set_thermostat* message is received from the *office*.

Reception of the *alarm_temp* message from the *heating* process allows the *domotic* process to call the user (via the *alarm_protocol* procedure) if the heating raises an alarm

The access to the actual temperature from the office is realised by calling the *get_actual_temp* procedure exported directly by the *heating* process.

Block *office* contains processes to handle the protocol to communicate with *domotic* and a user interface for the three functions: setting the value of the remote thermostat, getting the present value of the temperature at home, and signalling if there is a risk of frost at home.

6   Semantics of the SCADE/SDL Combination

The semantics of the GALS model is described here as a full SDL model. As the asynchronous part of the GALS model is already defined as an SDL model, only the behaviour of the synchronous processes (the heating process in the given example) needs to be expressed in terms of an SDL model.

It may appear as a paradox that SDL is proposed here, given the fact that we have just shown that it is not well-suited to safety critical control systems. But the communication scheme between the SDL part and SCADE which we propose preserves the good properties of the SCADE part: the SCADE loop has its own clock, and the SDL part neither blocks nor delays SCADE, nor interferes with its dataflow.

The model gives operational semantics of the GALS model. It is not intended to represent a way for a user to model a GALS with SDL only. The "SDL and control engineering" section of this paper gives every reason not to do it in such a way. The example presented in the previous section shows a nice way of modelling a GALS; this section has no more intentions than to define the meaning of such design.

6.1   Synchronous Model

Each synchronous process is semantically an SDL process composed of two SDL services as in Fig. 7. Instead of relying on the previous example to describe the semantics, we use a "general process", called synchronous. To simplify the description of the SDL model we suppose that the synchronous process designed in SCADE uses one instance, named rf, of the *receiveFromSDL* operator, one instance, named st, of the *sendToSDL* operator, and one instance, named et, of the *exportToSDL* operator, each with two parameters. The model can easily be extended for several instances of each kind of operators, and any number of parameters.

We also give a constraint on the usage of the "communicating" operators: we restrict their usage to be unique and only in the root node of the SCADE model. The reasons are:

methodologic: they are in fact (part of) the interface of the SCADE model;
practical: without the constraint, each operator may be called in several times and/or in several nodes. To define each individual behavior, it is necessary to duplicate the mechanisms we introduce below, shared variables and RPCs, leading to naming problems, etc.

The computation service implements the synchrononous computation while the communication service models the way data items are exchanged between the asynchronous and the synchronous parts of the GALS model. The two services share SDL variables. The features of the SDL services are very helpful to the definition of the synchronous process: first, a transition of one service must be completed before another transition of the other service can be executed; second, the SDL messages sent to the synchronous process are directed without additional information to the correct service.

The role of the *clock* signal appearing in that model is described in 6.3 - the computation service, after detailing the communication service.

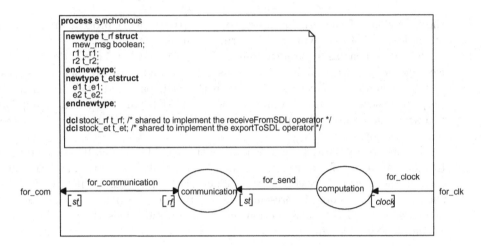

Fig. 7. SDL Process "Synchronous"

6.2   Communication Service

The communication service implements the three communication modes presented above between the asynchronous world and the synchronous computation. Several semantic choices are possible. One semantics choice is given here, and other possibilities are discussed afterwards:

ExportToSDL Operator: The *et* operator is realized by an SDL remote procedure exported by the communication service. The behavior of this procedure is simply to return the value of the *stock_et* variable. The fields of this variable are set by the *et* operator invoked from the *compute_one_step_procedure* (see 6.3 Computation Service).

ReceivedFromSDL operator: The *rf* operator is also realized thanks to a *variablestock_rf* shared between the two services. The chosen discipline is to overwrite this variable each time a message is received, so if several messages whose identifier is *rf* are received between two computations, only the last one will actually be taken into account.

   Another sampling discipline can easily be defined: by adding a condition *providednot*(*stock_rf*!*new_msg*) to the transition input *rf*($i1, i2$), we request that all messages received by the synchronous computation will serve at least once as output of the corresponding *rf* operator.

   The definition given here provides additional information to notify that a new message has been received since the previous computation. This information is stored in the *new_msg* field of the *stock_rf* variable, and is reset after each computation. This information is not used in the domotic example. To use it in a SCADE model it is sufficient to add a boolean ouput to the *rf* operator. But it is important to know that with the chosen semantics, a value true means that at least one message has been received. If it is important to know exactly the

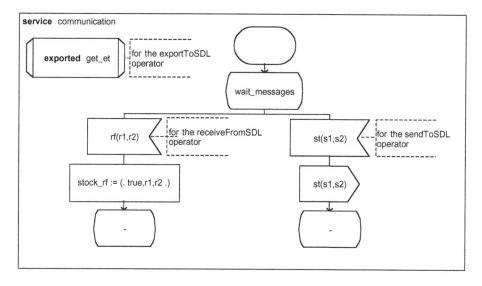

Fig. 8. SDL Service "Communication"

number of messages received, an integer can be used instead of a boolean value. If all the values carried by the different messages are important, then a better solution is to change the semantic to the one using the "provided" additional condition.

SendToSDL Operator: The *st* operator is not realized in the same way. If the communication between the two services were realized by a *stock_st* shared variable, a synchronization mechanism between the two services would be mandatory. Otherwise, the values stored during the execution of one cycle of the computation service may be overwritten during the next cycle before the communication service sends one *st* message carrying the value of a *stock_st* variable. Forcing the synchronous service to wait for something other than the clock would break the synchronous hypothesis.

The solution proposed here is to use an unbounded queue of data between the two services. This is naturally modelled by an SDL signalroute. The execution of the st operator sends an SDL message. The role of the communication service is simply to forward these messages to the asynchronous part of the GALS model. That way, the release of the output is made after the computation is entirely performed, as required by the synchronous hypothesis.

More on the ExportToSDL Operator: As this communication mode is modelled by an RPC, it seems that it is not a basic operator. Could it be modelled by a pair of *receiveFromSDL* and *sendToSDL* SCADE operators, as SDL RPCs are defined as SDL messages exchanges? For this to be possible, it is necessary for the *receiveFromSDL* operator to give the information that one message has actually been received during a cycle. This can be done thanks to the additional boolean ouput of the *receiveFromSDL* operator (value stored in the *new_msg* field of the variable). Now, it is mandatory that all messages from the asynchronous part of the GALS model are taken into account by the synchronous process. If one is "lost", the SDL process calling the RPC will block for ever. The sampling discipline chosen does not follow this property. In order for the *exportToSDL* operator not to be dependant on the sampling discipline chosen for the *receiveFromSDL* operator, it is sensible to define it as a basic operator.

6.3   Computation Service

The "basic clock" of the SCADE model defines the speed at which the synchronous execution is performed. The basic clocks of each synchronous process (there can be several of them in a GALS model) are independent. The semantics of that feature could be modelled in SDL using the spontaneous transition "input none". It would be correct semantics.

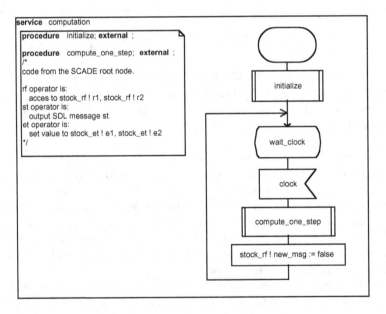

Fig. 9. SDL Process "Computation"

We propose here a better way to handle that behaviour: we have chosen to represent the basic clock explicitly by an SDL message *clock* (a different one per synchronous process). The environment provides this *clock* message. Without further information or constraint, this represents the inherent behavioural complexity of the whole GALS model, which is explosive. But these explicit clocks allow us to take into account hypotheses on the environment such as bounds on frequency of events. The restriction of the set of possible behaviours opens the door for validation of GALS systems under given hypotheses.

The *get_inputs* action introduced in the SCADE presentation section is realised thanks to the SDL services: the variables representing the communication
cannot be updated during the execution of the *compute_one_step* procedure.
Also, the *output_results* action is realised by the communication service without violation of the basic synchronous model: no message can be output from the
synchronous process before the transition of the computation service is entirely
performed.

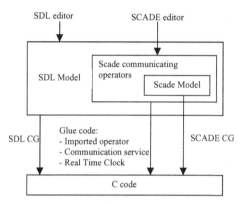

Fig. 10. Design and implementation process

7   Design and Implementation Process

The example presented above has been designed and implemented thanks to the
ObjectGeode and SCADE products. Fig. 10 pictures the process that has been
followed.

The model design is done with the ObjectGeode and SCADE graphical editors, apart from the Real Time Clock which is traditionally given as external
code (in our very simple example, it consists of a one second timer reset between
each cycle).

The code of the application is generated with the ObjectGeode and SCADE
code generator. In the current status of the products, the code of the communicating operators (SCADE imported operator) and the communication service (a
POSIX thread) are written by hand, but could easily be automatically generated
in the future.

8   Conclusion

In this paper, we have shown how we can address the specification and design of a distributed control system involving control units communicating asynchronously, by combining two of the most effective techniques currently available
for control systems and telecom systems.

We believe that this is a good starting point, as we base our approach on formalized languages, which have tool support and industrial acceptance.

A prototype of the above example has been implemented, using Object-Geode and SCADE code generators. Our interest now goes to in-depth formalization and formal verification of such systems. The ObjectGeode SDL simulator/validation tool, together with the semantics we presented, described in SDL and introducing an explicit clock, give a technical possibility to verify the GALS system defined here. Methodology to constrain the clock signal with the other SDL signals in order to restrict the set of possible behaviour is still lacking. Contributions from researchers would be welcome.

We also intend to explore the application of this hybrid technique to the telecom domain, for instance when mixing signal processing with event driven parts.

References

1. D. Brière, D. Ribot, D Pilaud, JL Camus, "Methods and specification tools for Airbus on-board systems" in *Avionics Conference and exhibition*, London, December 1994, ERA Technology.
2. G Le Goff, "Using synchronous languages for interlocking", *First International Conference on Computer Application in Transportation Systems*, 1996.
3. N Halbwachs, P Caspi, P Raymond and D Pilaud "The synchronous dataflow programming language Lustre". Proceedings of the IEEE, vol 79, n°9, pp 1305-1320, September 1991.
4. Modeling Reactive Systems With Statecharts: The Statemate Approach by David Harel and Michael Politi, McGraw Hill.
5. G Berry and G Gonthier, "The Esterel synchronous programming language, design, semantics, implementation", Science of Computer Programming, Vol 19, No 2, pp87-152, 1992.
6. P Le Guernic et al, Programming real-time applications with SIGNAL, another look at real time programming, special section of Proc. Of the IEEE, 79(9):1321-1336, Sept 1991.
7. ITU-T Recommendation Z.100 - Specification and Description Language (SDL).
8. Scheduling in SDL simulation. Application to Future Air Navigation Systems F. Boutet, G. Rieux, Y. Lejeune, E. Choveau, *2nd Workshop on SDL and MSC*, Grenoble, France June, 26-28 2000.
9. H. Kopetz and G. Grünsteidl. TTP A Protocol for Fault-Tolerant Real-Time Systems. COMPUTER, pages 14-23, January 1994.
10. Paul Caspi et Al. Formal design of distributed control systems with Lustre. In *Proc Safecomp'Safecomp99*, September 1999.
11. A Benveniste, B Caillaud, P Le Guernic, "From Synchrony to Asynchrony", CONCUR'99, Concurrency Theory,10th Int Conf, Lecture notes in Computer Science, Vol 1664, pp162-177, Springer Aug 1999.

Using Message Sequence Charts
to Accelerate Maintenance of Existing Systems

Nikolai Mansurov[1,2] and Djenana Campara[1]

[1] KLOCwork Solutions
1 Antares Drive, Nepean, K2E 8C4, Ontario, Canada
[2] Institute for System Programming
25 B. Kommunisticheskaya, Moscow 109004, Russia
nick@ispras.ru, djenana@nortelnetworks.com

Abstract. In this paper we describe our experiences in building tools for
accelerating maintenance of existing large telecommunications software.
We discuss how various maintenance activities can be accelerated by providing developers with the knowledge of the core scenarios of the system,
which approximate the intended use cases. We present a static approach
to extracting scenarios as source trajectories, by navigating through the
source code and capturing the source statements as events. We describe
our **PathFinder** tool for static capturing of scenarios. The possibility of
static capturing of the core scenarios and their representation as MSCs
have benefits in retaining expertise about existing software, in training
new personnel, in focusing understanding of legacy software, performing
architecture reviews, and in architecture analysis of existing systems. We
believe that this approach can contribute to closing the gap between tool
support for forward engineering in the so-called "green-field" projects,
and maintenance of existing software.

1 Introduction and Background

Recently, time-to-market has become the dominating factor for industrial success, placing enormous pressure on manufacturers and developers to spend less
effort on quality and performance, or to find much more efficient means of producing high-quality software for their products and services. Rapidly changing
economic conditions, new business paradigms and new technologies have produced a dynamic, global environment and a unique, unprecedented application
development challenge. The use of the Internet by both industries and consumers
is another dramatic force shaping the future of the global economy.

Traditional development methodologies and tools do not produce applications fast enough to keep up with customer demands and the competitive pres-
sures from more agile companies [9]. Therefore, there is an increasing demand
for new methodologies and tools that can speed up the time it takes to design,
construct, deploy and maintain applications. New methodologies should support
accurate, rapid development and increased developer productivity, enable easy
modification and customization of code, content, applications, and business rules
[9].

R. Reed and J. Reed (Eds.): SDL 2001, LNCS 2078, pp. 19–37, 2001.

Significant research and development efforts are being invested into acceleration of development and deployment of new systems. Companies like Rational and Telelogic demonstrate, that an up-front model can reap tremendous potential benefits in quality, productivity, reuse and maintenance. Object-oriented analysis, modeling, design and construction tools continue to grow more popular among development organizations [9].

Software components and the component-based development (CBD) paradigm are rapidly reaching a critical mass in terms of adoption by industry. Unlike the modeling-based paradigms, CBD approach relies on the ability to use leverage from a visual paradigm and automation to build and assemble components. Modeling tools can complement CBD tools by providing the means of understanding component internals and making it easier to change their functionality [9].

However, the issue of *accelerating the evolution of existing software* receives considerably less attention. To date, there exists no accepted maintenance methodology, although several such methodologies for development of new systems have already appeared. From the methodology perspective, maintenance is a very complex activity, because it involves a heterogeneous set of short-term processes, potentially with high concurrency. Accelerated maintenance requires a lot more "orchestration" than development of new systems.

Maintenance of existing software includes several activities:

1. corrective maintenance (or defect repair);
2. adaptation maintenance (or modifying the system to accommodate changes in the execution platform;
3. preventive maintenance (or improving the structure of the system to curb architecture erosion); and
4. evolution (or enhancement, adding new features) - designing a new feature of an existing system can be more difficult than designing an equivalent system from scratch, because all forward engineering decisions are constrained by legacy decisions.

Therefore, there are some problems, specific to maintenance, which are not present during development of new systems:

- high-level design decisions need to be rediscovered;
- higher volume of information that needs to be considered (understanding of legacy decisions in unfamiliar source code);
- "expertise" walks away as key developers change jobs;
- need for training new personnel
- multi-site collaborative environments, as for example maintenance is outsourced;
- change resistance due to architecture erosion (usually it is progressively more difficult to change the system).

Complexity of maintenance activity and the lack of methodologies have created a situation, where there exists a significant gap in tool support between the so-called "green-field" projects and maintenance of existing software. Maintenance of existing software has traditionally been the "lost world" for formal modeling methodologies and formal method based CASE tools. Up-front modeling of large existing systems is prohibitively expensive therefore the potential benefits of modern methodologies and tools are not attainable for programmers dealing with existing software. Tools for maintenance are usually built on an ad hoc basis, and have low adoption in industry.

On the other hand, pure maintenance activities (such as repair, adaptation and understanding legacy decisions) occupy a significantly bigger portion of the entire life-span of a software system, than pure design activities (such as understanding requirements for a feature, designing the feature and producing the source code). Also, according to [13], starting from 1990s, more programmers are involved in enhancements and repairs of existing software, than in "green-field" projects (see Table 1).

Table 1. Forecasts for distribution of programmers (in millions) per activities

Year	New projects	Enhancements	Repair	Total
1990	3 (43%)	3 (43%)	1 (14%)	7
2000	4 (40%)	4.5 (45%)	1.5 (15%)	10
2010	5 (35%)	7 (50%)	2 (14%)	14

Therefore, it is very important to produce *accelerated solutions*, suitable for maintenance. Our group is developing the so-called "software topography" methodology for the evolution of existing telecommunications software [1,2]. *"Software topography"* describes how maintenance activities can be accelerated using an integrated set of "topographic maps" of the system. Topographic maps represent all architecture views of the software, including both static and dynamic structures, and scenario views. The elements of these maps represent "real" entities and relations from the source code of the system. In our "topographic maps" architecture views are seamlessly integrated with each other and are traceable all the way down to the source code. "Software topography" addresses pure maintenance activities, such as defect repair, adaptation, preventive restructuring, as well as issues of understanding legacy software for the development of new features, and jump-starting new projects based on reuse. Software topography emphasizes concurrent development, multi-site teams, and issues of training new personnel [1].

We distinguish between the following four types of "topographic maps" of software (each serving a different purpose and therefore containing a different selection of elements): structure maps, execution maps, scenario maps and entity maps. *Structure maps* include Code Maps, Module Maps and Conceptual Maps. These maps correspond to well-known architecture views. *Code maps* represent physical software components (files, directories, etc.) and their relationships (import, generate, use, etc.). *Module maps* represent logical components (modules, subsystems, layers, etc.) and their interfaces. *Conceptual maps* represent high-level functional components. *Execution maps* represent run-time entities of the system. *Scenario maps* correspond to the use cases of the system. *Entity maps* include flowcharts of procedures, data structure diagrams, class diagrams and call graphs. Entity maps represent low-level code language-dependent entities (for example, procedures, variables, classes, etc.). Maps of the code below the physical components are essential for defect repair and adaptation maintenance. Paper [1] contains a high-level introduction into "software topography", and the process of building "topographic maps" of software in a cost effective way, combining automatic and manual steps.

The purpose of this paper is to describe our experiences in building and using *scenario maps* of existing software. We discuss how various maintenance activities can be significantly accelerated by providing developers with the knowledge of the *core scenarios* of the system, which *approximate the intended use cases* of that system. We present a *static approach* to extracting scenarios (as the so-called "source trajectories") by navigating through the source code and capturing the anchor locations in the source code of the program.

Scenario views of software are useful for understanding other views (Code View, Module View, Conceptual View and Execution View) [12,3]. Producing architecture views of existing software system is a capability that can bridge the gap between tool support for the so-called "green-field" projects and tool support for maintenance of existing software. The knowledge of the architecturally significant scenarios of an existing system can significantly accelerate maintenance activities by the means listed in Table 2, which are the main use cases for the approach suggested in this paper.

Table 2. Means to accelerate maintenance activities

- focused understanding of the architecture, especially during training of new personnel;
- focused impact analysis during bug localization;
- analysis of the architecture using scenarios [3];
- architecture reviews (walkthroughs);
- extraction of business rules;
- impact analysis at the level of business rules.

The rest of the paper has the following organization. In Sect. 2 we present an overview of the methodology steps, involved in approximating intended use cases of the system with "scenario maps". In Sect. 3 we discuss our experiments with various representations of scenarios, including Message Sequences Charts, UML Collaboration Diagrams, and Composite Code Flow Diagrams. We describe certain operations on scenarios (generalization, abstraction and projection), which provide a link between scenarios and architecture diagrams. In Sect. 4 we describe our **PathFinder** tool which supports manual capturing of the core scenarios in a cost effective way. In Sect. 5 we briefly discuss a case study, where we used our methodology and the **PathFinder** tool to capture the core scenarios of a small telecommunication-like system and used them for architecture reviews of the system. In Sect. 6 we compare our approach with related approaches and make some conclusions.

2   Approximating Use Cases with Source Trajectories

In this section we provide an outline of the methodology steps, involved in approximating the intended use cases of an existing system. The goal of the methodology is to build the set of the *core scenarios* of the system, which represent the architecturally significant paths through the system. We approximate the intended use cases as *source trajectories*. In our suggested approach, scenarios are captured *manually* by navigating through the corresponding trajectory in the source code. Later we automatically transform the initial model into Message Sequence Chart (MSC) notation [10]. Roadmaps between scenarios are represented as High-Level Message Sequence Charts (HMSC) [10]. The result is creation of a *formal scenario model* of the system.

Central to our scenario-based approach is the idea of extracting *core scenarios*, which approximate the intended use cases of the system. The objective of the extraction process is to obtain a *minimal set of scenarios*, which represent the original business objectives of the system (primary scenarios and key secondary scenarios). We believe that it is important to perform selection of architecturally significant events in the software system at the time, when scenarios are captured. Therefore, our approach involves *manual* selection of the key locations in the source code, which define the source trajectory.

Special-purpose tool support is essential to ensure effectiveness of this approach. The suggested solution is based on our inSight toolkit for maintenance of existing systems [1,2], the MOST Use Case Studio - the scenario-based forward engineering tool [7] and a special-purpose tool - the **PathFinder**, which is used as an integration platform between inSight and MOST. The **PathFinder** tool uses the extensibility APIs, provided by both tools. The **PathFinder** tool is used to capture, store, edit, animate and transform scenarios which are represented as sequences of the locations in the source code. Transformation of source trajectories into Message Sequence Charts is performed automatically. Our future plans involve automated selection of intermediate locations for a scenario guided by manual selection of fewer key locations, based on control flow analysis of the source code.

More precisely, our methodology consists of the following *steps*:

1. Analyze the code;
2. Collect known primary use cases;
3. Identify anchor locations;
4. Select modeling viewpoint;
   (For each use case)

5. Approximate the source trajectory for the selected use case;
6. Refine the trajectory;
7. Transform the scenario;
8. Update the HMSC roadmap;
9. Analyze Coverage.

Steps 5,6,7,8 are performed for one use case at a time. All steps are repeated iteratively. Iterations are controlled by termination conditions, which are checked at the last step. Below we describe each step in more detail.

Step 1. **Analyze the code**. The inSight toolkit [1] is used to perform architectural analysis of the system. This analysis is performed by the inSight Reverse Engineering Engine. The Engine extracts entities and relations from the source code of the existing system (and potentially, from some other sources) and builds the project repository. The result of this phase is the familiarity with the structure of the system, the main components and the call graph of the system.

Step 2. **Collect known primary scenarios**. There is much anecdotal evidence, that scenarios are used in a lot of industrial projects for design of applications, product families, architecture reviews and risk assessment [3], although there are quite few reports about the use of formal scenario notations. Informal knowledge of the primary scenarios can be very helpful in driving the static extraction of the core scenarios. In this case, the formal scenario model of the system is created by "anchoring" the known scenarios to the source code.

In a situation where there is no information about the primary scenarios of the system, they need to be re-discovered by analyzing the system. Our inSight tool and the "software topography" methodology can be used [1]. One of the claims of this paper is that organizing the exploration of an unfamiliar software system as the process of approximating the intended use cases of this system can significantly accelerate understanding of this model. Capturing the knowledge about the system in the form of the core scenarios focuses any further analysis of this system.

Step 3. **Identify anchor locations**. *Anchor locations* are certain architecturally significant statements in the source code of the system. They include certain *conditions* (where a choice between several scenarios is made), certain *loops* (where the system repeatedly performs a certain scenario), certain *function calls* (especially interfaces between components) as well as certain *operating system calls* (input and output, process creation and inter-process communication). Knowledge of such locations significantly accelerates the process of understanding the key trajectories of the system.

Step 4. **Select modeling viewpoint**. Extraction of scenarios is based on a certain modeling viewpoint. In [5,6] we identified three major kinds of modeling viewpoints:

1. *black-box* (interactions between the system and its environment)
2. *white-box* (core code), or
3. *grey-box* (collaborations between subsystems)

In the extreme case, source trajectories can be white-box (all events). However, in order to achieve a higher level of abstraction, and also to reduce the amount of work, it is useful to capture scenarios as grey-box (by cutting out segments of the trajectory, traversing through an architecturally insignificant subsystem). Selection of the modeling viewpoint is driven by the analysis of the structure of the system (see Step 1). Scenarios can be further changed using MSC tools by applying transformation techniques, described in Sect. 3.

Step 5. **Approximate the source trajectory for the selected use case**. This step is the key to our static extraction approach. First, a potentially architecturally significant scenario is identified. Then, navigation through the source code of the system is performed to identify the source trajectory for this scenario and to locate the anchor location for it. Navigation through source trajectories is performed using the inSight Flowchart tool [2]. Flowcharts are essential for understanding trajectories within the source code, since they visualize paths through the code. The inSight Flowchart tool uses the project repository to track procedure calls and inter-process communications.

When capturing a certain statement as an anchor location for the trajectory, the user selects the so-called *semantic tag* for this statement, specifying if this statement represents an architecturally significant task, condition or a communication point (send or receive).

The **PathFinder** tool is used to "record" interesting anchor locations as the trajectory is being navigated through, to store and edit scenarios. We suggest that at first only a few important anchor locations be captured for each scenario. Usually, the source trajectory is represented as a call-return skeleton. Then, the trajectory can be revisited and refined (see the next step).

Step 6. **Refine the trajectory**. Trajectories are *anchored* to locations in the source code (anchor locations). Integration with the inSight tool allows analysis of trajectories by displaying each step of the scenario in a Flowchart, or even at a certain architecture diagram. *Animation* of a trajectory can be performed by automatically displaying each location on a Flowchart in sequence. Our **PathFinder** tool has the capability to store and open trajectories in order to allow incremental refinement based on the modeling viewpoint.

In our opinion, the following manipulations are useful:

- Add/delete events (for example, after capturing the call-return skeleton of a scenario, the user can decide to capture some of the architecturally significant assignment, conditions and loops. Later, when the modeling viewpoint changes, certain events can be deleted from the trajectory);
- Redefine events (semantic tags of events can be changed to reflect changes in the modeling viewpoint);
- Hide interfaces, based on the selected modeling viewpoint (based on the modeling viewpoint, certain call-return segments of the trajectory can be folded, to cut architecturally insignificant parts of the trajectory)
- Capture a secondary scenario (certain alternative courses of action within the framework of the selected primary scenario can be additionally captured and saved).

Step 7. **Transform scenarios**. Source trajectory is the basic representation of a scenario. However, other representations are needed for the purposes of understanding scenarios, and using scenarios for analyzing the architecture of the system. The **PathFinder** tool is capable of transforming scenarios into the following representations:

- Message Sequence Charts [10];
- UML Collaboration Diagrams [8];
- Composite Code Flow Diagrams (a synthesized Flowchart representing a single flow across several procedures and/or processes) (see Sect. 3.4),

and the inSight tool is also used to visualize scenarios as UML Collaboration Diagrams and as Composite Code Flow Diagrams.

The MOST Use Case Studio [7] is used to visualize scenarios as Message Sequence Charts, to annotate scenarios and to publish scenario models on the Web. The MOST Use Case Studio is also used to maintain the roadmap of the derived scenarios in the HMSC notation (see the next step).

Step 8. **Update the HMSC roadmap.** Once scenarios are transformed into MSCs and imported into the MOST Use Case Studio, the HMSC roadmap is updated using the MOST Episode Sequence Editor tool [7]. For building an HMSC, a relation between scenarios is needed. Two source trajectories are related, if both include the same condition statement in the source code. Such a condition statement is transformed into a condition in MSC.

Step 9. **Analyze coverage.** We need to make sure that the extracted scenario model adequately captures the intended use cases of the legacy system. This may require several iterations of the extraction process. Coverage of the source code by the source trajectories can be used as the termination criterion. The coverage is defined as the ratio between statements, selected as anchor locations, to all statements in the system. Another useful form of coverage is procedure coverage. A procedure is covered by a certain source trajectory, if the trajectory includes locations from that procedure. Procedure coverage can be abstracted into architecture component coverage. All mentioned forms of coverage are computed and visualized by our insight toolkit [1,2].

3   Representations of Scenarios

In this section we discuss our experiments with various representations of scenarios, including Message Sequences Charts, UML Collaboration Diagrams and Composite Code Flow Diagrams.

3.1   Creating Scenario from the Source Trajectory

A *Scenario* corresponds to a certain *source trajectory* through the system, which includes all *anchor locations* from the scenario. A *Scenario* represents important (architecturally significant) *sequences of events* in the system. Each event is *anchored* to a certain *location* in the source code of the system. Each event contains information about the actual *source statement (the anchor location)*, the name of the so-called *key entity*, involved in the event and the so-called *identifier of the entity* (which is a unique representation of this entity in the project repository of the inSight toolkit. These identifiers are used to map scenarios onto architecture diagrams [1]. The key entity is selected by the user during event capturing. An *event* can correspond to a particular statement in the source code: a procedure call (connecting the procedure call statement and the start statement of the called procedure); or a return from a procedure (connecting the return statement with the call statement in the calling procedure); or an inter-process communication event (connecting two operating system calls). The semantic tag captures the type of an event.

Our **Pathfinder** tool allows the user to navigate through a certain *source trajectory* and builds the corresponding *scenario*. The inSight tool provides all information, required to build an event based on an anchor location in the source trajectory. The information, associated with an event, can be used to animate this trajectory in the inSight tool, to map it onto the architecture of the system, and to transform it into a number of other representations.

```
types
SemanticTag = tag_start | tar_stop | tag_task |
        tag_call | tag_entry | tag_return |
        tag_send | tag_receive |
        tag_condition_true | tag_condition_false |
        tag_loop | tag_end ;
SourceStatement = Text;
Location ::
        filename : FileName
        linenumber : nat1;
Event ::
        tag : SemanticTag
        anchor : Location
        name : EntityName
        eid : EntityId
        src : SourceStatement;
SourceTrajectory ::
        name : TrajectoryName
path : seq of Location;
Scenario ::
        name : ScenarioName
        events : seq of Event;
```

Scenarios should meet certain well-formedness conditions:

- the first event should have *start* (should have tag tag_start);
- the last event should have *stop*;
- each *call* event should be followed by an *entry* event;
- each *call* event should have a corresponding *return* event;
- *call* and *return* events should be balanced.

3.2   Representing Scenarios as MSCs

This section describes our approach to transforming scenarios into MSCs. Several mappings can be used to perform this transformation. Our basic mapping takes into account only calls, entry, return, send and receive events. More advanced mappings also take into account the following:

 − conditions and loops
 − tasks and parameters
 − objects (create, method call, delete)

We specify below the basic mapping of a scenario into an MSC using a simple algorithmic notation instead of full VDM-SL to conserve space. The algorithm shows, how traversing a Scenario structure can generate strings in MSC-PR (MSC textual notation) [10]. Constant parts of MSC strings are represented by underlined font. Variable parts of MSC strings are represented by *italic font*.

Algorithm 1: Basic MSC mapping

```
foreach event in events(scenario) do
    case tag(event):
    ( tag_start ) ->
        "msc name(scenario);"
        push call_stack, name(event)
        add proc_list, name(event)
        current = ''main"
        "main : instance;"
        idx = 1;
    (tag_stop) ->
        foreach proc in proc_list do
            "proc : endinstance;"
        endfor
        "endmsc;"
    (tag_call) ->
        "current: out name(event),idx to name(event);"
        push call_stack, name(event)
    (tag_entry) ->
        if ( ! name(event) in proc_list) then
            add proc_list, name(event)
            "name(event) : instance;"
        endif
        "name(event) : in name(event),idx from current;"
        idx = idx + 1
    (tag_return) ->
        pop call_stack, last_proc
        "current : out return_msg, idx to last_proc;"
        "last_proc : in return_msg, idx from current;"
        current = last_proc
        idx = idx + 1
    (tag_send) ->
```

```
    "current: out name(event),idx to ENV;"
    idx = idx + 1
(tag_receive) ->
    "current : in name(event),idx from ENV;"
    idx = idx + 1
endcase
endfor
```

Algorithm 1 is further illustrated at Figs. 1 and 2. Fig. 1 shows a simple system, consisting of four procedures ($P1,P2,P3$ and $P4$). The system has two components ($C1$ and $C2$). The trajectory contains 12 anchor points (represented as small circles) inside procedures. Procedure P1 receives a signal a from the environment (the corresponding source location contains an operating system call). Procedure $P3$ receives one more signal from the environment (signal b). Procedure P4 sends some signal to the environment (signal c).

The source trajectory at Fig. 1, is formalized by the following list of events (showing only tags and key entity names):

```
{ tag_start ' '' '';
tag_receive ' ''a"; tag_task ''p2"; tag_call ''p3";
tag_entry ''p3"; tag_receive ''b"; tag_return '' '';
tag_call ''p4";
tag_entry ''p4"; tag_send ''c"; tag_return " '';
tag_stop '' '' }
```

Applying Algorithm 1 to this scenario results in the following MSC:

```
msc example;
main : instance;
main : in a,1 from ENV;
main: out p3,2 to p3;
p3 : instance;
p3 : in p3,2 from main;
p3 : in b,3 from ENV;
p3 : out return_msg,4 to main;
main : in return_msg,4 textbffrom p3;
main : out p4,5 to p4;
p4 : instance;
p4 : in p4,5 from main;
p4 : out c,6 to ENV;
p4 : out return_msg,7 to main;
main : in return_msg,7 from p4;
p4 : endinstance;
p3 : endinstance;
main : endinstance;
endmsc
```

Fig. 2 shows the graphical representation of the MSC, which results by applying Algorithm 1 to the scenario.

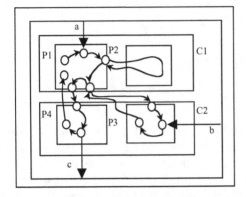 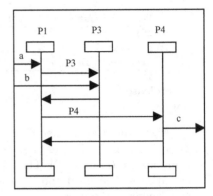

Fig. 1. Sample Source Trajectory   **Fig. 2.** MSC for sample

Note, that the trajectory visits procedure $P2$. However, the decision was to fold the corresponding part of the trajectory, by representing the corresponding anchor location as a *task*, rather than as a *call*.

3.3   Operations on Scenarios

In this section we define three operations on scenarios, represented as MSCs: generalization, abstraction and projection. These operations can be used to generate a variety of useful representations of scenarios. These operations formalize relations between scenario maps and architecture diagrams, which is essential for the "software topography" approach.

Definition 1. Generalization of a Scenario. Scenario B is *a generalization* of a scenario A with respect to a certain aggregation of instances $\{i_1, \ldots i_k -> i\}$, when the set of instances in scenario B is same as in scenario A, except that instead of instances $i_1, \ldots i_k$ there is a (composite) instance i. All call and return messages between instances $i_1, \ldots i_k$ are removed. All send (receive) messages from (to) instances $i_1, \ldots i_k$ are changed to send (receive) messages from (to) instance i. In other words, instance i is *a vertical composition* of instances $i_1, \ldots i_k$. The ordering of messages is unchanged.

Definition 2. Abstraction of a Scenario. Scenario B is *an abstraction* of scenario A with respect to a certain containment hierarchy $\{i -> c\}$, where i is an instance and c is a container (a mapping of instances to containers), when instances in scenario B are container names $c_1, \ldots c_k$, and each instance c_i is a vertical composition of the corresponding instances $i_1, \ldots i_n$, which are contained in c_i. The order of messaging is unchanged.

Abstraction uses a different input compared with generalization. Generalization is an operation on a scenario. Generalized scenario represents interaction between the original instances and a composite instance. Folding out a segment of the source trajectory inside an architecturally insignificant component is an example of a generalization. Abstraction uses architectural information (containment hierarchy) to perform generalization in a coordinated way. Abstracted scenarios represent interactions between containers.

Definition 3. Projection of a Scenario onto an Object Diagram. Object diagram B is *a projection* of scenario S onto an object diagram A, when B has same entities as A, and relationships on B are subset of relationships on A, such that B has a relationship between objects X and Y iff a generalization of scenario S has interaction between X and Y. Each relationship on B can be further annotated with its sequence number of the corresponding interaction.

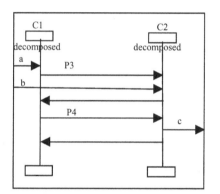 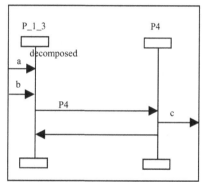

Fig. 3. Abstraction of sample   **Fig. 4.** Generalization of sample

Fig. 3 illustrates a generalization of scenario from Fig. 2, where instances $P1$ and $P3$ are composed into an instance P_1_3. Fig. 4 illustrates an abstraction of scenario from Fig. 2 with respect to the architecture at Fig. 1.

3.4   Alternative Representations of Scenarios

MSCs provide an intuitive representation of scenarios. In this section we discuss our experiments with various alternative representations of scenarios, including UML Collaboration Diagrams and Composite Code Flow Diagrams.

UML Collaboration Diagrams. A UML Collaboration Diagram (such as Fig. 5) can be built as a projection of a scenario onto a certain architecture diagram. The inSight tool under control of the PathFinder tool performs projection of scenarios. Unique identifiers of entities are received from the inSight tool as the user selects certain statements as anchor locations of some source trajectory. These identifiers are stored with the scenario. The inSight tool is

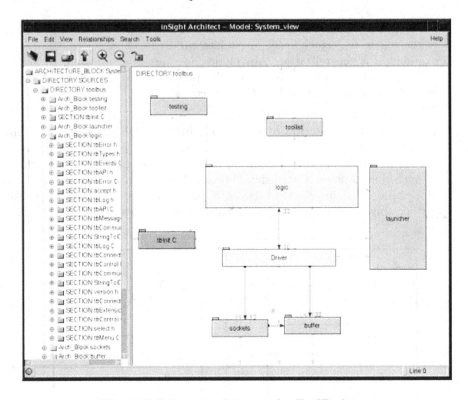

Fig. 5. Collaboration Diagram for ToolExchange

capable of projecting an individual entity onto any architecture diagram using unique identifier associated with that entity [1]. The inSight tool is also capable of highlighting relations on architecture diagrams. These APIs are used by our **PathFinder** tool to

1. project the entities associated with all entry points in the scenario onto an architecture diagram;
2. highlighting relations between the projected components.

The collaboration diagram for a scenario from the ToolExchange case study is presented in Fig. 5. Highlighted lines between blocks at Fig. 5 represent collaborations in the current scenario.

Composite Code Flow Diagrams   We have also experimented with other alternative representations of scenarios. Code Flow Diagrams use the inSight Flowchart tool to represent the complete scenario, providing visibility into the flow of control across procedure borders(see Fig. 6). The inSight Flowchart tool is capable of displaying a flowchart of a procedure in C/C++ or Java [2]. Our inSight Flowchart tool uses on-the-fly parsing and construction of the flowchart diagram using a patent-pending automatic layout algorithm. The **PathFinder**

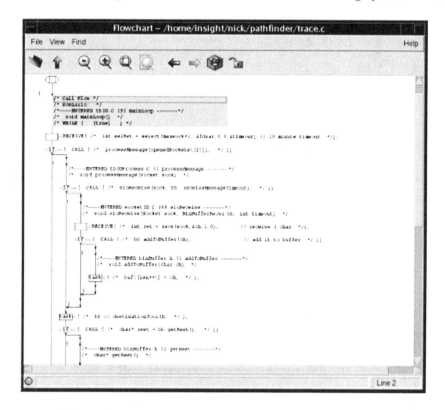

Fig. 6. Composite Code Flow Diagram for ToolExchange

tool synthesizes a certain C function and then uses an API to the inSight Flowchart tool to display this function. In the resulting flowchart diagram each flowline corresponds to a certain function. The indentation happens to represent the stack of function calls. Special comments and horizontal lines show the name of the procedure. All other events of the scenario are represented as TASK boxes at the Flowchart. The synthesized function represents a complete scenario.

A fragment of a Composite Code Flow Diagram for one of the scenarios from the ToolExchange case study is represented at Fig. 6.

4   The PathFinder Tool

The **PathFinder** tool (see Fig. 7) has the following capabilities:

- Navigate through source code; view the path; record anchor locations;
- Store/open/edit trajectory;
- Visualize anchor locations of the source trajectory using inSight Flowchart tool and inSight Architect tool;
- Animate (play) scenario back as the source trajectory using the inSight Flowchart tool and inSight Architect tool [1];

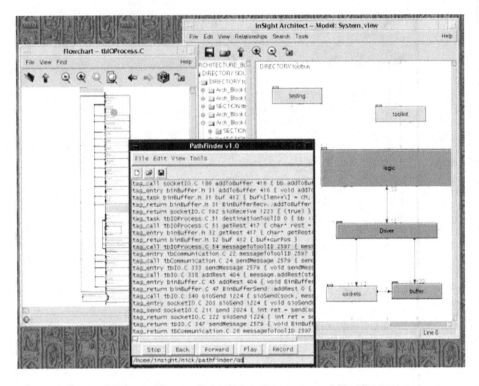

Fig. 7. Screenshot of the **PathFinder** Tool

- Check well-formedness of a scenario
- Maintain scenarios (check consistency between scenarios and the code)
- Automatically transform scenarios into MSCs, import them into the MOST Use Case Studio toolkit [7];
- Automatically transform scenarios into Collaboration Diagrams and Composite Code Flow Diagrams.

5   Case Study

We have applied our methodology to analyze a small-sized telecommunications-like system, called the ToolExchange described in [5].

Implementation of the ToolExchange system consists of 14 modules in C++ language. The implementation contains 110 functions. The total size of the implementation is 1128 LOC. There exists a small regression test suite, consisting of 15 test cases.

Considering the size of the system it was very easy to extract 11 scenarios, which exactly correspond to the original use cases of the system. These scenarios consist of 493 LOC of MSC-PR. The effort for the static extraction exercise was 2 days by a person with only superficial knowledge of the ToolExchange. This time also included getting familiar with the PathFinder and inSight tools.

6 Comparison to Related Work and Conclusions

The methodological contribution of this paper consists in the static approach to capturing source trajectories in order to approximate the intended use cases of the system. The objective of the extraction process is to obtain a *minimal set of scenarios*, which represent the original business objectives of the system (primary scenarios and key secondary scenarios). We believe that it is important to perform selection of architecturally significant events in the software system at the time when scenarios are captured. Therefore, our approach involves *manual* selection of the key locations in the source code, which define the source trajectory.

Special-purpose tool support is essential to ensure effectiveness of this approach. The suggested solution is based on our inSight toolkit for maintenance of existing systems [1,2], the MOST Use Case Studio - the scenario-based forward engineering tool [7] and a special-purpose tool - the **PathFinder**, which is used as an integration platform between inSight and MOST. The **PathFinder** tool uses the extensibility APIs, provided by both tools. The **PathFinder** tool is used to capture, store, edit, animate and transform scenarios which are represented as sequences of the locations in the source code. Transformation of source trajectories into Message Sequence Charts is performed automatically. Our future plans involve automated selection of intermediate locations for a scenario guided by manual selection of fewer key locations, based on control flow analysis of the source code.

Scenarios capture knowledge about the intended use cases and the key architectural patterns. The knowledge of the core scenarios of an existing system has significant benefits for acceleration maintenance activities. Core scenarios can be used for all the means listed in Table 2.

Kazman [3] claims, that although architecture reviews are done informally in almost every significant system development, the lack of formality of the process limits its usefulness. Scenarios do not get created and circulated throughout a large group of stakeholders, and they seldom form the basis of formal reviews of the software architecture [3]. Our **PathFinder** tool adds the required formality and supports the process of architecture reviews.

Several groups are developing scenario-based methodologies [3,4,5,6,7,11,12]. Scenarios are primarily used for eliciting and validating functional requirement [7].

The most complete set of literature on using scenarios for motivating software design centers around the practice of object oriented design. The major uses of scenarios and related notations in object-oriented design are:

- scenarios as a sequence of related events that represent dialogues between the user and the system;
- use cases, as employed by Jacobson, which include not only the notion of a scenario as a control flow, but also a set of pre- and post- conditions under which this use case applies [8];

- use case maps, as defined by Buhr and Casselman, which are a means of representing and reasoning about the flow of responsibility through a system [11].

Each of these uses of scenarios aids the architect in understanding how a software architecture supports its requirements; they are a means of mapping requirements onto an object-oriented design in a way that lets stakeholders "walk through" their execution [3].

Kruchten has gone one step further in the 4+1 model, proposing scenarios as the glue that aids in understanding a software architecture, in driving the various views needed to represent it, and in binding the views together [12]. Our "software topography" approach [1] leverages the so-called "scenario maps" of existing software using the methodology, presented in this paper, to focus investigation of other architectural views of the existing system.

It is interesting to note, that the suggested methodology works both for object-oriented as well as for procedural systems. In our experience, scenarios are quite useful for understanding non-object-oriented software, but they are almost indispensable for understanding object-oriented systems. On the other hand, structural architecture views are less insightful for understanding the functionality of an object-oriented system. The suggested approach handles conventional software, as well as concurrent systems, interacting asynchronously.

In [5,6] we described a dynamic scenario-based approach to the evolution of communications software, where we approximated the intended use cases by executing a suitable instrumented system using existing test cases. The dynamically extracted scenarios are represented by the so-called probe traces. We converted probe traces into MSCs. We used resulting scenarios to build an SDL model of the legacy system by applying the MOST Synthesizer Tool from the MOST toolkit to MSCs.

Several other groups suggested extracting scenarios dynamically from existing systems for the purpose of understanding the architecture, e.g. [4]. However, we used a very simple and powerful instrumentation technique, which allowed us to directly extract already abstracted MSCs. We identified functional interfaces of components we wanted to model. Then, for each function in such an interface, we created a macro with the same name. The macro emitted MSC events directly according to our call-return scheme, without capturing the actual probe trace. More precise instrumentation is of course possible, and is done by some authors [4], but led to huge information spaces which hinder understanding rather than provide any insight. According to [4], special tool support in the form of visualization was required to aid understanding of dynamically produced traces.

Representations and transformations of scenarios, presented in this paper, are applicable, no matter if a scenario was extracted statically as a *source trajectory*, or dynamically, as a *probe trace* of the instrumented system.

We believe the approach for extraction of core scenarios from existing systems and representing them as MSC presented in this paper helps bridge the gap between tool support for maintenance and tool support for "green-field" projects.

References

1. N. Mansurov, D. Campara, "Software Topography" approach to accelerating maintenance of existing telecommunications software, http://case.ispras.ru (submitted to IEEE Int. Conf. for Software Maintenance, ICSM'2001).
2. N. Mansurov, I.Ivanov, D. Campara, Using Entity Maps to accelerate maintenance of existing systems, http://case.ispras.ru (submitted to IEEE Int. Conf. for Software Maintenance, ICSM'2001).
3. R. Kazman, S.J. Carriere, S. Woods, Toward a Discipline of Scenario-based Architectural Engineering, in Annals of Software Engineering 9 (2000), pp. 5-33.
4. D. Jerding, S. Rugaber, Using Visualization for Architectural Localization and Extraction, in Proc. 4th Working Conf. On Reverse Engineering, 1997, Amsterdam.
5. N. Mansurov, R. Probert, Dynamic scenario-based approach to re-engineering of legacy telecommunications software, in Proc. 9th SDL Forum, Montreal, Canada, June 21-26, 1999, Elsevier Science Publishers B.V. (North-Holland), pp. 325-340.
6. N. Mansurov, R. Probert, Scenario-based approach to evolution of Communication Software, to be published in IEEE Communications, Special Edition, 2001.
7. N. Mansurov, Requirements capturing, validation and rapid prototyping in the MOST Use Case Studio, http://case.ispras.ru.
8. Jacobson, G. Booch, J. Rumbaugh, *The Unified Software Development Process*, Addison-Wesley, 1999.
9. IDC, *Application Design and Construction tools market forecast and analysis, 2000-2004*, May 2000.
10. Z.120 (11/99) Message Sequence Chart (MSC), ITU-T, Geneva 2001.
11. R. Buhr, R. Casselman, *Use Case Maps for Object-Oriented Systems*, Upper Saddle River, NJ, Prentice Hall, 1996.
12. P. Kruchten, The 4+1 View Model of Architecture, *IEEE Software*, pp 42-50, November, 1995.
13. A.van Deursen, P. Klint, C. Verhoef, Research issues in the Renovation of Legacy Systems, CWI research report P9902, April 1999.

From MSC-2000 to UML 2.0 –
The Future of Sequence Diagrams

Øystein Haugen

Ericsson Research NorARC, P.O. Box 34, N-1361 Billingstad, Norway
oystein.haugen@ericsson.no

Abstract. This paper discusses how MSC-2000 could influence the Sequence Diagrams within UML 2.0, and why the UML 1.x semantics is partly inadequate for what is needed in the area of sequence charts. Extracts of a possible UML meta-model is shown and this can be understood as a conceptual model for MSC-2000 as well and an indication of an approach to the future MSC-2000 semantics. UML Collaboration Diagrams have no direct counterpart in SDL/MSC and the difference between sequence diagrams and collaboration diagrams is analyzed.

1 Introduction

MSC-2000 [1] was finalized in 1999. The most updated version is on the SDL Forum Society web-site http://www.sdl-forum.org.

UML is becoming the predominant industry standard for modeling. We all know that UML 1.x [2] is not the ultimate language. Therefore experts of the SDL/MSC community are taking part in the standardization process of UML 2.0, which will probably emerge as a complete set of notations in 2002. In this context it has been recognized that UML 1.x is too limited with respect to sequence diagrams, and since MSC-2000 has been mentioned a number of times in the UML community [3] they are expecting to find inspiration from it.

It is not the situation that MSC-2000 as a whole will be included in UML 2.0. Firstly there is a need to recognize that UML 2.0 will have the same "look-and-feel" as UML 1.x. There will probably (as usual) be strong requests to keep the changes as backwards compatible as possible. Secondly while MSC is a standalone language, UML sequence diagrams are one notation out of about ten diagram types, and the basic concepts must be aligned across these notations.

The requirements for UML 2.0 are set down in the Request For Proposal [4] and contain the following

> *6.5.4.3 Interactions*
> – Proposals shall define mechanisms to describe the decomposition of a role in an interaction into an interaction of its constituent parts.
> – Proposals shall provide mechanisms to refer from one interaction to other interactions to support composition of interactions. It shall be possible to define, at least, sequences of interactions, alternative interactions based on conditions, parallel execution of interactions, and repetition of an interaction.

R. Reed and J. Reed (Eds.): SDL 2001, LNCS 2078, pp. 38–51, 2001.

It is easily seen that these requirements correspond well with mechanisms already found in MSC in 1996 [5].

In this paper we focus exclusively on interaction diagrams (sequence and collaboration diagrams). UML activity diagrams also have similarities with sequence diagrams, but are not considered in this paper. Furthermore we have concentrated on answering the requirements of the RFP cited above and have not found room to discuss the possible impact of introducing even more of MSC-2000 such as more formal data, improved time constraints and improved control thread description. These aspects of MSC-2000 are not excluded by the RFP, but are not the main focus. We expect that sequence diagrams of UML will follow a maturing process rather the same as that of MSC. As UML 1.x sequence diagrams are similar to MSC-92, it is reasonable that the most focused changes are those that appeared in MSC already in 1996.

The organization of this paper is to present and discuss their similarities and difference of the UML diagrams: Collaboration diagrams and Sequence diagrams. Then we show how the well-known MSC mechanisms could be applied in the context of UML 2.0 Sequence Diagrams. We present a suggestion for a UML meta-model. We then discuss Collaboration diagrams for UML 2.0. Finally we summarize.

2   The Conceptual Model of Collaboration and Sequence Diagrams

UML is a set of graphical notations, but there is no graphical grammar as we can find in SDL and MSC standards. They have an abstract grammar in the form of a "meta-model". This meta-model is described in subset of UML itself (though in principle it should be described in another notation called MOF, which is similar to a subset of UML). The subset used for the meta-model is basically the class diagram. Even though it is supposedly an abstract grammar, it is sometimes confused with a conceptual model, which is not always exactly the same.

In the UML 1.x meta-model and the corresponding semantics, Interaction diagrams and Collaboration diagrams are two different views of the same model. They both describe interactions between something (somebody).

There are two distinct issues here:

1. What happens when we want to describe more elaborate structures than simple method-call constructs?
2. What are the things that interact, send messages etc.? These are questions about "roles".

If we look at the two forms, collaboration and sequence diagrams, they are quite similar. Collaboration diagrams are Sequence diagrams where the time dimension has collapsed and been replaced by a numbering scheme. An example is shown in Fig. 1.

40   Ø. Haugen

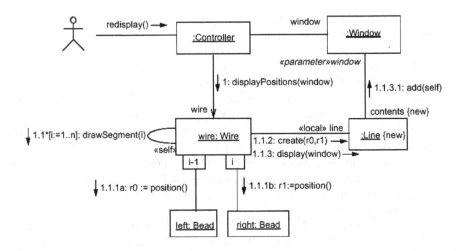

Fig. 1. Collaboration diagram example (from UML 1.4)

From the numbering scheme it is quite obvious that there is a clear limit to what can be described in a transparent way along the time axis. As it is the messages that are numbered, it is evident that only sequences of messages can be described. This understanding is strengthened by the UML 1.4 meta-model, an extract of which is shown in Fig. 2.

Because Sequence Diagrams and Collaboration diagrams are meant to represent the same kind of model the concepts of a sequence or collaboration diagram is not found in the UML meta-model. In fact no diagrams are found in the meta-model. They are described in the notation guide.

Thus both sequence diagrams and collaboration diagrams are meant to specify the abstract notion of an "Interaction". And as can be seen from the excerpts in Fig. 2, the meta-model can describe only sequences of Messages and not sequences of events such as sending a Message and receiving a Message. This means that the meta-model cannot describe asynchronous message overtaking.

This also reflects the most common use of sequence diagrams within the UML community. Sequence diagrams are very often used to describe situations where communication is done with procedure invocations. Typically there is only one thread of control as in a C++ program. Many of the entities depicted in the sequence diagram are in fact conceptually passive objects. Contrary to this, MSC has been used in situations of concurrency and asynchronous messaging. The merger of the two approaches is a challenge.

This leads to the conclusion that collaboration diagrams are mainly intended to give a rough view of a simple set of control threads.

On the other hand, sequence diagrams have their strength in that they have an explicit time dimension, and then collapse the topology into one dimension. From the experience with related notations such as MSC, what a sequence diagram describes is a partial order of Events. An Event is something that happens

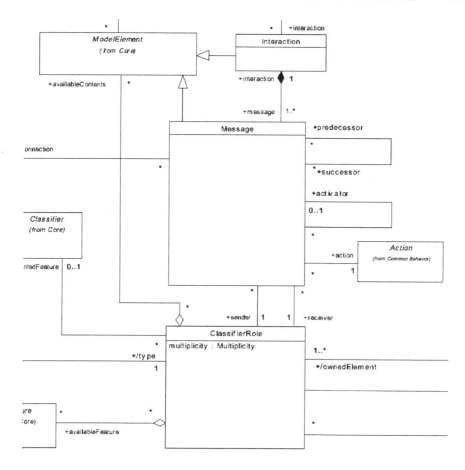

Fig. 2. Parts of UML 1.4 Meta-model for Collaborations

on an acting object that by itself has little duration. (Whether the duration of an Event has any significance is not considered in this paper).

The difference in perspective can be depicted through the conceptual model in Fig. 4. Collaboration diagrams describe a set of control threads based on sequences of messages. Sequence diagrams on the other hand describe partial orders of events. Events are related to Parts that own them. Messages have (normally) two Events, one output and one input. From this we see that collaboration diagrams can be transformed into sequence diagrams while the opposite is not always possible. Sequence diagrams are more expressive than collaboration diagrams, but collaboration diagrams may be more transparent when describing simple threads of control.

In Fig. 4 we have used Event as the atomic conceptual entity, but this may not be preferred term in UML, but the point is that it represents some atomic action associated with one active object.

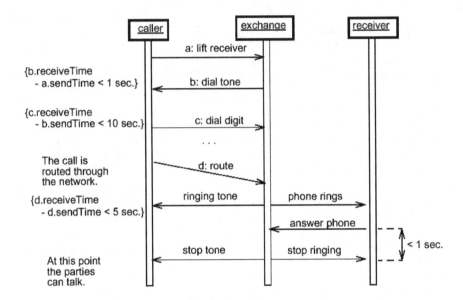

Fig. 3. Sequence Diagram example (from UML 1.4)

3   Structuring Sequence Diagrams

Sequence diagrams are used formally and informally to specify the behavior of parts of systems. The more popular these diagrams get the more the need for organizing them in a sensible way emerges.

From the work behind MSC-2000 we have come to a set of structuring principles that:

- should be equally applicable to sequence diagrams;
- correspond well with the requirements of the Request for Proposal.

3.1   Diagram References and Inline Expressions

Firstly we have the introduction of diagram references: a mechanism that allows a reference from within a diagram to another diagram. By introducing such references, sequence diagrams will have a feature as has been commonplace in programming languages for a long time: namely subroutines or procedures.

Secondly there is the ability to express variability and other expressions where elements of the diagram serve as operands.

We show in Fig. 5 how references to other diagrams look in the syntax of MSC (ITU Z.120). The two diagrams *EstablishAccess* and *OpenDoor* are referenced from *UserAccess*. The message *Mesg* and the *OpenDoor* reference are contained in an expression frame of the **optional**-expression.

The intuitive interpretation is obvious. A reference means that the description of the reference diagram can be substituted into the position of the reference

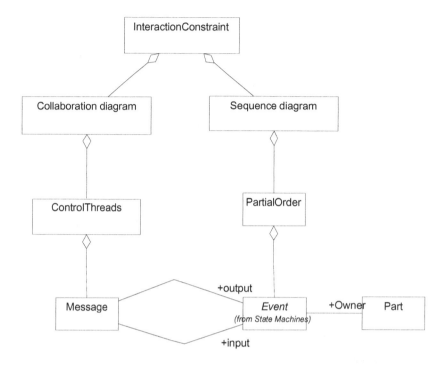

Fig. 4. Conceptual model for Collaboration diagrams and Sequence Diagrams

with event parameters bound. The interpretation of the expressions differs according to the kind of expression. The optional expression has one operand and the interpretation is that the traces described are both the ones including the operand and those excluding the operand.

As this is a sequence diagram we concentrate on the individual events and sequences of events. For a consistent and simple conceptual approach to the semantics of the diagram in Fig. 5, we reduce every event to a reference. Then we let the diagram be defined as an expression of such references restricted by the fact that output of a message must come before the input of that message. The places where messages go out of or into diagram references, are called "gates". There must be correspondence between the ("formal") definition of these gates (in the definition of the sequence diagram with that name) and the actual gates (on the diagram references).

Gates have no direct counterpart in UML 1.x. Gates have been discussed in the MSC community for several years. Simple gates have a trivial interpretation as representing connection points of messages. When the diagram references refer to more complicated expressions such as loops or alternative expressions, gates can no longer always be given such a simple interpretation. Hopefully the UML community may benefit from the MSC discussions to find a gate definition that is intuitive, but still precise. We will not cover this in greater detail in this paper.

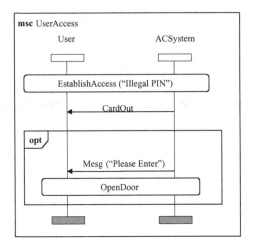

Fig. 5. References and inline expressions

On the right side of Fig. 6, we see how the diagram is transformed into having only references and connections through gates. On the left side, we see the corresponding expression tree of weak sequencing[1] [7] taking into account the restrictions on output coming before input of the same message.

We do not know what is in the referenced diagrams (e.g. *OpenDoor*), but to find the complete expression of *UserAccess*, the procedure is just to replace the *OpenDoor* reference by the expression graph of its referred sequence diagram. In the end we will have only references in the expression referring to atomic actions such as an output event or an input event. That expression is the canonic meaning of the sequence diagram.

3.2   Guarding the Expression Operand

The **opt**-expression in Fig. 5 defines two courses of action, either through the operand or outside it (empty operand). The expression defines two sets of traces. A common question that then arises is "when the option operand is executed and when it is not?" The expression as such defines both sets of traces as equally likely alternatives.

A statement that restricts when an operand is eligible is called a guard. We have guards in UML 1.x both for state machines and for sequence diagrams. Unfortunately these guards can only be applied to individual messages in the sequence diagrams, and that is not very powerful. It is simple to find scenarios where it is difficult to distinguish visually between the guarded alternatives, and again the underlying conceptual model is that of a sequence of messages (or method calls). By using notation from MSC and framing the alternative

---

[1] Weak sequencing informally means: if A **weakseq** B then for each message handling part, the events of A come before B, but for events on different message handling parts, there is no defined ordering.

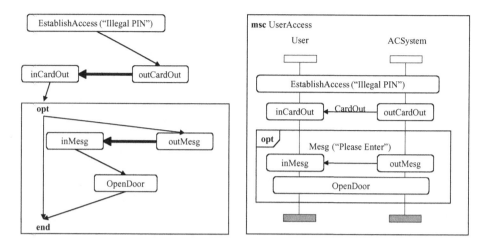

Fig. 6. Expression of interaction fragments

operands, we may reach a more powerful and more transparent syntax. Using a mixture of MSC and UML 1.x, the guards could look like the diagram in Fig. 7.

3.3   Decomposition

We assume that the internal structure of Parts defines an aggregate hierarchy of the substance. In Fig. 7 we have that the *ACSystem* consists of an *AccessPoint*, an Authorizer and a *Console*. It is also indicated that the *AccessPoint* consists of even smaller parts. The associated sequence diagrams focus on behavior of the contained parts. The behavior of one level is, of course, closely related to behavior on the next level. When *CardOut* is output from the *ACSystem*, this means that from somewhere within *ACSystem*, *CardOut* is output. This should therefore be visible if the *ACSystem* is decomposed.

Decomposing a part in a sequence diagram has both a substance dimension (the substance aggregate hierarchy) and a behavioral dimension (which scenario is decomposed). In Fig. 7 we see how MSC defines decomposition - at the top of a timeline. The *UserAccess* scenario is decomposed relative to the details of *ACSystem* in a sequence diagram *AC_UserAccess* also shown in Fig. 7.

It is necessary that decomposition is done in a way that preserves consistency. In particular this is important when diagram references cover parts that are decomposed as we see in the left part of Fig. 7 where *EstablishAccess* and *OpenDoor* cover *ACSystem*, which is decomposed. Since the decomposition can be understood as a diagram reference, there is intersection between references in this case. Consistency (called "commutative decomposition" [6]) means that it should be insignificant whether the decomposition is done before a reference or the reference is followed before decomposition. In our example in Fig. 7 the two different "roads" are:

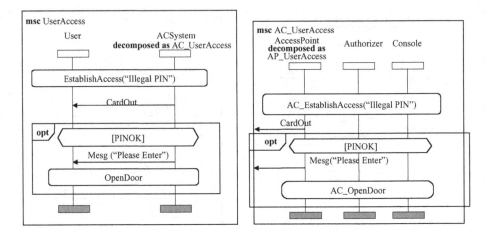

Fig. 7. Guards and decomposition

1. Decompose *ACSystem* within *UserAccess* (left), followed to *AC_UserAccess* (right). Then follow reference *AC_EstablishAccess* (that corresponds structurally to *EstablishAccess*).
2. Follow *EstablishAccess* reference to its diagram (not shown) and find in that diagram *ACSystem* decomposed as *AC_EstablishAccess*.

In MSC-2000 there are syntactic requirements to control that the decomposition is structurally similar to the origin. The definition of structurally similar is not interesting for this paper, but one can get the intuition from similarities in Fig. 7.

3.4 Meta-model for Sequence Diagram

We have summarized the approach to structuring Sequence diagrams in the meta-model depicted in Fig. 8.

4 Aggregate Hierarchy, Roles and Parameters

In the preceding section we have assumed a run-time structure specified for objects of a Class containing Parts and having an Interaction constraint defined by a set of sequence diagrams. These sequence diagrams have as their message handling components the Parts of the internal structure. This is very similar to procedures using the internal attributes of a class. It should be noted here that the concepts of "InteractionConstraint" and "Part" are not UML 1.x, but generic terms for concepts that most probably will appear in UML 2.0 as a result of response to other sections of the Request For Proposal relating to the "Runtime architectures".

MSC-2000 has the same concepts where MSC documents define the instance kinds (MSC terminology) and the internal instances are defined in the MSC document.

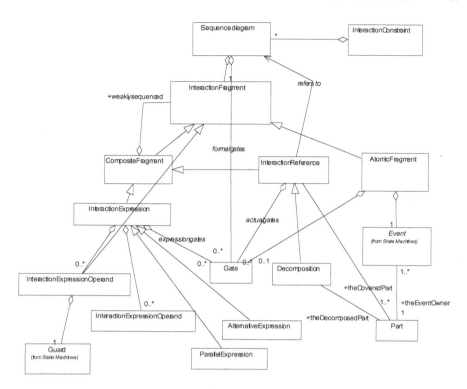

Fig. 8. Meta-model for Sequence Diagrams (some of it)

In UML 1.x the message handling entities are normally "roles" (Classifier-Roles). Roles can be seen as generic objects that are orthogonal to the internal aggregate structure. The sequence diagrams are behavioral fragments that can be applied to a number of different situations and the roles are subsequently synthesized into behavior of objects.

Our interpretation of roles is such that they are generic message handling objects, and as such are very similar to what is commonly known as parameters. Parameters are generic objects that can be bound to different actual contexts. MSC has instance parameters that would correspond to "part parameters" in a sequence diagram situation. We have attempted to model this in a meta-model in Fig. 9.

Seeing the roles as parameters helps in many ways because we need not relate to two different worlds: one world of roles and another of real objects. Parameters will of course be typed. Parameters are already found in other parts of the UML meta-model.

Expressed in words, the meta-model in Fig. 9 is intended to mean the following. InteractionConstraints (associated with Classifiers) consist of a set of BoundInteractions. These BoundInteractions are specializations of SequenceDiagrams containing no unbound formal Part parameters. Another class of Se-

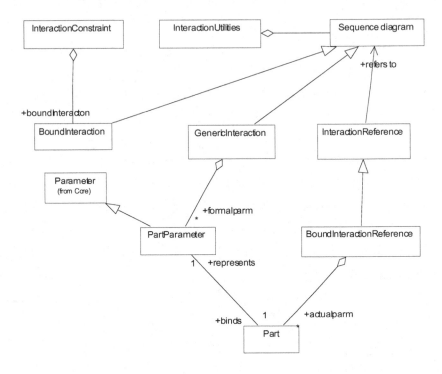

Fig. 9. Meta-model for generic part parameters

quence Diagrams is the GenericInteractions that contain PartParameters that
in turn can be bound by Parts. Parts are bound to parameters, for example in
InteractionReferences constituting a BoundInteractionReference.

UML also offers "interfaces" to classes, and it has been suggested that the
interacting entities in sequence diagrams should refer to such interfaces. The
concept of Interface is similar to a type, as it describes some (but not all) of the
capabilities of a class. Therefore it may be quite reasonable to let the types of interacting entities refer to interfaces (as well as classes). Typically such interface-
typed entities will be generic (that is parametric) parts that in the end must be
bound to real substance with a class as type.

4.1   Interpreting the Interaction Constraint

The sequence diagrams are parts of the interaction constraint. How one interprets
this is not obvious. The sequence diagram as it stands defines a set of traces (or
a partial order of actions if you like), but how is this set of traces related to the
internal structure and the other parts of the complete UML description?

Live Sequence Charts (LSC) by Damm and Harel [8] suggests to distinguish
between cold and hot diagrams designating whether the executable model should
be able to execute all the sequences of the sequence diagram, or possibly a subset
of the sequences.

There is also the possibility that the set of executable traces is disjoint from the set of traces defined by the interaction constraints, or that the two sets overlap without one being a subset of the other.

Another approach to describing an interpretation to sequence diagrams comes from TSC [9] where the purpose is to specify tests with their verdicts. The verdicts then represent an interpretation of the sequence. The TSC proposal has also a number of other creative syntaxes for structuring sequence diagrams.

5   Collaboration Diagrams

As shown in Fig. 4, the conceptual model for collaboration diagrams is not the same as the one for sequence diagrams. Collaboration diagrams are used to show simple sequences of Messages (or method calls, as they turn out to be more common).

Collaboration diagrams have roles, and these roles can be bound to classifiers as shown in Fig. 10.

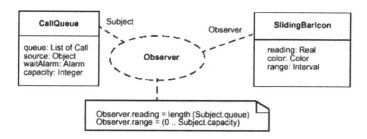

Fig. 10. Binding roles of a collaboration (from UML 1.4)

Unfortunately classifiers are bound to roles, while they should rather have been objects of classes because a collaboration may very well include communication and interaction between objects of the same class, but where the objects play different roles. Take for example a telephone call where there are two objects of Subscriber where one is an A subscriber and the other is a B subscriber.

Nevertheless this shows that binding of roles is quite similar to binding of parameters.

In Fig. 11 it is attempted to define how the *Component Framework* uses other Collaborations, and how the Roles of that collaboration are related to roles in the collaborations used by it. From this example, it is quite simple to state that these kinds of binding diagrams become messy very easily and that the precision of describing how one collaboration uses another one is not very precise. It would be impossible from Fig. 11 to reduce the *Component Framework* collaboration to a simple collaboration. Still it is an indication of how collaborations are intended to be reused. Fig. 11 shows more a static view of the roles and objects, than a dynamic behavioral view. This is also according to a statement found in UML

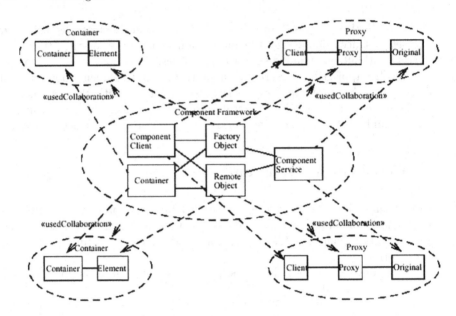

Fig. 11. Binding roles and using other Collaborations (from UML 1.4)

1.4: "Note that *patterns* as defined in *Design Patterns* by Gamma, Helm, Johnson, and Vlissides [10] include much more than structural descriptions. UML describes the structural aspects and some behavioral aspects of design patterns; however, UML notation does not include other important aspects of patterns, such as usage trade-offs or examples. These must be expressed by other means, such as in text or tables."

The problem with Collaboration diagrams is that there is no graphic time dimension. The sequencing is shown in the numbering on the arcs (the messages). This lends itself very poorly to structuring the behavior since describing a subsequence over the same set of message handling parts (or roles) cannot be shown graphically.

For simple sequences (meaning short sequences with little variability) the collaboration diagrams have been shown to have great appeal to designers. Normally this situation occurs very early in the project or in informal discussions from scratch. Later in the project, when it is of more importance to identify common behavioral patterns, the collaboration diagrams are not so helpful because it is hard to distinguish such similarities between collaboration diagrams.

The natural development for collaboration diagrams would be to concentrate on describing static aspects. This would make the collaboration diagrams one of the candidates for describing the internal structure of classifiers.

6   Summary

We have discussed how structuring mechanisms of MSC can be included in UML 2.0.

Our conclusion is that the structuring concepts of MSC should apply well in a UML context. Diagram references, decomposition and inline expressions with guards are mechanisms that lend themselves easily to being incorporated in UML.

There are, however, some fundamental issues that must be resolved. The conceptual model for asynchronous message interaction is that of a partial order of events and not a set of message sequences. Acknowledging this it is also necessary to realize that Collaboration diagrams and sequence diagrams focus on different conceptual aspects of the model, and that one cannot expect to describe them both with the same meta-model. We have given some suggestions for how such a meta-model could look with respect to sequence diagrams.

We have also discussed the concept of "role" in UML 1.x, and believe that UML 2.0 would achieve more by introducing parameterization of interacting parts. This would simplify the UML meta-model and make a smoother formal transition between the sequence diagrams and an executable model.

References

1. ITU (1999). Z.120. Message Sequence Charts (MSC). O. Haugen. Geneva, ITU-T: 126.
2. OMG (2000). OMG Unified Modeling Language, Object Management Group.
3. Haugen, Ø. (1999). Converging MSC and UML Sequence Diagrams. Beyond the Standard UML'99 - The Unified Modeling Language, Fort Collins, CO, USA.
4. OMG (2000). Request For Proposal. UML 2.0 Superstructure RFP. Needham, MA, USA, Object Management Group.RFP ad/2000-08-09
5. ITU (1996). Z.120. Message Sequence Charts (MSC). E. Rudolph. Geneva, ITU-T: 78.
6. Haugen, O. (1997). The MSC-96 Distillery. SDL '97 Time for Testing. SDL, MSC and Trends., Evry, Paris, France, North-Holland.
7. Rensink, A. and H. Wehrheim (1994). Weak sequential composition in process algebra. CONCUR '94, Proc. 5th Int. Conf. on Concurrency Theory, Uppsala, Sweden, Springer-Verlag.
8. Damm, W. and D. Harel (1999). LSCs: Breathing Life into Message Sequence Charts. FMOODS'99.
9. Schieferdecker, I., J. Grabowski, et al. (2000). TSC - Test Sequence Charts. Sophia Antipolis, ETSI.Draft recommendation.
10. Gamma, E., R. Helm, et al. (1994). Design Patterns. Elements of Reusable Object-Oriented Software, Addison-Wesley.
11. Reniers, M. A. (1998). Message Sequence Chart: Syntax and Semantics. PhD Thesis,. Departement of Computer Science. Eindhoven, Eindhoven University of Technology.

SDL and Layered Systems: Proposed Extensions to SDL to Better Support the Design of Layered Systems

Rodolphe Arthaud

Telelogic France, 150 Rue Nicolas Vauquelin - BP 1310
31106 Toulouse Cedex – France
rodolphe.arthaud@telelogic.com

Abstract. Designing complex systems as stacks of collaborating layers is a common practice in various domains, from operating systems to user interfaces. It proves to be particularly fruitful in the domain of telecom systems and, more generally, in distributed systems. After showing why, today, SDL is not well suited for the design of layered systems, we explore usual techniques available in programming languages. Then, we attempt giving SDL the power of expression necessary to view and manipulate signals at different abstraction levels while preserving the language spirit, staying at design level.

1 Introduction

Complex systems have been thought of as independent communicating layers relatively early in various domains, such as operating systems [1], user interfaces (think of Xlib, Xt and Motif as layers) [4,9], protocols or a generic architecture such as the OSI Reference Model [3,8]. The concept of layers was one of the early design patterns [5], long before the concept of pattern was born as such. It was also one of the earliest techniques to favor reusability, maintainability and extensibility. A successful one too, since it could never be replaced by the newer and more fashionable concepts in the area — inheritance, type genericity, dynamic binding and other varieties. On the contrary, all these techniques only made it easier to think, design and program in terms of layers.

Since its origins, SDL has been used mainly in the context of telecom applications and, more specifically, for the design of protocols. Surprisingly enough, SDL does not offer any support for the design of *several, communicating* layers. As we will see, it is even weaker in that respect than mere low-level, programming languages.

In this paper, we will make a set of proposals for additions to SDL, to make it better suited for the design of multi-layered systems.

Please note that these proposals are not final ones, in the sense that similar power of expression could be obtained by different means, possibly with better results regarding static typing or consistency with other mechanisms of the

R. Reed and J. Reed (Eds.): SDL 2001, LNCS 2078, pp. 52–71, 2001.

language. In particular, a complete proposal might consider improving signal description together with data types. In this paper, we have chosen to focus on one topic only for the sake of clarity.

Please note also that the author does not claim that these ideas are original. On the contrary, they were inspired by existing languages and by discussions with SDL users. In particular, the author remembers some discussions within ITU-T meetings. This paper is an attempt to capture good ideas that could now be inserted in SDL.

Though these proposals should be extensions to SDL 2000, we have used the syntax of SDL 96 in the discussion for the sake of understandability, since we thought most readers would be more familiar with it. This has no impact on the discussion.

2 Presentation of the Problem

2.1 What Are Layers?

A layer is a set of components offering a certain service to higher-level layers and using lower-level layers.

Layers are highly independent from each other. In other words, the design imposed to the clients of a layer allows them to run on top of any implementation of this (lower) layer, and the layer itself can be used in the implementation of various applications. This is true of most component-based design.

We will say that a system consists of *layers* when it is split into communicating subsystems that work at *different levels of abstraction*. In particular, such subsystems usually share no data types apart from the most primitive ones; one subsystem is seen as more abstract than the other, closer to the application, to the problem space, and the other closer to the implementation level, to the solution space. In the rest of this paper, we will refer to *application* and *implementation* layers, though layered systems often have many more — but two will suffice for the discussion.

Two adjacent layers see a given piece of data in different ways; typically, what the application layer sees as a well-structured data type, is a mere string of bytes to the implementation layer.

In the context of communication, a layer is more or less transparent to its clients: that is, communicating processes may ignore that the messages they exchange go through the implementation layer — just as the author of a letter may ignore the details of mail transport as long as he or she knows how to write the address.

The lower-layer applies the same transformation to all data structures from the upper-layer. The receiver, after applying the reverse transformation to a piece of data, does not know what type it instantiates and how to interpret it. Some wrapping will be necessary, such as additional bytes in a header or additional leading and/or ending signals, carrying information such as type and length.

We hope that the following examples will clarify this.

2.2   Presentation of the Layer Pattern

We have chosen the following communication examples to illustrate the concept of layers and various contexts in which it is currently used, where SDL would fail to capture the design elegantly. Our purpose is not to tell how these services are actually designed and implemented in actual systems.

The General Pattern. Let *foo* be a reversible transformation that applies to arbitrary data structures from the application layer and produces other data structures, as needed by the implementation layer. Let *defoo* be the reverse transformation. Data *foo*'ed on one end is *defoo*'ed at the other end.

The *fooing/defooing* algorithms could eventually depend on the bit representation of the data and not at all on its semantic, application-level structure. In other words, any piece of data could be *foo*'ed with the sole knowledge of the bitstream that represents it; this makes the *foo* service independent from any client application. For example,

```
class Scs {                    class Sb4 {
    char c;                        bool b[4];
    short s;                   };
};
```

To the *foo* service, both *Scs* and *Sb4* are only thirty-two bits (many of which are wasted due to the special attraction of computers for even numbers) that will be *foo*'ed just the same. However, the receiver of the foo'ed piece of data, after *defooing* it, only gets a decoded bitstream and does not know it should it be interpreted as *Scs* or as *Sb4*. To achieve this, the piece of data will be wrapped by some meta-information (such as type information, and length). For example:

```
class Wrapper {
    TYPE typeId; // sender/receiver specific
    short length; // split if > MAXSIZE
    char data[MAXSIZE];
    Foo(char[MAXSIZE]);
    char* DeFoo();
};
```

In this example, we see that before calling *Foo*, one must:

1. cast *Scs* or *Sb4* to the generic char[$MAXSIZE$] type,
2. explicitly set *typeId* with an arbitrary — but unique — constant standing for the type,
3. explicitly set *length*, using *sizeof* or something similar.

Conversely, after defooing the data, one must check its *typeId* and cast it accordingly to *Scs* or *Sb4* (and possibly set their size according to *length*).

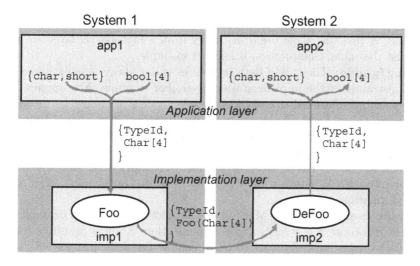

Fig. 1. *Foo*: a generic pattern for layered systems

Casting, *sizeof* and similar features like RTTI are quite characteristic of layered systems which, understanding the same piece of data at different abstraction levels, need to have access to some meta-information, that is information about the type itself: which is it? how large is it? etc. Such mechanisms are often characteristic of dirty programming too, but not always. In the particular case of layered systems, they are necessary to accomplish the very mission of the system. SDL lacks such mechanisms. We do not propose to introduce them in SDL, because they would allow "dirty tricks", and we would all like to preserve the relative cleanness of SDL. However, we will try to address the same areas in a different manner: the pattern, built in SDL, will allow a nice design for such systems.

2.3 Examples

The following examples follow the same pattern, and there is no need to go into similar details. The reader can imagine a combination of several of these layers, one running on top of another.

Routing: Probably the best example of layered systems. A layer (broker) is responsible for finding the recipient of a message and delivering it, independently from the nature of the message. Post mail is a particular instance of this pattern.

Portability, encoding/decoding: used to translate data structures into a format independent from the platform (programming language, compiler, operating system, hardware) to allow them to communicate and exchange data. In this example, note that there can be two different encoding and decoding functions at both ends. Following this pattern: RPCs [2], Corba [6], Microsoft's HAL (Hardware Abstraction Layer) [7].

Encrypting/decrypting: Data encrypted at one end is decrypted at the other end. A text file, a picture or a binary from a higher-level layer are treated just the same. Compression is another example.

Slicing/gluing: The size of data structures is not always compatible with the constraints of the implementation layer. For example, *Foo* applies only to structures of a fixed size of 512 bytes. Data must therefore be sliced before being *foo*'ed, and all chunks put back together after each is individually *defoo*'ed. Not only is each individual chunk wrapped, but the whole stream is preceded and followed by a leading and an ending message containing information such as an identity (necessary if chunks from various sources can be interleaved), number (to detect data loss, to sort chunks), data recovery information, etc.

Graphic user interfaces: Motif and X Windows offer a good example of layers in a context not related to communication. X offers primitive concepts, mechanisms and protocols such as window (actually just an active area), event, rectangle, resource, which Motif uses to build higher-level objects (shape and behavior) such as buttons, scrollbars, real windows... Thanks to the separation into layers, Motif can run on multiple implementations or emulations of X Windows, and many other window managers can run on top of X Windows. Finally, applications can be written using Motif elements.]

As one can see, layered design is quite common in distributed and communicating systems, which SDL addresses specifically[1]. We will now investigate and see what programming languages and SDL have to offer to support the pattern.

3   How SDL Is Weak and Lacking with Regard to Layers

3.1   Layered Systems in SDL

There are several ways to cope with layered systems in SDL. We will show the major ones and simplify them (to the level of a caricature) for the sake of clarity.

Explicit Switching

The model. We will now model in SDL the *Foo* example exposed earlier. *Block*1 receives *Scs* or *Sb*4 structures, embedded in signals *ScsSig* and *Sb*4*Sig*

---

[1] Another example pops to my mind that I did not dare mentioning but in a footnote: certain mechanisms of the human brain. Vision, for example. Frog eyes recognize certain aspects of reality very early in the information processing, such as shapes and movements. The brain receives pre-processed information such as "small, moving fast", "sharp and still" and interprets these as "insect", "fisherman" or "herb". The octopus's eye computes areas, that are later interpreted as "yellow submarine" or "Commandant Cousteau".

We believe that sophisticated information processing in general requires handling several abstraction levels at once and some access to meta-information (i.e. information about information) from the lower-layers.

respectively. These structures are carried to *Block2* after *Foo*-ing (for example encryption), wrapped in a *Wrapped* signal together with information needed for *DeFoo*-ing, a *TypeId* in our example.

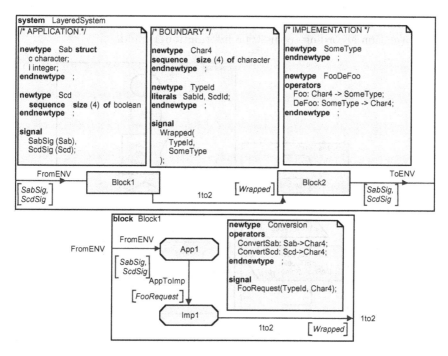

Block2 simply mirrors the structure of *Block1*, with two operators *RevScs* and *RevSb4* that respectively convert a *char4* back into the appropriate target type.

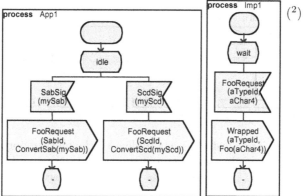

(2)

Down into *Block1*, *App1* must convert *Scs* or *Sb4* to *Char4* before transmitting them into the generic signal *FooRequest* for *Imp1* to process them.

---

[2] Two processes are used in the example to highlight layers, but of course, *Convertxxx* and *Foo* could be nested. the number of processes has no impact on the discussion.

Conversion operators *ConvertScs* and *ConvertSb4* allow a clean cast, but do not do any real transformation to the data. An actual implementation might simply use these SDL operators to embed (and hide) an actual C cast.

Imp2 is written in an application-independent fashion and simply *defoo*'s the data before forwarding it to *App2*. Once this is done, *App2* must read the information according to the type information TypeId.

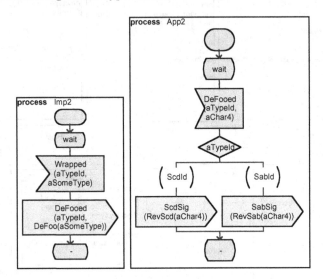

To sum up the global behavior in a simple example of the execution:

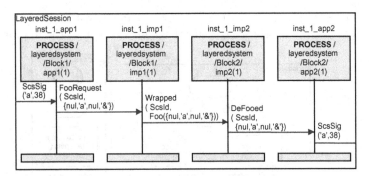

The problems. The most obvious problem — the most important, too — is that the design must evolve as the application changes. To have a new application use the *Foo* layer (that is *Imp1* and *Imp2*), one must:

- define *TypeId* accordingly, and associate a literal value to each type handled by *App*; this raises potential conflicts between several client applications, as each type must be identified uniquely;
- define appropriate converters in *Block1* and *Block2*, thus breaking modular design; indeed, these converters depend on the data structures defined in

client applications: ideally, no knowledge of client applications should be required in implementation layers;
- write an explicit switch to read the raw data, which is really bad, because the "decoder" *App2* may be shared by all client applications so that a change to any client application may have an impact on *App2*.

We may conclude that, in spite of a layered design, we could not achieve modularity and ensure maintainability and extendibility of our system, because any new application or any change to existing applications may require changes in our model.

Generic Signals. To avoid problems met in the first solution, one idea is to impose a fixed set of signals to all client applications and to hide data structures in the signal. This can be done by embedding C data structures in SDL signals, thus leaving it to C programmers to "cast and cheat", but we have chosen to use the ASN.1 Choice construct instead. In SDL–2000, we would have chosen the possibilities offered by the new object model. However, neither Choice nor inheritance can fully replace casting, as we will see.

The new solution uses only one signal *Sig* to carry all data structures (*Scs* and *Sb4* in our example), declaring one common "ancestor" *Scsb4* to both. Of course, adding more client applications would imply modifying this type explicitly; this is why we said that true casting "à la C" would be more powerful: a natural common ancestor for all types — including types not known yet — is available (say, *void*∗). If, for example, we wanted to add new applications dynamically, we would be forced to use C types embedded in our SDL system.

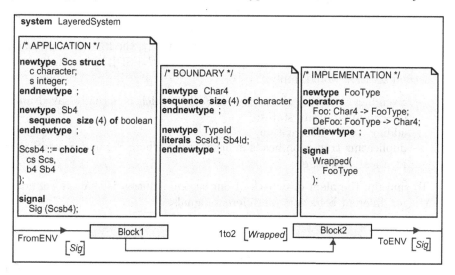

We can see the alternative as follows: either there is a switch at the control level (see *App2* in the previous solution), or the switch is hidden in the data, with type information.

Though this solution looks better, it is not perfect either. Apart from the fact that the ideally independent implementation layer cannot be modeled in plain C, we also had to avoid using explicit signals. Let us have a look at the recipient *App2*:

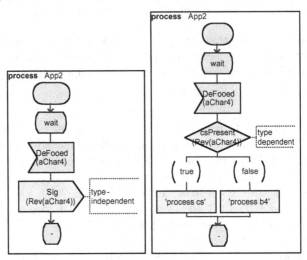

As one can see, one signal only triggers all actions, whatever data type is received. If different actions must be taken depending on the data type, it is possible to check it and to act accordingly (in the example, we use the **present** operator that supports ASN.1 Choice).

However, this makes it difficult to manage the input queue and to benefit from SDL instructions like **save** or **priority input**. For example, to accept certain data types as immediate input, to discard some other types, and to save one for future use in appropriate states. Comparatively, the first solution was much better in this respect. For example, simulating the SDL save construct in our second solution would imply:

- resending to **self** a *SavedSig(myScsb4)* which is actually saved and used later in appropriate states;
- adding sender information;
- duplicating transition heads in other states where *Sig* and *SavedSig* trigger identical actions, etc.

To sum up: the main drawback of our second solution is that we cannot easily trigger different behaviors on different signals.

Modeling One Layer Only. A common practice is simply to model only one layer at a time in SDL, considering that lower layers are implementation details of the communication means — implementing the SDL **input** and **output** statements. It is actually a correct assumption. However, one might want to model interacting layers and to check their compatibility as well as their correct cooperation.

It could indeed be a good idea for protocol environments to support the modeling of layers in independent models, and to allow their co-execution, using one (implementation) model as the definition of Input and Output in the other (application) model.

Sect. 5 will explore further related modeling issues.

Context Parameters. It may appear at first glance that context parameters solve the problem. Indeed, one may define a component using a generic signal S defined as a parameter, later instantiated with an actual signal.

A closer look shows it is not the case. Context parameters are instantiated statically. Our implementation layer should be able to handle any set of signals without requiring anything apart from conformance with the general structure of signals. The *same dynamic instance* will accept requests from *several different clients* using their *own set of signals*.

3.2   Common Mechanisms Used for Layered Systems in Programming Languages

Most often, generic components of all kinds (operating systems, drivers, and more) are implemented in C and (to a lower extent) in C++. Let us explore some of the most common mechanisms used to implement a layered design. We will also see what other languages have to offer.

Casting. Casting usually occurs twice in a layered system, when crossing the boundary between layers one way or the other. Down in the implementation layer, *Foo* functions see structured data at a lower abstraction level (such as as bytes). *DeFoo* will conversely return a mere unstructured string of bytes that clients must interpret as structured data. Casting is not an actual conversion: no transformation is applied to data at that stage; casting is only a trick used in some typed languages to bypass type control and allow the very same piece of data to be seen as an instance of two different types by different actors.

Casting can occur in different ways: explicitly by the use of cast or conversion operators, or implicitly by using untyped data structures, or by overlaying such as COMMON sections in Fortran or union in C.

SDL–96 offers no such mechanism.

Untyped or Weakly-Typed Data. In C, one would typically use *void*∗ to pass data to a lower layer. Inheritance in Object-Oriented languages (like SDL–2000) offers a new possibility: some languages introduce one or more data types that are an implicit common ancestor to all other data types, such as:
Any, Variant, Object.

At the opposite extreme are languages with no types at all.

Type Information. As shown earlier, one must associate some information about the original data type to the raw data, so that it can be interpreted correctly by the receiver. The most common information is an *Id* that identifies the type and some information about the actual length to read — if the data structure does not have a fixed size. A language like C++ has built-in mechanisms such as RTTI (dynamic_cast, type_info, etc.) More elaborate techniques can be found in languages such as Smalltalk, Lisp or Prolog. They allow metaprogramming: all information about the structure of data types is available to the program, and new data structures can be built dynamically, as needed. Unfortunately, such powerful features are reserved to interpreted languages, at the expense of efficiency.

3.3   Dangers of Usual Techniques

No type safety: in C, casting is usually error-prone, since it is possible to cast to the wrong type and to interpret data incorrectly. Consequences can be terrible.

Finding errors at runtime: operators like C++ *dynamic_cast type_info* and others allow safer casting. The remaining problem is that type correctness cannot be checked statically.

Not working at design-level: a more methodological issue is that we have been forced to write design-level SDL operators that merely wrap "dirty cast" operations written in C — or to renounce writing an implementation layer that is fully application-independent. Wrapping implementation details is fine, but in the case of layered systems, *writing a generic, signal-independent component is not an implementation detail: it is the essential mission of the system and the key aspect of the design.*

Our proposals aim at supporting the design of such systems in plain SDL, without having to use a lower-level programming language. We will also try to avoid the inconvenience of weak typing, at least for simple problems.

4   A Proposal for Generic Signal Handling

4.1   The Type Tsignal

Tsignal Variables. We propose a new predefined data type **Tsignal**[3]. Contrary to other variables, a **Tsignal** variable can be used in an **input** or in an **output**. The instruction **input** *myTsignal* behaves exactly like **input** *[4]but,

---

[3] Proposals for a better name are welcome.

[4] Actually, a better proposal is to allow the syntax **input** $x := a, b$, where a and b are declared signals and x a **Tsignal** variable (possibly declared implicitly for transition-scope); indeed, this narrows the **input** to a and b, while allowing to discard or save other signals. In particular, this seems necessary since there could be some specific management messages addressed by clients to the low layer itself (letters to the Post administration, phone calls the phone company...), that should be recognized from messages that are to be processed and forwarded.

The syntax exposed in this paper would be a short notation for **input** $x := *$.

in addition, gives access to *myTsignal* for future use. Among other things, it is possible to execute output *myTsignal*. Unlike a construct that would only give access to the signal that triggered the current transition, the signal is accessible at all times.

Let *myTsignal* be a variable of type **Tsignal**.

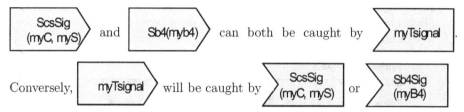

depending on its runtime content. It is possible to output a signal without actually knowing what signal it is or what values it carries.

A Simple Example: Permuting Messages. This is quite a silly and useless example in real systems, but we believe it illustrates ideally the kind of operations that cannot be specified in SDL in a way that is both simple and generic. Our "silly layer" must wait until it receives two messages and then forward them in the reverse order.

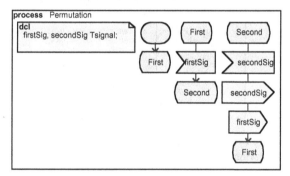

Because *firstSig* and *secondSig* are **Tsignal** variables, we are able to write the permutation service in a generic, signal-independent way. No "dirty" or dangerous code is needed there. No constraint is put on the application layer: no need to wrap data in one generic signal, no need to encode all signals in a different set of signals understandable to the implementation layer.

However, note that forwarding the message could alter its content: should the sender of the message be the initial sender of the message, or should it become the Permutation process? SDL would suggest that Permutation is the new sender. On the other hand, it would be useful in layered systems to have the possibility to forward a message in a transparent way, preserving sender information. When you receive a letter, you know the postman did not write it, don't you? You may even ignore totally how the letter traveled to your place.

We will not answer right now and will try to come up with something satisfactory later, in Sect. 5. For the time being, let us just propose that what looks like standard SDL behaves like it (that is, the sender of a forwarded message is altered) and introduce a new keyword from to explicitly set the sender to something different than **self**:

> **output** *aSig* **from** *aSig!sender*;

Redesigning Foo with Tsignals. The top level is simplified because no explicit machinery is necessary to translate data between levels. We do not have to exhibit the *char*4 implementation type, nor any converters back and forth.

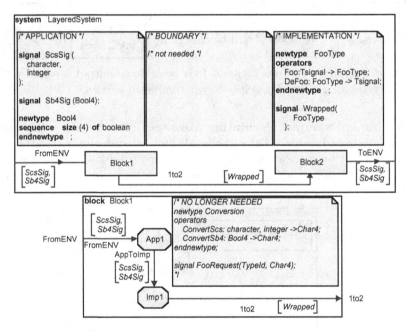

Process *App*1 has actually become useless, since all it does is now to forward all signals, unchanged, to the next layer (its sole purpose was to highlight the transformations that occurred at the boundaries between layers). Note that it is possible to write a new, simpler, generic version of this useless *App*1. Similar simplifications occur in *Block*2.

The most remarkable improvement is that $Imp1$ does not depend in any way on signals received from the upper layer. Since **Tsignal** is also a simple data type, it can be used as a parameter for an operator (Foo in the example), as a field of a more complex structure... or of another signal.

The same simplification occurs at the other end of the system: $DeFoo$ returns a **Tsignal** that can directly be output from $Imp2$ to $App2$.

Whereas $Imp2$ simply outputs whatever signal results from $DeFoo$, $App2$ can handle each signal as appropriate. The switch on the signal is implicit. We were neither forced to switch explicitly depending on some type information to restore the correct signal, nor were we forced to wrap all signals, thus loosing the expressive power of SDL and possibilities to use **save, priority input**... or implicit discard.

4.2   Benefits and Drawbacks

We hope that the benefits of the construct are obvious enough at that point. In short: we are able to describe components or protocols that are:

- more generic components or protocols;
- simpler, more readable;
- keeping the possibilities of SDL, discriminating in a given state which signals to input, discard, save or prioritize.

We have not fully explored the dark side and leave it to the reader. As all languages that allow the use of a weak type — as $void*$, **any** or **Tsignal** — it does not allow much static checking. In other words, there is no way to check statically that an output of a **Tsignal** targets an appropriate process... just like in SDL–96 in fact, before typed Pids were introduced in SDL–2000.

4.3   Metadata in Tsignals

The Need for Metadata. We have not completely addressed the issue of layered systems. Indeed, we can only cope, so far, with behaviors that do not do any transformation to the signal itself. It was easy to describe signal permutation. In the Foo example though, we have only pushed the difficulty one step further by hiding it into the operator (but simplifying the SDL design, and which is interesting in itslef) and the operator had to be written in plain C.

The Foo example uses two operators to transform any **Tsignal** to a $FooType$ and to reverse the operation. How can we write anything useful if do not know anything about **Tsignals**? For example, can we decide to split a large message into smaller one without information about the size? Can we encode data without knowing its type?

We will now explore two possible directions to solve such problems.

"True" Metadata. One example of what we call true metadata can be found in languages such as Lisp, Prolog or SmallTalk. In such languages, a data structure is itself an instance of another (meta) data structure and, as such, can be queried about its properties or even built at runtime.

Applied to **Tsignals**, the concept would allow us to ask how many fields the signal has, what is the type of each field and so on, recursively, until the level of basic types that can be worked on directly is reached. One could think of **Tsignals** as "interpreted" rather than "compiled".

This could for example be useful for the encoding of signals. In systems where efficiency is an issue, a drawback might be that interpreted traversal is time-consuming, and that embedding structure information in protocols is byte consuming. However, more elaborate solutions could be imagined. We do not wish to explore these possibilities further, only to show possible extensions of the concept. Besides, we will see that true, "clean" metadata might not be sufficient.

"Raw" Metadata. By contrast with true metadata, raw metadata does not give any access to the structure of data, only to some basic information such as the number of bytes needed or actually used, or the marker that ends the sequence. The idea is that almost nothing of the data structure itself is known to the implementation layer.

Though this is usually thought of as "dirty", we believe it cannot be avoided in some layered systems. For example, let us consider an application whose function is to cut messages into slices of five bytes at one end of a pipe, to wrap and send these chunks through the pipe and to rebuild messages at the other end. To do this, we need to know a few things about the generic arrangement of all signals in the upper layer.

Byte	0	1	2	3	4	5	6	7	8	9	10	11	12	...
	SId	sndr	target	Size	data									...
	Str	sndr	target	length	char1	char2	char3	char4	char5	char6	char7	char8	char9	...
	2flts	sndr	target	8	float1				float2					

Chunk 1	length	app sender	app target	imp target	freight 1-5				
					Chunk 2	length (e.g.3)	initial sender	initial target	target	freight 6-10

In this example, the first four bytes (0-3) contain information necessary to the implementation layer, the following bytes contain the actual data to be split. As illustrated, the separation between two chunks might well fall between two bytes of an "atomic" type like float. Only the "raw" approach can address this.

Telecom systems are full of such applications and most of the examples we used in Sect. 2.3 fall into this category. Therefore, in spite of the potential dangers, we propose to provide such mechanisms in SDL. A possible structure for **Tsignals** that would allow their instances to be handled, accessed, modified or built as needed:

```
newtype Tsignal
    struct
    id SId;
    sender Pid;/* SDL sender */
```

```
    target Pid;/* if 'to Pid' used, NULL otherwise */
    data bitstring; /* string includes length info */
endnewtype;
```

4.4   Data and Signals

It is quite obvious from what precedes that signals are just ordinary data structures that, in addition, can be sent around and implicitly wrapped with routing
information — the Pid of the sender and the designation of some target. Could
we not declare that signals and data types are the same thing?[5] Could we not
allow to output any data type?

We believe this is something to work on, and have left that out to focus on
our main topic. However, a solution that would consistently integrate data and
signal structures would be more elegant (and probably more powerful and even
simpler) than our partial proposal.

5   The (Somewhat Unexpected) Comeback of Channel Substructures

In the previous sections, we have added to SDL some concepts that make it
possible to describe several interacting layers of a system. However, a drawback
of our approach so far is that different views of the same data structure are
presented together, mixed. We would like, on the contrary, a clean and clear
separation between layers. In addition, we would like an implementation layer to
be *transparent* to application layers. Ideally, an application could almost ignore
its dependence on the implementation layer for certain things — just as you may
ignore how letters travel when you post one.

5.1   Reminder on SDL-88 Channel Substructures

A Brief Description. A Channel Substructure, as the name suggests, refines
a channel with a lower-level structure.

Inside the substructure, we may have an arbitrarily deep decomposition describing for example the actual implementation of the communication medium.

One could expect, from the figure below, that the substructure is transparent,
i.e. removing it from the model does not alter the behavior of the system.

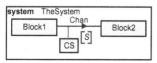

The channel *Chan* is refined by a substructure CS...

---

[5] We do not own this idea. We heard, at least, Anders Ek from Telelogic discussing
this possibility. We believe we heard a few words about it at ITU meetings too.

... which is itself further refined. Pseudo external channels indicate which block, connected to *Chan*, is actually connected to the local channel. In this example, signals sent by *Block*1 are actually sent to *Imp*1, whereas signals received by *Block*2 actually come from *Imp*2.

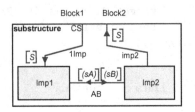

As one can see, we have a nice separation between the application layer, in *TheSystem*, and the implementation layer of the communication medium *CS*.

Limitations of Channel Substructures. Unfortunately, Channel Substructures as defined in SDL-88 failed to add real support for design of layered systems.

Altering Pids. The substructure *CS* described previously is not very different in essence from the one described here.

The major (the only?) benefits of Channel Substructures, are to hide the lower-layer, to highlight it is a refinement of the channel *Chan*, to show than *Chan*1 and *Chan*2 are halves of the same channel Chan and to constrain them to have consistent signal lists.

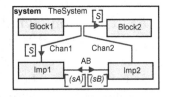

However, in SDL–96, *CS* is not transparent: for example, if *Block*1 uses the construct output to *Pid*, then the *Pid* must identify a process instance within *Imp*1, not *Block*2; and, when *Block*2 queries the sender of a signal, it obtains the *Pid* of a process instance within *Imp*2 rather than *Block*1. This is contrary to intuition: the address you write on an envelope is not that of the post office; and at the back of the envelope, what you read is not the address of the postman[6].

Other limitations. There are other limitations that we will not explore now in depth. Among others:

– *Only point-to-point communication is supported*: since Channel Substructures are attached to a channel, they are not suitable for the representation of architectures in which several logical one-to-one links share a common implementation.
– *No type-level definitions*: the concept was introduced in SDL–88 and never updated since then. Neither SDL–92, nor SDL–96 defined anything like a Channel Substructure Type. For SDL–2000, the concept was dropped together with the complementary concept of signal refinement.

---

[6] ...though the postman may act as a client of mailing services. The same happens in layered systems, since a layer and its clients may exchange information (e.g. to configure future communications or to check status before or after actual use). In Telecommunications Network Management (TMN), the application that manages the network requires a network for its own implementation.

5.2   Transparency

We suggest a form of Channel Substructures that are transparent, so that it is possible to remove them without any impact on the semantics of the model — unless they are explicitly addressed in some way. In other words, a Channel Substructure defines some aspect of the implementation of communication, but does not affect the semantics of **input** and **output**.

Therefore, communication through a Channel Substructure should not affect the *Pid*. In the previous example, it should be possible to address a process of *Block*2 from *Block*1 by its *Pid*. Conversely, the **sender** of such a signal should identify a process of *Block*1.

Within the substructure, it is possible to access signal information thanks to the mechanisms described in 4.3, *"Raw" metadata*.

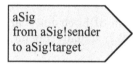

This example illustrates how to write a simple forward. The **to** clause uses $Tsignal!target$ to find where the signal was initially sent. The **from** clause is used to preserve the initial **sender** of the message.

5.3   Miscellaneous

The proposals we make here are not mature (even less than proposals we made until now) and just show some directions that we should explore before the whole idea can be turned into a usable proposal ready for evaluation by ITU for integration into SDL.

Separation Between Application and Implementation Layers. We think that it would be a good thing if models did not mix two abstraction levels in the same agent or in siblings: that is, if the same signal could not be seen as two different things in the same place. Channel Substructures may give us the opportunity to solve the problem elegantly.

This modeling rule could be enforced by formal rules. For example, we could consider allowing **Tsignal**s only in very specific situations:

1. in Channel Substructures on the pseudo-channels between sub-blocks and the environment;
2. in sub-agents of a Channel Substructure, and only to handle messages from or to the upper-channel.
3. in variables or fields assigned by a signal complying with the aforementioned situations. For example, it should be possible to forward a message from the upper layer, or to wrap it as a field of a local signal.

Such rules can probably not be checked statically, but tools could detect at simulation time if the rule is violated, e.g. if an explicit output is received as a **Tsignal** in the same layer.

The spirit of these rules would be to ensure that **Tsignal**s are used for the sole purpose of clean layering, and not for tricky programming.

Communication Between Application and Implementation Layers. If needed, how can the application and implementation layers communicate?

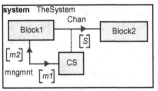

A first idea would be to allow ordinary channels between blocks and substructures, like **mngmnt** in the example.

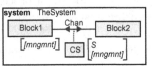

Another proposal is to have *two* signallists on each channel direction: one to the opposite end, and an optional one to the substructure.

Substructures Shared by Several Channels. In SDL, channels are logical communication paths. They do not imply that there is actually a point-to-point connection only. For example,

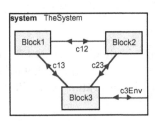

may very well run on the following infrastructure:

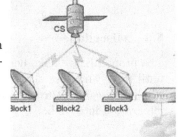

To cope with such situations, we would need several channels to share a common Channel Substructure, as in the following diagram. Since the same medium could be used in different blocks, not belonging to the same diagram, an alternate textual syntax would be helpful (e.g. c12[cs])

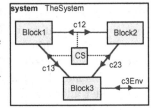

Channel Substructure Types, Channel Types. Finally, Channel Substructures should be made homogeneous with other kinds of agents as they were introduced in SDL–2000. For example, we would expect that several Channel Substructures with similar content can be declared as instances of a type, possibly parameterized[7].

---

[7] This must not be confused with the proposal made in Substructures shared by several channels: we talk here of sharing *definitions* (e.g. a communication protocol implementation) between several instances, whereas we previously discussed sharing one *instance* (e.g. one communication resource) between several channels.

6   Conclusion

In this paper we have tried to highlight that although layering is a good design practice and widely used (specifically in telecommmunication architectures), SDL does not assist designers in describing layered systems.

We have made a few proposals that in our opinion could be of some help. The essential idea is that SDL should make it possible to view signals at two different abstraction levels, so that designers are not forced to cheat in some way (such as embedding signals, programming some parts in plain C. . .) thus loosing some major benefits of SDL and, sometimes, some of the power of SDL tools too.

Besides, SDL is often used as a programming (though abstract) language. Maybe as an "implementable design language"? At least, it should be powerful enough compared to basic, general-purpose programming languages. Ideally, designing a layered system should be *simpler* in SDL than it is in C, not the contrary. We believe it is possible and desirable to give SDL such power while maintaining its integrity and "design spirit".

References

1. Maurice J. Bach, *Design of the Unix Operating System*, Prentice Hall (even the cover shows layers).
2. John Bloomer, *Power Programming With RPC*, O'Reilly& Associates, 1992.
3. J. D. Day and H. Zimmermann, *The OSI Reference Model*, Proc. of IEEE, vol. 71, pp. 1334-1340, dec. 1983.
4. M. Ferguson, David Brennan, Motif Reference Manual (The Definitive Guides to the X Window System, Vol. 6B), O'Reilly & Associates, Inc..
5. Erich Gamma, Richard Helm, Ralph Johnson, and John Vlissides, *Design Patterns*, Addison-Wesley.
6. Object Management Group, *The Common Object Request Broker: Architecture and Specification*, v. 2.4, October 2000.
7. J. Richter, *Advanced Windows*, 3rd edition,MS Press, 1997.
8. Andrew S. Tanenbaum, *Computer networks*, 3rd ed.,Prentice Hall, New Jersey, 1996.
9. Douglas A. Young, *X Window System, The Programming and Applications with Xt*,OSF Motif Edition.

Collaboration-Based Design of SDL Systems*

Frank Roessler[1], Birgit Geppert[1], and Reinhard Gotzhein[2]

[1] Avaya Inc., Software Technology Research
600-700 Mountain Ave., P.O. Box 636, Murray Hill, NJ 07974-0636, USA
{roessler,bgeppert}@avaya.com
[2] Computer Networks Group, Computer Science Department
University of Kaiserslautern
P.O. Box 3049, D-67653 Kaiserslautern, Germany
gotzhein@informatik.uni-kl.de

Abstract. The concept of *collaborations* capturing dynamic aspects of a distributed system across agent boundaries is elaborated in the context of SDL-2000. Several ways of composing collaborations are introduced, with collaborations being implicitly represented as SDL fragments. A new language for their *explicit* formal description, called CoSDL (**Co**llaborations in **SDL** systems) is then introduced and illustrated.

1 Introduction

The flexibility of protocol architectures is becoming more and more important as the pressure to deliver converged communication services to every corner of the globe grows. Due to ever growing demand of customers for new features, organisations are seeking for ways to reduce the time it takes to evolve their communications platforms and therefore reduce time to get new features to market. In this paper, we describe a new concept for structuring communicating state machines, which can help overcome these problems. Though the general approach works for the whole spectrum of automata based description techniques, we have yet focused on elaborating the details for the description and specification language SDL-2000 [17].

The research work underlying this paper originally aimed at formalising the SDL pattern approach [4,5,6] and providing adequate tool support. SDL patterns were introduced as reusable SDL artefacts accompanied with an incremental, scenario-driven design process. Though the relevance and usefulness of SDL patterns was demonstrated in many case studies, there was still an essential element missing: in order to be a true *construction set of protocol building blocks* (as envisioned in [5]), the approach needed a proper compositional framework for its reuse artefacts. When tackling the research problems regarding SDL pattern composition, soon it turned out that much broader architectural questions of an SDL system were involved.

---

* The work reported here was conducted while all authors were working at the University of Kaiserslautern, Germany.

R. Reed and J. Reed (Eds.): SDL 2001, LNCS 2078, pp. 72–89, 2001.

Many of the SDL patterns defined in [5] address interaction behaviour between two SDL agents. At first, we tried to treat such SDL patterns as components. This approach was not feasible, because SDL does not offer adequate language constructs for defining the subtle interface between an SDL pattern and its context specification. However, one can assemble individual SDL patterns to larger aggregates that have a much simpler interface. We call these aggregates *collaborations* and demonstrate how collaborations can be composed to complete SDL systems. Collaborations enable very flexible protocol architectures and are an important design principle for the development of SDL systems.

There are also two other viewpoints motivating the concept of collaborations independently from SDL patterns and showing that collaboration-based design is in fact a new approach to SDL system development that is independent of SDL patterns:

In general, interactions between communicating agents can always be described in two complementary ways: one of them centred on individual agents and the other focusing on a set of cooperating agents. Agents in SDL (as in other automata based description techniques such as Estelle [8], StateCharts [7], or ROOM [15]) are behavioural views that are deep but local. An SDL agent is precisely specified and immediately leads to executable code. However, it can be quite difficult to understand the overall functioning of an SDL system, because the behaviour of many agents must be analysed and combined for determining the entire system behaviour. To overcome this problem, a more holistic view on the behaviour of a collection of agents should be provided, and as shown in this paper, the concept of collaborations plays a key role here. Note that UML also supports this general notion of collaborations [2] and suggests class-, sequence-, and collaboration diagrams for their description. However, this paper develops the concept much further (in the context of SDL), as a much tighter integration of scenario and automata modelling is supported.

An important and complicated phase in distributed system development lies in the transition from system behaviour to the behaviour of interacting components. While the former is generally specified using scenario modelling techniques such as use case maps (UCM) [3], use cases [9], or message sequence charts (MSC) [16], the latter is typically captured by description techniques based on communicating extended finite state machines. It is commonly accepted that a systematic approach is required for this transition. One observation about these approaches is that scenario models are deliberately incomplete. They typically capture main system behaviour and important exceptional cases. While this adequately reflects and supports the process of understanding a distributed system from the perspective of a developer and maintainer, there is no reason other than complexity for not capturing all possible scenarios. In contrast to the scenario model, the automata model must eventually capture all possible scenarios. The problem is that these are not explicitly represented and that automata do not provide a natural way for designing scenario structures. We have explained this in the last paragraph and suggest that collaborations is a good solution for this.

The approach will then allow a smooth transition from example system traces to collaborations and finally to complete SDL systems.

In Sect. 2, the concept of collaborations is motivated and developed, and several ways of composing collaborations are introduced, where collaborations are *implicitly* represented as SDL fragments. Sect. 3 introduces a new language for the *explicit* description of collaborations, called CoSDL (**Co**llaborations in **SDL** systems). Conclusions are drawn in Sect. 4.

2   Collaboration-Based Design

This Sect. explains the basic principles of collaboration-based design. We will introduce the notion of collaboration and outline the collaboration-based design process. In this Sect., collaborations will be *implicitly* represented by SDL fragments.

2.1   The Concept of Collaborations

Collaboration-based design integrates smoothly with modern principles of software architecture. Thus, we will first clarify our notion of software architecture and then show how collaborations fit into this framework and what they contribute.

Architectural Structures. We share the common view that there is no single architecture of a software system but a set of structures that together describe a system's architecture [1,10,14]. Each software structure consists of components, their externally visible properties, and relationships among them. The semantics of components and relationships differ for each structure defined. Some of the most common software structures are:

- *Module structure*: large projects are typically partitioned into components that serve as work assignments to development teams. Components usually comprise programs and data that define a publicly accessible interface and a hidden, non-accessible part. Components may be subdivided for assignment to sub teams, which defines a submodule-of relationship between components.
- *Conceptual structure*: for mature domains, reference models are defined, such as the ISO/OSI model for the communication systems domain. These models decompose a problem domain into functionalities together with connecting data flows, so that the pieces cooperatively solve the problem. Reference models are typical examples of conceptual structures, which have abstractions of functional system requirements as components and data flows as relationships.
- *Dynamic structure*: this structure deals with communication, synchronisation, and concurrency aspects of software systems. The components are processes or threads, while relationships represent synchronises-with or communicates-with links. This is the kind of software structure handled by description techniques such as SDL.

– *Physical structure*: for distributed systems the mapping from the design model to physical components becomes an issue. Components are hardware resources that are capable of executing the system. Communication pathways define the relationships.

Multiple structures clearly help to separate concerns, as they address different sets of system quality attributes. The module structure, for instance, is engineered to produce changeable systems [11], it shapes project structure and plan. The conceptual structure assists in understanding the problem domain. The dynamic structure is important for scheduling and performance issues. The physical structure is helpful for analysing system availability or security. There are other useful structures not listed above. The *uses* structure [12], for instance, is engineered such that a system can be easily replaced or extended in the sense of an incremental build approach. In essence, each structure tends to be a system abstraction with respect to different quality attributes. As such, it can have its own notation, components & relationships, and validation methods.

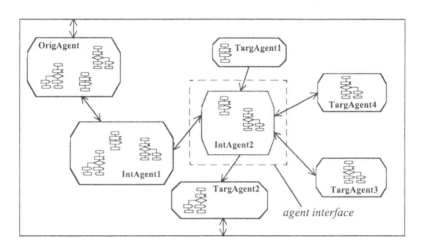

Fig. 1. Agent View

Collaborations as a Complementary View on Dynamic Structures. In this paper, we elaborate on the dynamic structure of a software architecture as introduced in the previous Sect.. As we have seen, each architectural structure implies a semantically different set of components and relationships. Collaborations do not define new kinds of components or relationships and can therefore not be considered a new architectural structure. In fact, collaborations describe the dynamic structure from a different perspective than is usual in the literature. While the traditional view on the dynamic structure emphasises its components, collaborations emphasise the relationships. We call the component-centric view the *agent view* of the dynamic structure as opposed to the *collaboration view*. We consider both views of equal importance, but focusing on different aspects.

As it turns out, there is a clearly defined mapping between both views. Note that we cannot expect the same between the different structures of a software architecture.

Before we explain what collaboration really is, let us briefly review the main characteristics and modelling guidelines of the traditional agent view. The dominating semantic model - as in the case of SDL - is a set of communicating extended finite state machines. Fig. 1 represents an SDL system with the graphics inside the process symbols representing a couple of state transitions. The agents communicate with each other and the environment via signal exchange. Each agent essentially models four things: the events to which it can respond, the responses to these events, the impact of the state machine's history on current behaviour, and allowable sequences of events and responses.

Usually, SDL agents are developed around their interface (which is the set of possible events and responses) and lifetime. Starting from an initial state, the designer lays out intermediate states until he reaches the final state. These states are connected by transitions, which are triggered by appropriate events and lead to corresponding responses. An SDL agent is usually developed in the context of its neighbouring agents. The designer makes sure that events from the environment can be adequately processed and that responses to the environment will be consumed. Checking an agent's behaviour against expected event/response scenarios or searching for unreachable states and deadlocks are common validation activities.

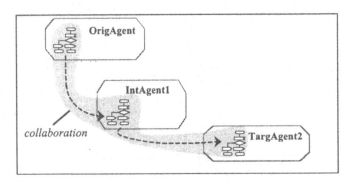

Fig. 2. Single collaboration

SDL agents are well suited for specifying distributed and reactive systems, favouring the traditional agent view as illustrated in Fig. 1. As mentioned earlier, this results in deep but local behavioural descriptions, which make it hard to get an overall understanding of the system functionality.

To overcome this insufficiency, a more holistic view on the behaviour of a collection of SDL agents must be provided in addition. Fig. 1 shows that *OrigAgent* has two interfaces: one to *IntAgent1* and another to its environment. In reality, *OrigAgent*, however, also collaborates with all other agents. One of

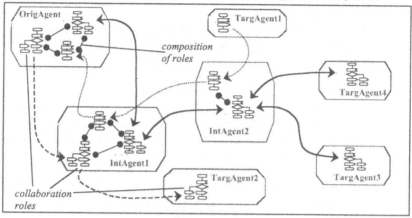

Fig. 3. Collaboration view: composition of roles

these collaborations is illustrated in Fig. 2, where *OrigAgent*, *IntAgent1*, and *TargAgent2* work together to realise a specific system functionality. The purpose of collaboration-based design is to equally focus the development process on collaborating agents in addition to the agent view.

As a matter of fact, the same system agent often participates in several collaborations. Thus, the state machine of an agent needs to be split off between these different collaborations, which leads us to the concept of *collaboration roles*. In Fig. 3, other collaborations between the agents are shown, which means that *OrigAgent* and *IntAgent1* now participate in two additional collaborations. As a consequence, the agents comprise additional transitions that represent the new roles and are composed with the already existing ones. Fig. 3 reflects the collaboration view of Fig. 1. In total, we have three collaborations, which are scattered across the seven agents. There are agents participating in all, two, or only one collaboration, with the corresponding state machines decomposed into roles accordingly. Note that the collaboration view still preserves agent boundaries and that state machine modelling guidelines still apply to each role of an agent. However, they will be applied from a holistic viewpoint, and a state machine's lifetime will also be mapped onto roles that characterise different phases of an agent's behaviour.

2.2   Collaboration-Based Design Process

The previous Sect. made it apparent that an SDL system has an inherent structure, namely the collaboration view, which is not adequately reflected - neither by means of the language itself nor in the development process. Nevertheless, SDL agents could be structured accordingly, so that they consist of mainly two parts: code for implementing collaboration roles and glue code for realising compositional relationships. As we have to consider that SDL only offers limited language support for collaboration implementation, the final SDL design, however, will not make this distinction explicit. We therefore propose a collaboration description language (called CoSDL, see Sect. 3.2) for explicit specification of single collaborations, and a systematic transformation procedure of collaboration diagrams into SDL.

It is relatively straightforward to specify an SDL design for realising an individual collaboration, because then, each SDL agent accommodates only one role. However, managing several roles within an SDL agent needs clear understanding of the collaborations' compositional relationships. Sect. 2.3 defines different types of compositional operators and illustrates mechanisms for implementation. However, the question remains what the criteria are for applying compositional operators in different situations. Technically, two collaborations can be composed in several ways. It is their interrelation in terms of shared data, which ultimately determines the correct compositional relationship.

The observations above suggest a basic, three-step process for collaboration-based design of SDL systems:

Step 1: Collaboration design. First, a set of individual collaborations is designed. Starting from system requirements, we extract those functionalities that will be realised in a distributed way and describe them as collaborations. It is necessary to identify participating agents and the message sequences between them, which will implement the intended functionality. We abstract from internal agent logic and focus only on the interactions that make them work together.

Step 2: Collaboration analysis. Second, collaborations are analysed according to their data dependencies. Though collaborations are designed individually at first, they may interact at run-time. One collaboration may establish some agent state that is used by another collaboration later. For a correct functioning of the whole system, data dependencies imply certain causal relationships between collaborations. We need to make sure that collaborations are composed accordingly, so that those causal relationships are enforced. Collaboration analysis must also decide on a set of *system agents* that can accommodate the collaboration roles defined in Step 1. Each collaboration role must be assigned to one of those system agents.

Step 3: Collaboration implementation. Finally, the collaboration model is transformed into an SDL system. In [13], we have described guidelines and mechanisms for implementing CoSDL roles and compositional operators in SDL. By means of these mapping rules, single collaboration roles will first be translated into SDL fragments and then composed according to the analysis results of Step 2. Several forms of composition are presented in the following Section.

2.3   Composition of Collaborations

As we have seen in the previous Sections, collaborations consist of a set of roles each located in a different SDL agent. Though composition of collaborations is finally expressed as a composition of roles in the SDL agents, the whole collaboration must be taken into account. We have defined three composition categories that each split into individual compositional operators. Each operator takes two collaborations and puts them in a causal relationship within the SDL agent structure. The operators also apply to composites of collaborations.

The composition categories and operators reflect the different sorts of data sharing and control dependency that are needed between collaborations. Though collaborations can be developed independently from each other, they definitely do not act in isolation. Each collaboration establishes some state within its agents that may be used or manipulated by others. Thus, access to common data often needs to be synchronised, while competing collaborations are in progress. It may be sufficient to use mutual exclusion on the basis of individual agents, or it may be necessary to globally lock the state of all involved agents at once. Depending on the situation, different compositional operators allow different degrees of data integrity.

Regarding flow of control, the most important aspect is to make sure that roles are executable whenever this is required by an ongoing collaboration. We have to consider that several roles belonging to different collaborations are accommodated by the same agent. Proper composition mechanisms must therefore orchestrate collaboration roles so that both data integrity and role executability are ensured.

Sequential Composition. In the following, we consider two collaborations C_1 and C_2. If the execution of C_2 is preceded by C_1, the compositional category is called *sequential*. We distinguish between *globally* and *locally sequential* composition. Globally sequential composition means that C_1 needs to be completely finished before C_2 even starts. Locally sequential composition enforces this condition only on a per agent basis: within each agent instance of a collaboration, the role belonging to C_1 is completed before the role belonging to C_2 starts. For sequential composition, it is a necessary precondition that C_1 eventually stops. In the global case, we additionally need a agent that can decide the completion of C_1. This agent will then trigger the beginning of C_2.

We illustrate the compositional operators by an example shown in Fig. 4. We assume to have two collaborations: one for connection establishment and

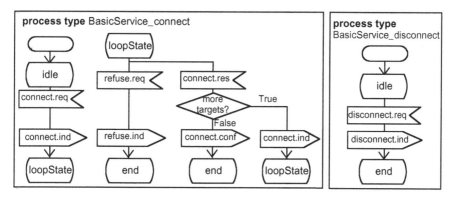

Fig. 4. Example collaborations

the other for connection tear down. To keep things simple, each collaboration involves only three agent types. A calling application that is capable of initiating a connection, a called application that accepts connections, and the basic service that forwards connection requests from the caller to a callee. Furthermore, we assume that each collaboration comprises one instance of the calling application, one instance of the basic service, but possibly many instances of the called application. That is, we allow multicast connections to be established.

Fig. 4 shows a partial implementation of each of the two collaborations in SDL. The process type **BasicService_connect** implements the role of the basic service for connection establishment, while the process type **BasicService_disconnect** implements the role of the basic service for the connection tear down collaboration. An SDL implementation of the caller/callee would send and receive connection messages to and from the basic service accordingly. Note that at this stage both roles of the basic service are implemented by separate agents. For each compositional operator, we will show how to compose both roles into one agent. For role composition, only the major states of the agents will be used.

In Fig. 5, the two roles are composed sequentially. An adequate implementation mechanism for sequential composition is to have disjoint state sets for both roles except for designated "transfer" states that explicitly transfer control from one role to the other. In the example, the state *connected* has this responsibility. We want to make sure that the preceding role has completed before it transfers control, and that the succeeding role keeps control once it has obtained it. The mechanism exemplified in Fig. 5 must be applied uniformly to all agents accommodating roles of the connection establishment or tear down collaboration. That is, the SDL implementations of the caller/callee roles (which are not shown) must be treated analogous to **BasicService_connect** and **BasicService_disconnect**, in order to achieve a sequential composition of the entire collaborations.

So far, implementation of locally and globally sequential composition can proceed equally. There is, however, a basic difference, because for global composition, an agent that can decide on the global completion of the preceding collaboration needs to trigger the succeeding collaboration when this has hap-

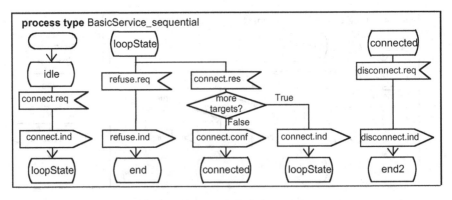

Fig. 5. Sequential Composition

pened. The succeeding collaboration can either be triggered remotely by a newly introduced trigger message, or locally by a particular state transition. Note that the agent receiving the trigger needs also to be in the state where the succeeding collaboration is finally started.

Exclusive Composition. Sequential composition ensures mutual exclusion between two collaborations and furthermore imposes a particular order for executing the collaborations. If execution order is irrelevant, it is possible to merely guarantee mutual exclusion by so-called *exclusive composition*. Exclusive composition again comes in two versions: *global* and *local*. Globally exclusive composition of two collaborations C_1 and C_2 guarantees for entire collaborations that either C_1 or C_2 is in progress, while locally exclusive composition limits this to individual agents. That is, for each agent, the agent's portion of either C_1 or C_2 is in progress at the same time. Of course, both compositional operators require the collaborations to eventually stop in order to avoid starving each other.

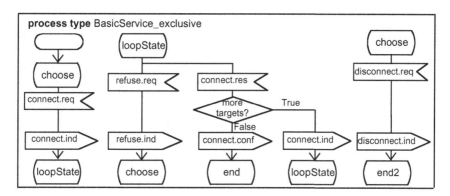

Fig. 6. Exclusive composition

Figure 6 shows how our example collaborations are exclusively composed. The applied implementation mechanism for exclusive composition is to have disjoint state sets for both roles except for certain transfer states that explicitly transfer control from one role to the other. In the example, the state choose has this responsibility. We need to make sure that a role has completed before it transfers control. The mechanism exemplified in Fig. 6 must be applied uniformly to all agents accommodating roles of the connection establishment or tear down collaboration, in order to achieve an exclusive composition of the entire collaborations.

So far, implementation of locally and globally exclusive composition can proceed equally. There is, however, a difference, because in the global case an agent that determines global completion of one collaboration needs to trigger the other one and vice versa. The other collaboration can either be triggered remotely by a newly introduced trigger message or locally by a particular state transition.

Note that the agent receiving the trigger needs also to be in the state where the next collaboration is started.

Parallel Composition. Sequential and exclusive composition offer either no or a rather low level of concurrency. If joint access to common data can be handled on the SDL transition level or if there is no common data at all, parallel composition provides better performance. There are two subtypes of parallel composition: *ordinary* and *dynamic*. Ordinary parallel composition takes two collaborations and makes them execute concurrently: that is, both collaborations can proceed at the same time and independently from each other. It is implemented by having disjoint state sets for each role and constructing the product automaton as illustrated in Fig. 7. As mentioned above, the illustrations of this section have only shown the composition of those roles that are played by the basic service when establishing or closing a connection. In order to have a complete composition of the entire collaborations, the roles of the caller/callee must be composed also.

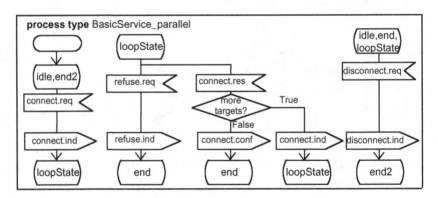

Fig. 7. Ordinary parallel composition

Dynamic parallel composition represents some kind of self-composition of a collaboration. In certain situations it may be necessary to have the same collaboration running multiple times. For this purpose, each involved agent has to manage not only one role, but an active set of roles. This is done by a role administrator, which dynamically creates a new role and adds it to the set, whenever a new collaboration instance is needed. In SDL, we can implement this concept via dynamic creation of SDL agents, so that each agent represents only the role in question. An administrator agent is responsible for agent creation and for forwarding incoming messages to the right agent instance. Note, that this is actually an application of the "DynamicEntitySet" pattern that was first introduced in [4].

3   Explicit Specification of Collaborations

We recall from Sect. 2 that SDL was designed to support the traditional agent view of a software architecture, where collaborations are described implicitly. While Sect. 2 has motivated and introduced the concepts, it is not clear yet how a collaboration can be described explicitly. However, we seek a collaboration notation that is as close as possible to SDL, so that we do not introduce any semantic gaps or prevent the definition of a mapping between a collaboration description and its representation in SDL.

3.1   Describing Collaborations in MSC

Given the fact that collaborations consist of roles which exchange messages, we certainly need language constructs for describing send and receive events. These events need to be assigned to roles, and corresponding send and receive events need to be linked somehow. Furthermore, control structures are necessary for building simple event sequences, as well as alternative, iterated, and parallel event sequences. These are all elements that can be expressed with basic MSC and inline expressions. However, readability of MSC diagrams decreases as the number of inline expressions increases. This is mainly due to the geometrical restrictions that "swim lanes" impose on an MSC diagram. By swim lanes, we refer to the convention that all events of the same instance are attached to a vertically arranged instance axis. Swim lanes are a convenient concept, but in the context of collaborations - which make intense use of "inline expressions" - the geometrical constraints of swim lanes seem to outweigh their benefits. One may argue that High-level MSC (HMSC) could be used instead, but then we would disconnect control structure from message events: we would not be able to show events and control structures in the same diagram, which we think is essential for collaborations. To some extent, a collaboration description can be compared to a sequential algorithm, where one would not describe if_then_else- or while-constructs on a different level than assignments.

With send-/receive events and the control structures mentioned above, we are only able to describe a fixed set of roles. This does not adequately reflect the nature of collaborations, which often are more generic. Assume, for instance, a web browser requesting a new page. The collaboration that describes interaction behaviour between the client, server, and possible proxies should be considered the same, irrespective of the number of proxies involved. Within the collaboration, the role of each proxy is identical, but the number of roles for a concrete deployment can vary. We want to be able to express this by a collaboration description. Thus, a possible collaboration description language needs some kind of type concept and a way to describe allowable configurations of roles for collaboration instantiation. These requirements are not met by MSC, because it focuses on interaction behaviour of instances only.

We think that the requirements above justify the design of a new language instead of extending MSC, before the concepts of collaboration-based design become more widespread in use.

3.2   CoSDL

We have designed a graphical language called *CoSDL* (**Co**llaborations in **SDL** systems) for explicit description of single collaborations. CoSDL has a formal syntactical and semantic basis and serves as a proof of concept for the idea of collaboration-based design.

We informally introduce the language by an example specification shown in Fig. 8. Actually, there are two collaborations illustrated, one for connection establishment and one for connection tear down. This example corresponds to the SDL specification shown in Fig. 4. However, in Fig. 4 we did only show the roles that are played by the basic service. In Fig. 8, both collaborations are completely specified. In the following explanation, we elaborate on the left diagram. The meaning of the collaboration on the right follows accordingly.

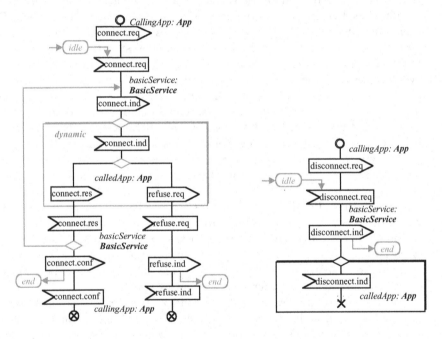

Fig. 8. CoSDL diagrams

The specification shows a set of causally related messages (written as plain text, e.g., connect.res) together with the corresponding sending and receiving roles (written in italics, e.g., *calledApp*). If we ignore the parts of the specification that are coloured in grey, the following collaboration is described: if the calling application (*callingApp*) invokes a connect.req, the basic service (*basicService*) will issue a connect.ind for the called application (*calledApp*). The decision symbol ◇ states that reception of a connect.ind at the called application either causes a connect.res or a refuse.req signal sent back to the basic service. This makes the basic service report a connect.conf or refuse.ind, respectively, back to the calling application.

We expect a system implementing this collaboration to execute one of the two alternative message sequences, when triggered by a connect.req and synchronisation conditions with respect to other collaborations hold.

It turns out that the specification has a flow of control from top to bottom. The straight line prescribes this sequencing order for the operations attached to it. Whenever necessary, arrows can additionally be used to make the order of operations clear. In Fig. 8, four kinds of operations are used. First of all, we have the send ⊐⟩ and receive ⟩⊐ operation. The corresponding symbols always occur in pairs with the same signal name inside, which represents a message transmission from the corresponding sending to the corresponding receiving role. For instance, the connect.req message in Fig. 8 is sent from the calling application to the basic service.

Furthermore, there are symbols start ○ and stop ⊗. The start operation can be performed at any time without restriction and is followed by the operations specified by the collaboration. In case of Fig. 8, it should be expressed that the collaboration starts to execute, if and only if the calling application sends a connect.req. We do not want to determine under what circumstances the application triggers the collaboration. In particular, this implies that several collaboration threads can be active at the same time, i.e., the calling application can issue further connect.req signals before the current thread has been completed. The different threads execute concurrently and can overtake each other. If such behaviour should not be allowed, we need control structures coloured grey in Fig. 8.

At a stop symbol, a collaboration thread terminates. This does not mean that the involved role is terminated, because a parallel thread of the collaboration may still be active. The semantics of a stop symbol is: stop here but continue somewhere else, if possible. Note that most operation symbols are borrowed from SDL and have analogous meaning.

We have just introduced the new term *collaboration thread*. Let us briefly clarify its relation to other concepts before we proceed with the discussion of CoSDL. A collaboration as, for instance, specified in Fig. 8 describes all possible message flows that can occur between participating agents at run-time in order to realise an intended functionality. The message flows that are triggered by an external stimulus is called a collaboration thread. As a consequence, many threads can traverse the agents at the same time, depending on how many stimuli enter the system and whether the collaboration allows to start a new thread before others have terminated. As mentioned earlier, the behaviour specification of a single agent within a collaboration is called a role. At run-time a role is executed by its agent and therefore extends a collaboration thread as soon as it reaches the agent. In an SDL system, roles are executed by SDL agents, which means that a role as a behaviour description is part of the agent description. One of the essential ideas of this paper is that the state machine of an SDL agent can be viewed as a composition of smaller state machines, each describing the behaviour of a different but single role. With these clarifications in mind we can continue our discussion of CoSDL.

In spite of its simplicity, Fig. 8 demonstrates one of the most important features of CoSDL. It is shown that operations of all involved roles are explicitly put in a causal order. Compared with current approaches, we do not restrict the causal order to individual agents and then define agent composition rules for building more complex systems. Rather, the collaboration-based approach considers collaborations as the basic architectural units that can be composed to form more complex systems. Each collaboration handles only one specific role of an agent, so that CoSDL specifications provide a clear view on the overall system functioning.

We call a consecutive sequence of operations of the same role a *behaviour* segment. The basic service in Fig. 8, for instance, has three behaviour segments: if it consumes a connect.req from the calling application, it sends a connect.ind to the called application. Additionally, there are the two alternative behaviour segments for a successful connection establishment and a rejection, respectively. It turns out that the segments follow strictly one after the other and are interleaved by a behaviour segment of the called application.

As behaviour segments of individual roles are scattered across the specification, it is important to attach the instance and type name of the corresponding role as a label to each behaviour segment. The label syntax is: the instance name in italics is followed by a bold type name, where instance and type name are separated by a colon.

Let us now take a closer look at the parts of the specification that are coloured grey. The colour actually has no semantic meaning, we only use it for the purpose of explanation. In case of a successful connection establishment, an additional connection to another communicating peer can be requested. This is done via a loop. A loop is established when a decision branch is directed back into a preceding behaviour segment. It is required that the backward branch originates and ends in a behaviour segment of the same role. As a consequence, the behaviour segments in between are iterated a nondeterministic number of times. As CoSDL abstracts from data aspects, the concrete number of iterations cannot be derived from the specification.

The loop in Fig. 8 involves only one additional role (*calledApp*) per iteration, though in general, any number of different roles is possible, even none. For the current example, however, the more important question is whether individual iterations address the same or different role instances. The shadowed rectangle around the called application's behaviour segment in combination with the keyword *dynamic* denotes that each iteration employs a different instance of the role type **App**. For each iteration, the involved role type keeps the same, but the instances change, so that several called applications with identical behaviour segments participate in the multicast collaboration. CoSDL offers several language constructs for generating concurrency: at start symbols, new message flows can be created at any time. In addition, forks split off concurrent message flows.

With concurrent message flows in place, there is a need for adequate synchronisation mechanisms. CoSDL offers the concept of a join where different message flows merge. Message flows can also be terminated by the stop and termination symbol. However, with these constructs alone, a role can hardly control the order of message flows traversing through its behaviour segments. Whenever a message is received, the reaction can theoretically follow promptly, as all causal dependencies are met (CoSDL leaves timing issues open). In case of Fig. 8, for instance, this means that whenever *callingApp* decides to send a connect.req, a specific set of called applications will be connected. That is, several of these message flows may be processed concurrently.

For handling the synchronisation of message flows within a role, the concept of *states* is applied. A state is a static condition that a role instance meets between processing steps. States are preconditions for processing steps and can also change afterwards. In order not to clutter a graphical CoSDL specification too much, an operation by default turns the state of a role into an unique state (*default state*) if an incoming state is attached. The default state must not be used otherwise. Syntactically, states ⟨*idle*⟩ are used as shown in Fig. 8. States that act as a precondition have an outgoing arrow that points onto the conditional operation. In Fig. 8, *basicService* will only process incoming connect.req signals, if it is in state *idle*. After processing a connect.req signal, *basicService* enters the default state, as no other output state is specified. This differs when *basicService* is sending a connect.conf or a refuse.ind signal. Then the state will change explicitly from *default* to *end*. Note that operations with no state attached to them do not change the agent's state at all. They can be triggered by any state and do remain in this state after completing the operation.

There are two additional issues to be clarified: what initial state does a role instance enter when created, and how are incoming signals handled by conditional receive operations? The state at the top of Fig. 8 has an incoming arrow. Such an arrow must only be used once for each role type because it denotes the initial state of a role instance. There remains the question of what happens to signals at a conditional receive operation, if the precondition does not hold? There, we have two possibilities: with the *save* attribute stated, the incoming signals will be stored. Otherwise, i.e., without any attributes, the signals will be simply discarded. It follows that in Fig. 8, only the very first connection request of the calling application will be processed, because the basic service never returns to the idle state. If the basic service returned to the idle state, subsequent connection requests could be handled again.

4   Conclusions and Future Work

We have introduced the concept of collaboration-based design and applied it to the description language SDL. This resulted in the definition of a set of compositional operators for collaborations in SDL and a collaboration description language called CoSDL. CoSDL is a graphical language to exploit human visual perception. This is especially beneficial for collaborations, because CoSDL visualises the engineer's conceptual model of the interaction relationships in a distributed system. A strong syntactical correspondence to SDL helps with keeping the gap between CoSDL and SDL small, so that traceability during the design process is intuitive. This allows the definition of a systematic process for the transition from sample system execution traces to collaborations and finally to complete SDL systems.

The complete syntax and semantics definition of CoSDL and a description of the collaboration-based design process is given in [13]. We have defined the semantics of CoSDL in two steps. First, the token string derived by applying the production rules of the CoSDL grammar is transformed into a directed graph (CoSDL graph), which is then interpreted by the abstract CoSDL machine, yielding the set of possible execution traces described by a CoSDL specification. The collaboration-based design process integrates smoothly with SDL pattern based design as defined in [6]. As a matter of fact, the SDL pattern pool provides adequate building blocks for collaborations, and SDL pattern selection is greatly simplified through the individual development of collaborations. Thus, both approaches benefit from each other and are best applied jointly. One implication of a further integration of the two approaches could be the definition of specific SDL patterns for the different compositional operators introduced in Sect. 2.3.

The set of compositional operators for collaborations that were introduced in this paper should be extended and formalised. We basically gave a flavour of what is possible for future work. For instance, we focused our discussion on compositional operators that describe collaboration interactions on common data. We could also define operators for collaboration dependencies in the sense of "uses" and "refines" relationships typically defined for use cases [9].

References

1. L. Bass, P. Clemens, R. Kazman: Software Architecture in Practice. SEI Series in Software Engineering, Addison Wesley, 1998.
2. G. Booch, J. Rumbaugh, I. Jacobson: The Unified Modeling Language User Guide. Addison Wesley, 1999.
3. R. Buhr, R. Casselman: Use Case Maps for Object-Oriented Systems, Prentice Hall, 1996.
4. B. Geppert, R. Gotzhein, F. Rößler: Configuring Communication Protocols Using SDL Patterns. 8th SDL Forum (SDL'97), France, 1997.
5. B. Geppert: The SDL Pattern Approach - A Reuse-driven SDL Methodology for Designing Communication Software Systems. PhD thesis, Univ. of Kaiserslautern, Germany, 2001.
6. B. Geppert, F. Rößler: The SDL-Pattern Approach - A Reuse-driven SDL Design Methodology. To appear in Special Issue of the Computer Networks Journal, 2001,
7. D. Harel, StateCharts: A Visual Formalism for Complex Systems. Science of Computer Programming, Vol. 8, pp. 231-274, 1987.
8. ISO 9074, Information technology - Open Systems Interconnection - Estelle: A Formal Description Technique Based on an Extended State Transition Model, 1989.
9. I. Jacobson, M. Christerson, P. Jonsson, G. Övergaard: Object-Oriented Software Engineering - A Use Case Driven Approach. Addison-Wesley, 1995.
10. P. Kruchten: The 4+1 View Model of Architecture. IEEE Software, 11/1995, 12(6).
11. D.L.Parnas: On the Criteria for Decomposing Systems into Modules. CACM, 15(12), 1972.

12. D.L.Parnas: Designing Software for Ease of Extension and Contraction. IEEE Transactions on Software Engineering, 5(2), 1979.
13. F. Rößler: Collaboration-based Design of Communicating Systems with SDL. PhD thesis, University of Kaiserslautern, Germany, submitted for review.
14. J.Rumbaugh, I.Jacobson, G.Booch: The Unified Modeling Language Reference Manual, Addison-Wesley, 1999.
15. B. Selic, G. Gullickson, P.T. Ward: Real-Time Object-Oriented Modeling. Wiley, 1994.
16. ITU-T Recommendation Z.120 (11/99) – Message Sequence Chart (MSC), International Telecommunication Union (ITU), to be published.
17. ITU-T Recommendation Z.100 (11/99) – Specification and Description Language (SDL), International Telecommunication Union (ITU).

Using UML for Implementation Design
of SDL Systems

Jacqueline Floch[1], Richard Sanders[1], Ulrik Johansen[1], and Rolv Bræk[2]

[1] SINTEF Telecom and Informatics, Trondheim, Norway
{Jacqueline.Floch, Richard.Sanders, Ulrik.Johansen}@informatics.sintef.no
[2] Department of Telematics,
Norwegian University of Science and Technology, Trondheim, Norway
Rolv.Braek@item.ntnu.no

Abstract. The purpose of Implementation Design is to bridge the gap between an abstract system, e.g. in SDL, and its implementation in hardware and software. Expressing the Implementation Design decisions is a general challenge in systems engineering. Notations have been defined to describe abstract models and implementations, but little work has been done to define a notation for Implementation Design. In this paper, we discuss UML solutions and extend them. Our work is intended as a contribution to a common SDL approach, and an inspiration to ITU-T Study Group 10 question 11 on deployment and configuration language DCL. We also present how the ProgGen tool was extended in order to generate code controlled by the Implementation Design model.

1 Introduction

There exist several differences between SDL [1] systems and real systems. While an SDL system consists of abstract entities such as processes and channels, a real system is composed of concrete entities such as computers, software processes, buses, and communication networks. [2] discusses how real systems differ from SDL. There are:

- *fundamental* differences originating from the nature of real components and their "imperfections". They encompass processing time, errors developed by physical components and noise, physical distribution and limitation of resources.
- *conceptual* differences originating from the way that components function. They encompass concurrency, communication modes (e.g. stream vs. message), synchronisation, and data abstractions.

When realising an SDL system, one must decide how to implement the abstract functions. These decisions and the concrete system should be documented. In Open Distributed Processing [3] parlance this means describing the Engineering Viewpoint and the Technology Viewpoint, while SDL contributes to the Computational and Information Viewpoint.

R. Reed and J. Reed (Eds.): SDL 2001, LNCS 2078, pp. 90–107, 2001.

The term *deployment* has recently been introduced to denote the Engineering and the Technology Viewpoints [4]. We prefer the more general term *Implementation Design*. While the deployment model describes a particular configuration of run-time processing elements and the software components that live on them, the Implementation Design additionally describes the implementation principles in terms of generic types that can be instantiated in a given deployment. This enables reuse of implementation principles. When applying an Implementation Design model to an SDL system, a concrete system will result.

In addition to defining implementation principles, there is a need to define specific configurations, including aspects such as priorities, capacity parameters and dynamic reconfiguration. These issues are the specific concern of configuration languages, and are not fully treated in this paper.

Although notations have been defined to describe functionality (for example SDL, UML) and implementation (for example C++, Java), little work has been done to define a notation for describing Implementation Design. Some of the few attempts hitherto have been:

- SOON [2], which represents hardware and software structures and the mappings from the SDL entities to the concrete system components. Although we have found it to be very useful, it suffers from lack of tool support. Moreover, SOON lacks support for describing other non-functional properties of the system, such as information that is needed during code generation and installation.
- UML Implementation Diagrams [4], which describe the system software structure and the deployment of software components on a hardware platform. These solutions, although as yet immature and incomplete, have gained our attention.

In this paper, we discuss how appropriate the UML solutions are, and suggest extensions to them. We believe that SDL users would benefit from a common approach to Implementation Design. This paper is intended as a contribution to such an approach and an inspiration to the ITU-T Study Group 10 question on deployment and configuration language DCL [5], to which other projects also contribute [6,7].

In the following, we first recall the main purposes of Implementation Design. Then we describe how UML basic concepts and extension mechanisms can be used for Implementation Design. Examples based on the Access Control system presented in [2,8] are given. Finally we illustrate, using ProgGen [9], how the Implementation Design can be used to control code generation.

2   Implementation Design

Every system development must bridge the gap between the abstract and the concrete system [10]. The Implementation Design model is an important document for the system developers, and for tools such as code generators, configuration and building tools. The model should describe implementation aspects of the concrete system, and relate it to the abstract SDL system. It should model:

- the hardware structure identifying at least all physical nodes and interconnections needed to implement the abstract system;
- the software structure identifying at least all software components, software communications and relations needed to implement the abstract system (in terms of processes, procedures, and data). Additional aspects include process priorities, queues and semaphores, interrupt handlers, timers, watchdogs, scheduling, etc;
- the mapping from the abstract SDL system (and other abstract representations, such as UML) to components of the concrete system, and identify legacy components; and
- the principles of installation, activation, run-time reconfiguration, removal, etc.

In order to support implementation design in the wide sense and not only deployment of instances, the hardware and software structures should be capable of expressing generic and reusable entities that describe common aspects of hardware and software instances. It should model generic types and object structures, support inheritance, and enable flexible configuration of system instances.

The implementation design description is an important document for the system developers and system users, and also for tools. The Implementation Design should be expressed formally, so that tools such as code generators can interpret it. In our work on automatic code generation from SDL, we give detailed examples of implementation design information elements that may be relevant during code generation [11]. We classify this information as follows:

- Information related to the implementation of the SDL application. This includes runtime priorities, timer units, static vs. dynamic allocation of variables, optimisation requirements (for example, for size or speed), response time requirements, and security requirements.
- Information related to the execution target. This includes distribution of the software into files, software processes, target language (for example, C++ or Java), external import of code, and interfaces with the environment.
- Information independent of any model. This includes the identification of the SDL code generator and compilers used to produce the concrete system.

3   UML Concepts for Implementation Design

UML [4] is a general-purpose family of visual modelling languages that can be used to specify and document functionality, implementation and other aspects of a system. In addition to the basic UML model elements, UML introduces extension mechanisms that enable one to customise and extend model elements. These extension mechanisms give UML great power of expression, albeit with the main drawback of reducing the universality of the models. An extensive use of these mechanisms may lead to models that are not portable among tools, and that are difficult to understand even by UML experts. One solution to this is to define extensions as a UML Profile that is commonly agreed within a user community.

In this section, we investigate how UML basic elements and extension mechanisms can be used in order to represent the Implementation Design information. We believe that SDL users should avoid developing their own UML extensions, but rather agree on common extensions for Implementation Design. Our approach may be a step towards a common solution. It is an input to the graphical form of DCL [5].

UML implementation diagrams set restrictions with respect to types and instances, and these restrictions make it difficult to model hardware/software types, compared with such modeling in SOON. We believe that the concepts of the implementation diagrams should be aligned with those of the class and object diagrams (that is allowing type and instance diagrams for the hardware and software structures). We propose to extend the UML deployment diagram to represent hardware and software structure types [12].

A main drawback of UML is the lack of formality of the language specification[1]. Some concepts in the specification (especially the concepts related to the implementation diagrams) are defined in a vague manner. Some concepts are not completely defined in the UML Semantics (as they should be), but instead are mentioned in the UML Notation Guide.

In the following, we assume that the reader has a basic knowledge of the concepts of class, object, association and attribute in UML.

3.1   UML Components and Nodes

UML defines components and nodes that represent software and hardware units. The component diagram is used to show the overall software structure, while the deployment diagram is used to show the run-time environment. Note that UML focuses on deployment, meaning a configuration of run-time systems, rather than a generic hardware structure.

Components and Component Diagrams. A component in UML represents a physical piece of the implementation of a system, and includes code (source, binary, executable), or equivalent items such as scripts or files. We suggest using components in order to represent software processes[2], procedures and data blocks. A component may conform to and provide a set of interfaces. The purpose of the component diagram is to describe components and show the dependencies among software components. A UML component diagram has only a type form, not an instance form. To show component instances, a deployment diagram is used. The deployment diagram may possibly be a degenerate[3] deployment diagram (i.e. one without nodes).

---

[1] By the UML specification we mean both the UML Notation Guide and the UML Semantics.

[2] The term software process is used here to denote a software component operating concurrently with other software processes, normally scheduled by an operating system.

[3] "degenerate" is the term used in the UML specification.

The UML specification does not mention that relationships may be defined between a component and other classifiers (for example classes or nodes). According to the UML meta-model and semantics, such relationships are possible. Thus, we may specify (for example) that a component realises classes, which is useful information to capture. The Notation Guide does not explicitly state that the component diagram can be used to describe composite components or other relations among components. However, the Notation Guide does give examples where some component instances consist of objects.

As shown in Fig. 1, we represent composition in component diagrams (in a type form showing components that consist of other components or classes). When further details about a component have to be specified, it may be wise to split the information into several diagrams, and possibly to use the attribute notation to save space. The nested elements may have a role name (for example "co") within the composition. The multiplicity is shown in the upper right hand corner of the nested component symbol, where a specific number (here "1" for each nested component); a range (that is a minimum and maximum) or a set of values may also be specified.

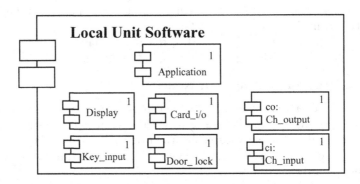

Fig. 1. An example of a composite component diagram in UML

Note that several notations are defined in UML in order to represent composition (that is attributes, associations with filled diamonds, graphically nested elements). Any of them could be used. We have chosen the nested graphical representation since it is close to SDL and SOON. However, we are aware that most UML tools do not support the nested representation.

Composition is a form of aggregation with strong ownership and coincident lifetime of part with the whole. When lifetimes do not coincide, open aggregation should be used instead.

Nodes. A node represents a run-time computational resource, and generally has at least memory and often has a processing capability, where components can be deployed. Thus, a node is a piece of hardware. We use nodes to represent hardware blocks.

The UML specification does not explicitly state that a node may have its own features such as attributes and operations, and that it can realise interfaces. According to the UML meta-model and semantics, this is allowed. It is also possible to define relationships between a node and other classifiers (for example classes or components). We suggest specifying node properties such as memory size, CPU frequency, etc, and describing relationships between nodes and other classifiers. For example, one may specify that a node is composed of other nodes or that a node realises some classes.

Deployment Diagram. The purpose of the deployment diagram is to show the configuration of run-time processing elements and the software components that live on them. The deployment diagram represents run-time manifestations of code units (instances of software components), and not source code. Components that do not exist as run-time entities are shown in component diagrams, instead of deployment diagrams.

The UML specification does not clearly state which diagram should be used to model node types. Although the examples given only illustrate node instances, the UML Notation Guide explains that node types can be represented. In their UML textbook [13], Booch and others give some examples where the deployment diagram is used to show the overall hardware structure of a system in a type form.

Figure 2 shows how nodes and deployment diagrams can be used to model the run-time configuration for a hardware sub-system, *CentralHW*. As the diagram has an instance form, we are not allowed to show a variable multiplicity. In the figure, the links between the node instance *Computer* and the two instances of *Operatorterminal* represent physical connections between the nodes. UML extension mechanisms can represent the direction of signaling (not shown here).

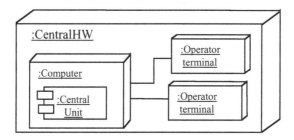

Fig. 2. An example of a deployment diagram in UML

Representing Hardware/Software Structure Types - Extension to UML. One of the shortcomings of UML is that component types are not allowed in deployment diagrams. This prevents us from modelling the overall structure of software and hardware in a type form (showing both hardware blocks, block properties and relations between hardware block types and software component types). In order to represent a hybrid software/hardware structure, we propose an extension of the UML deployment diagram, that allows node and component types to be represented in a deployment diagram. We wish to model two main relations between nodes and components:

- the *execute* relation means that a hardware node is able to execute a software component. The *execute* relation can be represented using composition.
- the *implement* relation means that an abstract component (or a class) is realised by a hardware node of software component. The *implement* relation is represented using the UML stereotype «realize».

Figure 3 shows a hybrid software/hardware structure. The node *Computer* in *CentralHW* executes the software component *CentralUnit*, which implements the class *CentralUnit*. The node *Controller* in *LocalControl* implements the class *AccessPoint*. The predefined UML icon for the stereotype «realize», i.e. the dashed line with a closed triangular arrowhead, is used to represent the *implement* relation.

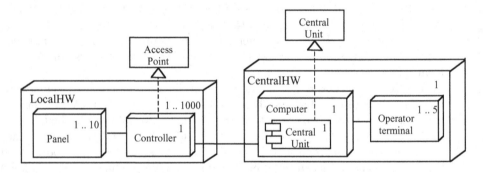

Fig. 3. Using an extension of the UML deployment diagram for hardware/software structure

3.2   UML Extension Mechanisms

The basic model elements defined in the UML specification can be customised and extended with new semantics. Three extension mechanisms are defined in UML: stereotypes, tagged values and constraints. Applied to model elements they enable new kinds of elements to be added to the modeller's repertoire. They represent extensions to the modelling language, and are used to create UML profiles, such as the UML profile for DCL/GR suggested here.

Tagged Values. A tagged value enables arbitrary information to be attached to any model element. A tagged value is expressed by a (name, value) pair. The interpretation of the tagged value is beyond the scope of UML; it is determined by the user or a tool. The UML specification gives some examples of possible use, such as code generation options, model management information, etc.

Properties may be attached to an element using attributes, associations, or tagged values. Attributes and tagged values are quite similar concepts. A question thus arises: when should we use tagged values rather than attributes? We suggest viewing tagged values as meta-data, because its values apply to a model element, not to its instances. We suggest using attributes to describe properties of instances of the concrete entities in the model, while we will use tagged values to qualify elements (i.e. group elements that share some characteristic). Tagged values are particularly of interest when describing stereotypes, and this is the only practical use of tagged values we have found. We describe the code generation information using attributes when this information varies for the different instances of a (user-defined) model element.

Constraints. The UML constraint concept allows a constraint (or semantic restriction) to be specified for a model element or a group of model elements. The constraint can be written as an expression in a constraint language chosen by the user (or accepted by the tools used by the user). The UML specification suggests OCL (Object Constraint Language) for doing this. We suggest using UML constraints to express constraints on the concrete system, for example optimisation aspects, response time requirements, security requirements, and so on.

Figure 4 shows how UML constraints can be used. Here constraints are attached to single model elements (a component, a node or a connection). It is also possible to represent constraints between elements, and to refer to attributes of the model elements that the constraint is attached to.

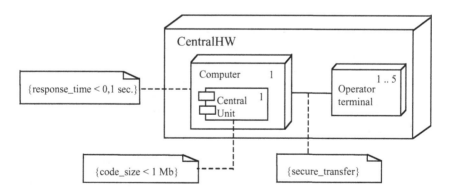

Fig. 4. Using constraints in an implementation diagram

Stereotypes. The stereotype concept provides a way of classifying elements in order to add some new meaning to a basic model element. The stereotype may specify constraints and tagged values that apply to instances. The UML specification itself defines several stereotypes.

The general UML presentation of a stereotype is to use the symbol of the basic model element where the name of the stereotype is added within guillemets « ».

For example, the class stereotype *process* is defined in Z.109 [14] and the syntax
«**process**» is defined in Z.100 [1] to denote the SDL process class. A graphic
icon can be defined to represent the stereotype. Z.100 [1] (and the Telelogic
TAU deployment editor [15]) use the SDL process symbol as an alternative for
«**process**».

Declaration of Stereotypes. The classification of stereotypes themselves can
be shown on a class diagram. This meta-model diagram usually is not developed
by the system developer, but by the methodology or tool provider. The UML
specification does not clearly specify how the properties of the stereotype can be
defined. We believe that the exact semantic of the stereotypes should be specified
by the methodology or tool provider, and that the specification should be made
available to the system developer.

Based on our positive experiences with the classification of software components in SOON, we propose to define the corresponding stereotypes in UML. The
SOON software components are defined in Fig. 5. We have retained the SOON
icons. Note that when defining new stereotypes, guillemets (« ») are added to
the keyword **stereotype**, but not to the names of the stereotypes. This notation
is used in [16]. When using stereotypes during modelling, the stereotype name
is written within guillemets, for example «procedure».

Using Stereotypes. In Fig. 6, we assume definition of the «connection» stereotype (connection between hardware blocks), and we use the «process» stereotype (representing an SDL process) defined in Z.109 [14]. Note that a «process»
stereotype is also defined in UML for representing a heavyweight flow of control[4].

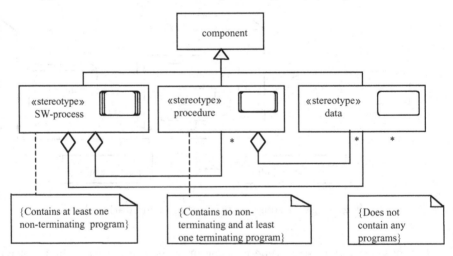

Fig. 5. Stereotype definitions for software entities

---

[4] SDL processes (stereotype «process» in Z.109) have their own flow of control (stereotype «process» in UML), but the inverse is not true. In order to avoid ambiguity,
Z.109 should have introduced different names, e.g. «SDL-process».

The UML specification lacks to disclose how a reference to a particular UML profile should be specified. We indicate a reference to Z.109 profile by using a comment in Fig. 6. Stereotypes can be defined for any UML model elements, also for attributes. In this figure, we represent some properties of the variables of an SDL process using the stereotypes «static» and «dynamic».

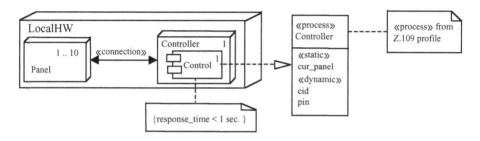

Fig. 6. Using stereotypes

Stereotypes Defined in UML. UML predefines several stereotypes for the basic model elements. Some of them are relevant for Implementation Design. We present them here:

- «process» and «thread» stereotypes specify that classifiers represent respectively a heavyweight flow of control, or a lightweight flow of control.
- «realize» specifies a relationship between a specification model element(s) and a model element(s) that implements it. This stereotype can be used to show that a hardware or software entity realises a class (e.g. an SDL process).
- The «file» stereotype denotes a document containing a program while the «executable» stereotype denotes a program that may run on a node. Both stereotypes are useful for system building.
- «derive» specifies a derivation dependency among model elements. This stereotype can be used to show that one entity is computed from another.
- «become» specifies a flow relationship between the same instance at different points of time. It can be used to show the migration of components from node to node, or objects from component to component as shown in Fig. 7.
- «create» specifies a creation relationship between two classifier instances. It can be used to represent creation of component instances, either in sequence and collaboration diagrams, or in deployment diagrams (extension to UML) as shown in Fig. 7.
- «call» specifies that an operation in the source class invokes an operation in the target class. We can use this stereotype in order to show activation or call relations between software components.

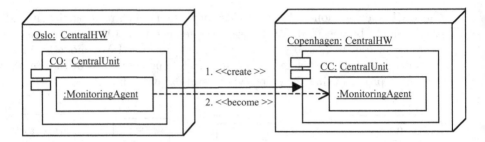

Fig. 7. Showing creation and migration of components

4   Implementation Design and Code Generation

A primary interest for system developers is how the Implementation Design is used during system generation and configuration: Is it "just another nice drawing", or can it be used to control the code generation? If the latter were true, the model not only depicts the intentions, but also in fact documents the actual system implementation. This is particularly important in situations where alternative system realisations are derived due to different non-functional requirements.

4.1   Description of the Implementation Design Model

As explained earlier in this paper, we classify implementation design information elements that may be relevant during code generation as follows [11]:

- Information related to the implementation of the SDL application;
- Information related to the execution target;
- Information independent of any model.

Each type of information is optional in the Implementation Design model. In the following we give examples on how information of each type can be modelled in UML.

Design Information Related to the Implementation of the SDL Application. In the Implementation Design model, the SDL entities are defined as stereotyped classes. UML constraints are used to describe optimisation aspects, response time requirements and security requirements.

General properties related to types of entities (for example, type definition expansion) are represented using stereotypes (constraints or tagged values may be attached to stereotypes). Properties may be related to some properties of an entity type (for example, for a process variable properties stating if the variable should be placed in static or dynamic memory - see Fig. 6); in that case the properties are represented using stereotypes.

Implementation properties related to SDL entity types and instances (such as priority, SDL timer units) are represented using attributes. In Fig. 8, a priority attribute is defined for the SDL process type *Validation*. Default values may be specified for the attributes. Note how the realisation of *Validation* by *CentralUnit* is represented in the figure. The realisation association indicates a difference of abstraction between the entities.

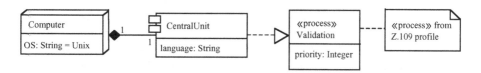

Fig. 8. Using attributes to represent implementation information in a type diagram (UML extension)

Design Information Related to the Execution Target. This information is represented by components and nodes, and by the properties of these nodes and components (using attributes or stereotypes). Interfaces to the environment are described using UML interfaces. It is also possible to indicate which file contains the software that realises the interface using a dependency relation. This is shown in Fig. 9.

Note that the use of «realize» is overloaded; it is necessary to take into account the model element it is applied to (object, class, component, interface). The rule is that an entity is realized by a more concrete model element.

Design Information Independent of any Model. Here we may want to indicate the type of tool that is used to produce the implementation (for example code generator, compiler). This tool information can either be described in a generic manner using stereotypes (for example, all components attached the stereotype «java» are compiled using a Java compiler) or for specific components using the «derive» stereotype. In the latter case, the tool used for derivation could be specified using a constraint attached to the derivation, as shown in Fig. 10. It is also possible to define new subtype stereotypes of «derive».

Coupling Models. It is necessary to tie the SDL model and the Implementation Design model together. The coupling is done using stereotypes from Z.109 profile representing SDL entities such as «process», «block», etc. and using the same entity names as in SDL. Figure 11 shows an example of such coupling.

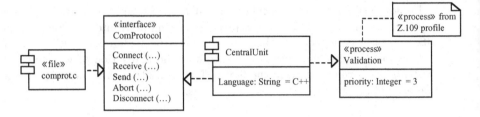

Fig. 9. Description of an interface to the environment

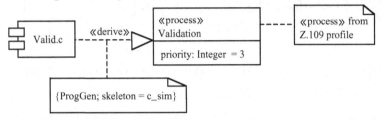

Fig. 10. System building information using «derive» and tagged values

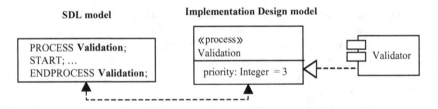

Fig. 11. Coupling SDL and Implementation Design models

4.2   A Solution for a Code Generator Controlled by Implementation Design

As we have seen in the previous sections, UML can be used to describe an Implementation Design for SDL systems. The question now is: can this be used to control the code generation process? We describe here a solution and explain how the code generation tool ProgGen [9,17] has been extended for this purpose [11].

The particularity of our approach is that we keep the functional aspects separate from the implementation aspects. Some code generation solutions support implementation specific information within the SDL model, e.g. inline code in [15]. We believe that the tight integration of functional and implementation models makes the SDL model more complex and reduces its portability and ease of configuration.

The Code Generator. ProgGen is a generic tool that can be used to produce translators. ProgGen is based on the idea that the structure of most programs

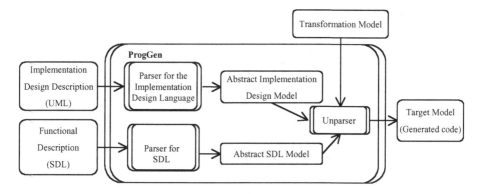

Fig. 12. ProgGen's architecture – Implementation Design model example

follows some general scheme [18]. These schemes are described in a transformation model that is interpreted by ProgGen during code generation.

Using ProgGen, it is possible to generate code from any textual language. Several parsers can be defined and used simultaneously. An SDL/PR parser has traditionally been integrated in ProgGen. We have defined a new parser for the Implementation Design. Figure 12 shows ProgGen parsing two models:

- The SDL model containing the abstract functional model in SDL/PR.
- The Implementation Design model, containing implementation specific information expressed in UML, as defined earlier. No textual representation of UML has yet been defined (unlike SDL/PR). Hence it has been necessary for us to define a textual linear grammar that is specific to the purpose.

The transformation model, shown in Fig. 12 using the icons defined in Fig. 5, defines how the SDL model and Implementation Design model is transformed to another representation (the target model or implementation code). The transformation model refers to ProgGen's internal abstract representations of the SDL and Implementation Design models, and generates the target model by applying a set of transformation rules. The Implementation Design model may describe which of several alternative transformation models should be used during program generation.

In the following, we present an example to demonstrate the relationship between the SDL model and the Implementation Design model, see Fig. 13. The example also shows how information that is not related to the SDL model can be expressed in the Implementation Design model. Note that the model presented here is not complete.

Here follows a brief explanation of the Implementation Design model:

- We specify new general implementation properties of SDL processes and procedures in an UML profile. The attribute *priority* is defined for all SDL processes and the attribute *code* for SDL procedures.

New profile

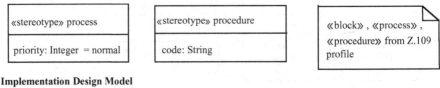

Implementation Design Model

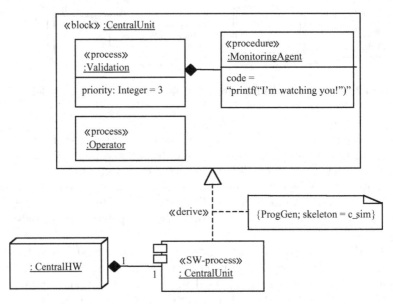

Fig. 13. SDL model and the Implementation Design model

- In the application model dependent part, the *priority* attribute of the process *Validation* SDL block *CentralUnit* is set to a specific value. The SDL procedure *MonitoringAgent* is given a specific content in the chosen target language. The transformation shall ignore the procedure content defined in SDL model and use the content specified here instead. The change of contents of the SDL procedure shows that the designers can specify implementation information according to their own special needs.
- In the execution target dependent part, we specifies that the software process *Validator* realizes the SDL block processes *CentralUnit*. This software process executes on the hardware node *CentralHW*.
- In the application and target independent part we specify which transformation model(s) to use. Here the c_sim transformation model is used.

[11] gives more examples, e.g. an SDL model may be used together with two different Implementation Design models, or an Implementation Design model may be used together with two different SDL models. It should be noted that running code was produced by ProgGen in this experiment. The Implementation

Design model was expressed in a textual language of our own making; the textual model was manually derived from UML.

5   Related Work

As the Z.109 recommendation [14] focuses on the transition from domain analysis with UML to application design with SDL, we believe that Z.109 and our work describing Implementation Design for SDL with UML should be unified.

The SDT deployment editor (or DP editor) [15] is a tool in Telelogic Tau SDL Suite. The tool allows the developer to describe graphically, using the UML notation, the partitioning of the SDL entities in the run-time environment. Build-scripts can be generated automatically. The DP editor is a step in the right direction for capturing Implementation Design with UML. However, the focus is on the distribution of the SDL entities on nodes and on concurrency aspects, and the tool does not make it possible to completely describe the overall hardware and software architecture, nor provide a control point for tuning the code generation process. The set of properties of the concrete entities that can be represented is limited. The DP editor does not make use of the full UML semantics for classifiers.

The deployment and configuration language DCL has recently been added to the list of future ITU-T languages [5]. Several projects work with contributions to this [6,7].

6   Conclusion

In this paper we have outlined a solution for using UML to express Implementation Design for systems described using SDL. We have pointed out opportunities and drawbacks with UML in this respect, and suggested extensions to UML. We believe that the solutions that we propose are general and may also be applied when notations other than SDL and UML are used for modelling the application system. Our work may serve as a contribution to a common SDL and UML approach, and to the ITU-T Study Group 10 question 11 on deployment and configuration language, DCL/PR.

In our approach, the Implementation Design information is specified separately from the specification of the SDL application. We believe that the separation of the application and Implementation Design models is an improvement over today's practice of cluttering up the SDL model with implementation details. We have shown, by extending ProgGen, that the automatic code generation can be controlled by the Implementation Design. It should be possible to apply the principles in our solution in any code generator tool.

References

1. Z.100 - Specification and Description Language (SDL). ITU-T Recommendation Z.100. November 1999.

2. Bræk, Rolv & Haugen, Øystein. Engineering Real Time Systems. Prentice Hall, 1993. ISBN 0-13-034448-6.
3. X.902 - Open Distributed Processing – Reference Model, ITU-T Recommendation X.902.
4. OMG Unified Modeling Language Specification. Version 1.3, June 1999.
5. ITU-T Study Group 10: DCL: Deployment and Configuration Language. Information available at http://www.itu.int/ITU-T/com10/questions.html
6. Eurescom: Deployment and Configuration Support for Distributed Services at http://www.eurescom.de/~pub/seminars/past/2000/TelecomIT2000/17Andreas Hoffmann/17aAndreasHoffmann/
7. Eurescom: Support for Distribution and Configuration of Distributed Applications http://www.eurescom.de/~pub/seminars/past/2000/TelecomIT2000 /10Efremidis/10aEfremidis/
8. TIMe: The Integrated Method. Information available at http://www.informatics.sintef.no/projects/time/
9. ProgGen User's Guide. Available at http://www.informatics.sintef.no/projects/proggen/
10. Gorman, Joe and Johansen, Ulrik. Engineering the Implementation of SDL Specifications, Proceedings of the Fifth SDL Forum, North Holland, 1991.
11. Johansen, Ulrik. Design controlled code generation from SDL. (In Norwegian). SINTEF rapport STF 40 A99040, 1999. ISBN 82-14-01729-7.
12. Floch, Jacqueline. Using UML for architectural design of SDL systems. SINTEF rapport STF40 A00009, 2000. ISBN 82-14-01927-3.
13. Booch, Grady, Rumbaugh, James & Jacobson, Ivar. The Unified Modeling Language User Guide. Addison Wesley, 1998. ISBN 0-201-57168-4.
14. Z.109 - SDL Combined with UML. ITU-T Recommendation Z.109. November 1999.
15. Telelogic TAU SDL Suite. Information available at http://www.telelogic.se/
16. Booch, Grady, Rumbaugh, James & Jacobson, Ivar. The Unified Modeling Language Reference Manual. Addison Wesley, 1999. ISBN 0-201-30998-X.
17. ProgGen Skeleton Author's Guide. Available at http://www.informatics.sintef.no/projects/proggen/
18. Floch, Jacqueline. Supporting evolution and maintenance by using flexible automatic code generator. Proc. of the 17'th International Conference on Software Engineering, 1995.

Deployment of SDL Systems Using UML

Niclas Bauer

Telelogic Technologies Malmö
P.O. Box 4128, SE-20312 Malmö, Sweden
niclas.bauer@telelogic.com

Abstract. The increasing complexity of SDL software (for example, as required for distributed system architectures and advanced operating system integrations) has generated a need for a powerful notation for deployment. The notation should be used for modeling the run-time configuration of SDL applications and the communication between these. This paper formulates requirements on a deployment notation to be used with SDL. Using the mapping between SDL and UML in Z.109, it is investigated how the UML implementation diagrams can be used for deployment of SDL systems, and how well the diagrams conform to the requirements. It is found that the UML deployment diagram can be used for showing deployment of SDL agent instance sets and static instances. Agent instances dynamically created and destroyed require certain mapping rules in order to be modeled. The interface concept in UML is found to be adequate for modeling communication. Using the UML extension construct "tagged value", information at arbitrary detail levels can be shown. This information can be used by a variety of targeting tools such as code generators.

1 Introduction

1.1 Background

The ever-increasing interest in SDL [1] continuously produces new application areas where SDL is used for system development. SDL is today used for implementation of both hardware and software. Code can be generated in a variety of programming languages, both compiled and interpreted. Integrations exist with a great variety of target platforms, such as different real-time operating systems (RTOSes).

It is well known that platform-independent specification languages like SDL can be used to specify functionality independently of the target implementation. The rise of embedded systems, where customized software and hardware is often developed, emphasizes the need for simple co-design of hardware and software.

The rise and merge of the Internet and mobile communication put new requirements on existing systems. Integration between existing software/hardware systems and Internet services are common. New protocols, such as WAP and Bluetooth, create new possibilities for interaction. Implementing communication between systems has become an essential part of system development.

R. Reed and J. Reed (Eds.): SDL 2001, LNCS 2078, pp. 107–122, 2001.

The rising popularity of distributed system architectures leads to complex interaction models between distributed units. For systems developed using SDL, it is also relevant to specify the execution model of a system or a part of it. Organization of SDL entities into operating system threads or tasks is a common scenario.

The high complexity of systems, with regard to the above aspects, implies that architectural design today is a substantial part of system development. An effective notation for *targeting* is needed. Targeting is referred to as the transition from a logical functional model to a concrete system. The notation should be used for controlling code generation from SDL models as well as configuration of communication and integration with target platforms.

1.2 Objectives

The objective of this paper is to discuss requirements on SDL systems deployment. It is described how the UML component and deployment diagrams can be used for deployment of SDL systems. Based on this description, a discussion is held on the usability of the UML diagrams for SDL deployment purposes.

1.3 Scope

This paper focuses on the modeling aspects of SDL system deployment. The presented requirements focus on static deployment of SDL objects at run-time.

The reader is assumed to have basic SDL and UML knowledge. Where necessary, the UML notation is explained.

1.4 Organization

The paper is organized as follows: In Sect. 2, requirements for deployment of SDL systems are discussed. In order to illustrate the requirements, some typical SDL deployment scenarios are presented. In Sect. 3, it is described how the UML component and deployment diagrams can be used for SDL deployment. The deployment scenarios from Sect. 2 are modeled using the UML diagrams. The usability of UML for SDL deployment purposes is discussed in Sect. 4. Finally, a summary is given in Sect. 5.

2 Deployment Requirements

2.1 Requirements on a Deployment Notation for SDL

The following requirements apply for a notation for SDL system deployment:

- The notation should support targeting of SDL systems;
- The notation should be unambiguous and possible to use as input to code generators and makefile generators;

- The notation should describe the organization of deployed SDL systems at run-time;
- The notation should describe static deployment of SDL systems: it should not support run-time migration of objects;
- The notation should not describe deployment of code generated from an SDL model, as this does not add any useful information. Only deployment of SDL entities onto target is relevant;
- It should be possible to partition an SDL system into many separate executable entities;
- It should be possible to deploy SDL agents onto parallel execution points. Depending on the underlying operating system, the execution points are represented using *threads* (lightweight processes with shared memory-space) or *tasks* (heavyweight processes with separate memory spaces);
- It should be possible to select different granularity levels when deploying SDL agents onto threads or tasks: it should be possible to deploy onto a single thread or task both single instances and instance sets;
- The deployment notation should be applicable for deployment of SDL agents that are translated into code that is to be either compiled or interpreted (that is through a virtual machine);
- The notation should support both modeling of SDL agents executing on hardware and SDL agents implementing hardware;
- The deployment notation should be general enough to support any code generator using SDL as input language;
- The notation should support all languages into which SDL models are translated;
- It should be possible to make instance-specific settings: both for executable entities and the platforms on which they execute;
- The notation should support communication between distributed SDL agents deployed in separate executables.
- The notation should support communication between SDL agents and external entities, e.g. libraries, processes, and scripts.

These requirements build the base for the discussions in this paper. The requirements are optimized for deployment of applications that are suited for development using SDL: that is, communication-intensive applications with real-time constraints that may execute in embedded environments.

In the following section, some typical deployment scenarios are presented in order to illustrate the requirements. Each scenario is illustrated using simple structural drawings. An ad-hoc graphical notation is used.

2.2 Deployment Scenarios

Deployment of Agents onto Threads of Execution. An SDL system should execute in several threads (lightweight operating system processes) on an RTOS. In order to optimize performance, experimentation with different thread settings, e.g. priority and stack size, is needed. There is also a requirement on simple mappings between SDL agents and the threads that they should execute in. The desired mapping is shown schematically in Fig. 1.

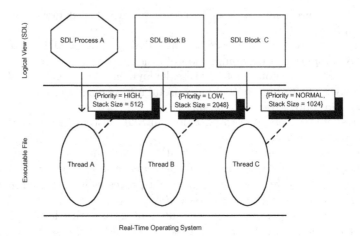

Fig. 1. SDL blocks and processes mapped onto RTOS threads

Deployment with Communication with External Software Components. An SDL system that should communicate with external software via a TCP/IP connection is developed. Data received by the SDL system is parsed and translated into SDL signals. In order for the executable files to find each other, routing tables with IP address and port number for each executable file must be created. An architectural overview is shown in Fig. 2. A notation should provide an overview of the systems and their dependencies and support simple configuration of communication parameters.

Deployment of a Client-Server Architecture. Client-server architectures are common in communication-intensive applications. When building client-

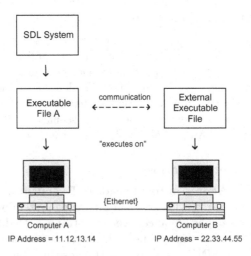

Fig. 2. An SDL system communicating with external software via a network

server applications, it is common that the designer must consider an arbitrary number of clients communicating with the server side. Modeling an arbitrary number of clients, instance-specific settings are irrelevant on the client side. For a server application, however, instance specific settings are highly relevant. This scenario is shown in Fig. 3. An arbitrary number of clients running SDL client systems communicate with a server running a dedicated SDL system. The systems are deployed in a computer network, but the principles are applicable for other scenarios, such as GSM mobile phones communicating with a base station.

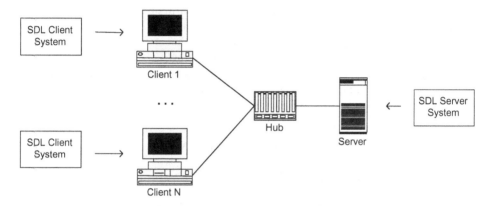

Fig. 3. Deployed SDL systems with a client/server architecture

Deployment of an SDL System that Implements Hardware. The state-machine properties of SDL enable translations from SDL into hardware description languages. A deployment notation should support both SDL models implementing hardware and SDL models implementing software executing on hardware. A simple drawing of this scenario is shown in Fig. 4.

Fig. 4. An SDL system deployed onto hardware

3   Using UML for Deployment of SDL Systems

3.1   Architectural Design

Architectural Design is defined in [2] as the mapping from a logical functional model to a concrete system. This definition is used throughout the paper. Deployment can be considered as the part of architectural design that describes the run-time configuration of a system.

[2] and [3] recognize architectural design as a general problem in software development. Little work has been done on notations in this area.

SDL is a language focusing on the object- and state machine views of a system. Its semantics are complete for those views. SDL relies on other notations for displaying complementary views. For instance, Message Sequence Charts (MSCs) [4] are used for displaying interaction. SDL contains no notation for architectural design. No complementary notation exists to date.

The most widespread notation for architectural design is the UML implementation diagrams, which consist of the component diagram and the deployment diagram. SDL and UML are currently being harmonized, as in [5]. Using the compatible subsets of the languages, it is interesting to evaluate how UML can be used as a complementary notation in areas where SDL lacks notation. This suggests an evaluation of how the UML implementation diagrams can be used for modeling deployment of SDL systems.

3.2   The Mapping between SDL and UML

Mappings between SDL-2000 and UML 1.3 [6] are described in [5] using the extension mechanisms of the UML to adapt it [7]. Z.109 is a mapping between a subset of UML and a subset of SDL. On the UML side, the mapping concerns the UML object and state machine diagrams. The MSC and the UML sequence diagram notations are similar, but no explicit mapping has been presented. The UML use case, collaboration and activity diagrams have no equivalents in SDL. The use case diagram is used for modeling roles and tasks. The collaboration diagram is used for showing interaction and is complementary to the sequence diagram. The activity diagram is used for showing a state-machine view of activities. None of these three diagrams are considered relevant for deployment.

As stated in Sect. 2.1, a notation for deployment of SDL systems should show how SDL instances and instance sets execute. The notation should show the run-time organization, and should support mapping between SDL instances/instance sets and OS threads/tasks.

In the following sections, it is discussed how the UML implementation diagrams can be used for modeling deployment of SDL systems. The discussion is based on the mapping between SDL and UML in [5] and is held from the perspective of the requirements presented in Sect. 2.1. The deployment is modeled using the UML extension concepts tagged value, stereotype, and constraint.

3.3   Combining SDL and the UML Component Diagram

The UML Component Diagram. The UML component diagram represents a type-based view. The types modeled in a component diagram are instantiated in a deployment diagram. The notation is based on the concept of the component, which is a container for software.

A component encapsulates the operations of the classifiers it contains. The most common scenario is where classes are contained within a component, but a component can also contain other components [7].

The contents of a component can be modeled by using either containment: drawing the contents inside a component; or by using the dependency relation. The latter is used in this paper.

One or many different *artifacts* implement a component. An artifact is a classifier that is described in [7] as the outcome of a software development process. Stereotyping is used to specify the type of an artifact. Typical artifact stereotypes are: executable, library and file. The artifact for a component is only shown in the component diagram.

Using the UML extension mechanisms, detailed information can be specified for a component. This information is modeled using *constraints* and *tagged values*. The information needed varies with the target platform. For a component that contains SDL agents, information about the code generation from the SDL model is relevant, and tagged values can be used for the generated code language and the target directory.

An example of a component with tagged values and an artifact is shown in Fig. 5. The component has no contents. Two tagged values are shown within braces inside the component.

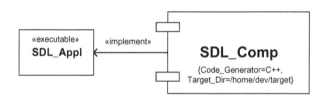

Fig. 5. A Component type implemented by an artifact

Mapping Agents to Components. The most common form of the UML component diagram is where classes are connected to components. This can be used for showing SDL agent types in component diagrams. [5] shows how the UML class can be mapped to the SDL agent type. This implies that UML objects map to SDL agent instances. The stereotype extension mechanism of the UML is used to specify the sort of SDL agent type that a class maps to. A class is given *process*, *block* or *system* as stereotype. Implicit agent types have no UML equivalents, and cannot be mapped. In order to show implicit agent types in a component diagram, a tagged value can be used on the implicit agent type. The suggested tagged value is `agent_type = implicit`. In the component diagram, the class symbol is then given the same name as the instance set of the implicit type.

An example of how agent types are connected to a component is shown in Fig. 6 and 7. Parts of an SDL system controlling a car are modeled in a component diagram. The component contains two SDL agents taken from the

114   N. Bauer

CarControlSystem. As seen in Fig. 6, the ESP block is an implicit type. This is indicated by the tagged value in Fig. 7. The component *Control_Comp* is implemented by the artifact *SDL_Appl*, which has executable as stereotype.

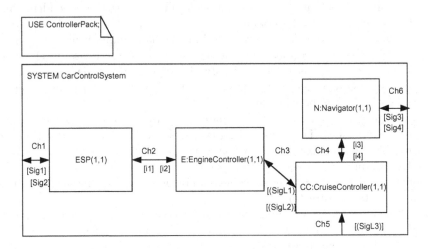

Fig. 6. An SDL system with an imported block type

In Fig. 6, the channel *Ch*1 is used for sending the signals *Sig*1 and *Sig*2. *Ch*2 contains the interfaces *i*1 and *i*2. *Ch*3 contains two signallists, called *SigL*1 and *SigL*2. Channel *Ch*4 has interfaces, *Ch*5 has a signallist and *Ch*6 has two signals.

Organizing Agents into Threads and Tasks. The UML meta-model contains some predefined stereotypes, which can be used for any classifier. Two stereotypes, thread and process, are of particular interest when modeling systems with multiple execution points and real-time constraints. In [7], thread is defined as a flow of control. In this paper, thread is more specifically considered a lightweight process, having shared memory space. For a thread, the tagged value `memory_space = shared` is used to describe its properties. A component with thread as a stereotype is displayed in the diagram as a rectangle containing its name. The stereotype is shown within guillemets.

Process is defined in [7] as a heavyweight flow of control. However, the process notion is already used in Z.109 as a stereotype for a class that is mapped to an SDL process type. The stereotype task is introduced to denote a heavyweight flow of control, having a separate memory space. For a task, the tagged value `memory_space = separate` is used. A task is displayed in the diagram as a rectangle containing its name. The stereotype is shown within guillemets.

Threads and tasks are connected to components using *residence*: an abstraction that has the stereotype `reside`. The residence arrow is drawn from the component to the thread or task.

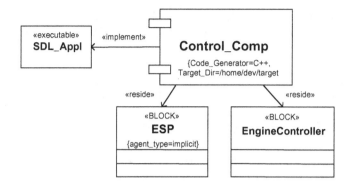

Fig. 7. A component, implemented by an artifact, containing two SDL agents

Using components having *thread* or *task* as stereotype, agents can be organized as they are intended to execute within the component. Threads can be contained within tasks, but the opposite is not allowed. If a class is connected directly to a top component, a default execution model is assumed. The default execution model is determined by the target platform that the instantiated component will reside on.

Tagged values are used to set thread and task properties such as priority, stack size and signal queue size. The tagged values are visible both in the component diagram and the deployment diagram.

An example of the notation for threads and tasks is given in Fig. 8. Two agents from the SDL system in Fig. 6 are deployed onto one thread each. Each thread has a tagged value for expressing its priority. *Control_Comp* is implemented by the artifact *SDL_Appl*, which has `executable` as stereotype.

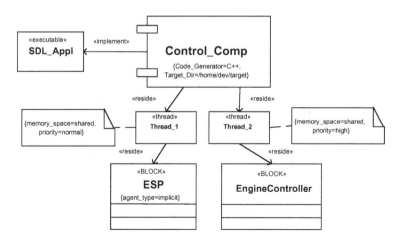

Fig. 8. A component with two threads, each connected to an agent

Mapping the UML Interface to the SDL Interface. In the component diagram, the interface notion is used to specify operations of a component that are externally visible. The component realizes interfaces through the classifiers it contains. The UML interface is visualized in the component diagram as an empty circle that is connected to a component via an association. External components that use an interface can be connected to the interface using a dependency arrow: a dashed arrow directed towards the interface.

The interface notion exists in SDL-2000 as well. An SDL interface defines the set of remote procedure calls (RPCs), signals, remote variables and exceptions that are exposed by an agent. It has two forms. *Implemented interface* is the type of interface that describes what requests an agent should respond to. *Required interface* is the set of signals that may be sent and the RPCs that may be called from an agent.

A UML interface maps to an implemented SDL interface. SDL signals and exceptions are modeled using stereotyped operations. Access to remote variables in an agent is modeled in the component diagram using GET and SET operations in the agent.

If an agent contains a channel where no interface is defined, an implicit UML interface is set up on the component. Signals received by an agent on one channel produce an implicit interface. Similarly, a set of signals sent over a single channel from an agent without an interface also produces an implicit interface.

Implemented interfaces are modeled in the component diagram using the interface notation. Required interfaces are modeled in the component diagram using dependency arrows. The UML interfaces are given the name of the implemented SDL interface. If the interface is implicit, the name of the signallist should be used.

A component that contains agent types should expose all interfaces of the agent types that are necessary for inter-component communication. Each implemented interface at an agent should produce a separate interface on the component. A bi-directional channel produces one interface at each component with dependency arrows directed towards each interface. If agents that communicate with each other reside within the same component, no interfaces should be set up for the communication between these.

Figure 9 shows two agents from the SDL system in Fig. 6 deployed onto two components. The two blocks communicate in both directions, as can be seen in Fig. 6. The interfaces $i1$ and $i2$ are implemented interfaces. Dependencies are drawn from *Engine_Control* to $i1$ and from *ESP_Control* to $i2$. Two interfaces are not connected, as the agents using these are not modeled in the diagram. The interfaces are given names from the signallist and the signal, respectively, as these are implicit.

3.4   Combining SDL and the UML Deployment Diagram

The Deployment Diagram. The UML Deployment Diagram represents a view of the run-time configuration of components that are deployed on nodes. The UML deployment diagram has a descriptor form and an instance form. The descriptor form shows the mapping between component types and node types. The instance form shows a run-time configuration where component instances map to node instances. In this paper, the instance-based deployment view is described.

As the deployment diagram shows instances, it represents a snapshot of a system configuration at a specific time instance. Dynamic creation and destruction of instances cannot be shown in a deployment diagram.

The component diagram represents a view of the structural, static organization of a piece of software. The deployment diagram shows instantiated components with platform-specific settings. The most common action when doing UML deployment is to create a component diagram and one or more deployment diagrams where the components are instantiated.

A node is a hardware platform having processing power and probably some memory. A node can be stereotyped to reflect the type of platform it represents. Virtual machines (RTOS soft kernels or other run-time environments) can be modeled using stereotyped nodes.

Component instances can either reside and execute, on nodes, or implement nodes. The former case is modeled using an abstraction arrow with reside as the stereotype. The latter is modeled using an abstraction with the stereotype implement.

Component and node instances are named using the "Instance:Type" UML notation. If a node or component is instantiated without having a type, only ":Type" is used.

Modeling Agent Instance Sets. In the deployment diagram, it is common practice that only component and node instances are shown. The component is considered a black box. This implies that the names of objects, etc. inside component instances are not specified. If a class is contained in a component in a component diagram, this practice implies that all instances of the class should execute within an instance of the component, no matter at how many locations the class is instantiated or the number of instances.

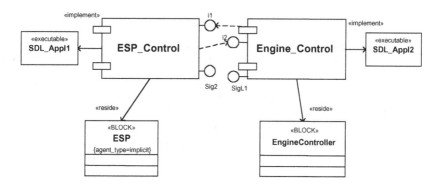

Fig. 9. Two components communicating through interfaces

The deployment diagram should show all instances of threads and all instance sets of agents that reside within a component. It is considered relevant to explicitly specify the names of the agent instance sets, because the focus of SDL system deployment is real-time behavior and partitioning into threads or tasks. As agent types may be instantiated at many different places in an SDL system, it can not be implicitly assumed that all instance sets of an agent type should execute together. The name of an agent instance set should be retrieved from the SDL specification to get a consistent mapping. This makes it possible to partition an SDL system on how it should execute, rather than partitioning by the agent type specification.

An agent instance set is named using the "`Instance:Type`" UML naming convention. An agent instance in an agent instance set has its index as complementary information, such as "`Instance[Index]:Type`". For agents with implicit types, only "`Instance:`" is used.

Referring to the requirements in Sect. 2, it should be possible to map a single instance of an agent to a thread. It is only possible to show at a specific instant in time an SDL agent instance that is created and destroyed at run-time.

The deployment diagram shows instances of threads. As threads are created and destroyed dynamically, the snapshot principle applies also for thread instances in a deployment diagram.

This implies that only static threads can be shown. If dynamic threads are to be shown, this can only be done at single time instants.

Modeling Migration. The Deployment Diagram contains notation for run-time migration. Objects can migrate between components, and components can migrate between nodes. Migration is modeled using a dashed arrow with the stereotype become. In this article, run-time migration is considered out of scope and is not discussed further.

Modeling Communication. Communication is modeled in a Deployment Diagram using associations and dependencies. Associations are drawn between nodes and are used to model physical connections such as a computer network. Associations can be stereotyped to specify the type of connection: for example, Ethernet. Stereotyped nodes can be used for modeling network parts such as hubs or switches.

Dependencies are used for modeling logical connections. They are displayed the same way as in the component diagram: using a dashed arrow that points from a component to an interface at another component. For a component that is instantiated from a component diagram, all interfaces for that component are available in the deployment diagram as well. Interfaces for an agent type are instantiated in a deployment diagram as the agent type is instantiated.

For an instantiated interface, it is relevant to specify the communication mechanism that should be used, such as inter-process communication or a protocol stack. It is modeled using a constraint on the interface.

Using protocol stacks for communication requires addresses on node and component level. This data can be modeled using tagged values. For an interface using TCP/IP, a tagged value for the IP address is needed on node level and one for the TCP port number is needed on the interface. The usage of TCP/IP is modeled as a constraint on the interface.

The communication mechanisms specified at an interface are used both by the component implementing the interface and all components using the interface. This can be illustrated by the TCP/IP example above, where the component implementing the interface acts as a TCP/IP server and the components that use the interface act as TCP/IP clients. The interface construct works for both communication between components containing SDL Agents and for communication between a component with SDL Agents and components containing external code.

Figure 10 shows a deployment diagram using interface communication. The diagram corresponds to the component diagram in Fig. 9. The components are instantiated and deployed onto node instances. An association having Ethernet as a constraint connects the nodes. Each node instance has an IP address as a tagged value. The interfaces $i1$ and $i2$ have TCP/IP as a constraint and TCP port numbers as tagged values. This information is sufficient for addressing components using TCP/IP at run-time.

In the following section, it is described how the scenarios in Sect. 2.2 can be modeled using component and deployment diagrams.

3.5   Deployment Scenarios Modeled Using UML

Deployment of Agents onto Threads of Execution. If the SDL process A is static, this case is easy to model using a deployment diagram. The instances of the threads and the agents can be displayed in a diagram that is considered a valid description during the whole execution.

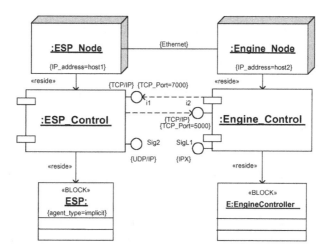

Fig. 10. A deployment diagram with communication through interfaces

If Process A is a process instance set with dynamic creation and destruction of process instances, a deployment diagram of the SDL system on instance level can only show the configuration at a single time instant.

Deployment with Communication with External Software Components. Using the UML interface concept, the communication can be modeled both between SDL Agents in different components and between an SDL Agent and external software. Tagged values can be used as illustrated in Fig. 10 to specify communication parameters. The deployment diagram should contain two components, each having an interface. Each component is deployed on a node using the reside stereotype.

Deployment of a Client-Server Architecture. The core of the client-server scenario is the modeling of an unspecified number of clients communicating with a server. The deployment diagram can show a snapshot of a scenario where a specific number of clients are connected to the server. The component diagram provides a type-based overview of the communication, without any possibility to specify any instance settings. Here, the two diagrams are used in a complementary way. Each diagram provides information not contained in the other.

Deployment of an SDL System that Implements Hardware. This deployment scenario is easy to model using a deployment diagram. The component instance containing the SDL system is connected to a node instance using the implement stereotype.

4   The Usability of UML for SDL Deployment

The requirements presented in Sect. 2.1 focus on modeling of run-time properties. Diagrams drawn using a notation that fulfills the requirements should minimize manual work when SDL systems are targeted. After having presented how the UML implementation diagrams can be used for deployment of SDL systems, the usability and the conformance to the requirements should be evaluated.

Regarding communication, the mapping between the SDL interface and the UML interface works well for deployment. For an agent type having interfaces, each instance renders interface instances at the enclosing component. Tagged values and constraints can be used to describe communication properties at an arbitrary level of detail.

An important detail to consider is that the behavioral part of SDL is built on the concept of processes. This emphasizes the need for a simple mapping between SDL process instances and operating system threads or tasks.

Modeling threads and tasks using components works fairly well. By modeling thread settings using tagged values, thread parameters can be passed on to code generation tools. If desired, new thread stereotypes with platform-specific tagged values and constraints can be created and reused.

As long as only static instances are considered, an instance view alone is adequate for modeling deployment. When dynamic process creation and destruction is considered, however, an instance view does not suffice. A diagram with an instance view cannot display dynamic behavior. Neither can a diagram with a type view, because agent types are not considered relevant to deploy: only instances and instance sets are interesting to deploy.

The fact that instances are the focus of SDL deployment questions the relevance in using the component diagram for SDL deployment. For applications without real-time constraints, having one single execution point, the component diagram provides sufficient information. The targeting process can manage with type information only. When real-time issues are considered, however, the complexity rises. Deployment of instances onto threads or tasks provides better possibilities for optimization. For applications having real-time constraints, that mapping is crucial.

A view is needed which contains mapping rules for dynamic creation and destruction of instances. It should be possible to map an instance set to a thread type and state the policy for creation of thread instances. At the same time, it should allow direct mapping of static instances to thread or task instances. This produces a view that is instance-based with an exception for dynamic instances. Using the mapping rules, a diagram containing sufficient information for targeting is created. The diagram can be regarded as an instance-based diagram containing complementary information. The information in the diagram can be used as input to a code generator.

The level of detail in component and deployment diagrams is dependent on the target platform and the tools that are used for targeting. Tagged values provide a powerful extension mechanism.

Throughout Sect. 3, some examples on tagged values were given. For a node, typical tagged values are the operating system, the processor, and size of memory. For a component, the programming language of the generated code and the names of the make utility, compiler and linker are useful tagged values. Other relevant data is the target directory for the generated code and directories for libraries and object files. For a thread or task, tagged values for priority and stack size are relevant.

Much is left to be done before complete semantics for deployment and targeting can be realized. The UML implementation diagrams provide many ideas for further research in this area.

5   Summary

In this article, requirements on a possible notation for deployment of SDL systems have been formulated. The most important requirements are the support for targeting from a deployment model, the support for mapping SDL agents to OS threads or tasks and the support for communication between applications. Some typical targeting scenarios, where SDL systems execute in target environments, were presented.

The most widespread notation for deployment is the UML implementation diagrams. Using the mapping between SDL and UML in [7], it was investigated how the UML component and deployment diagrams can be used for deployment of SDL systems. Solutions to the targeting scenarios were discussed using notation from the UML implementation diagrams.

Finally, a discussion was held on the usability of the UML implementation diagrams for deployment of SDL systems. The interface primitive for communication provides an effective notation. It was found that the type view in the component diagram adds little value in SDL system deployment, where multiple execution points and real-time constraints are important issues. SDL deployment should be performed using an instance view. When dynamic creation and destruction of agent instances are considered, an instance view can only provide a snapshot of the system run-time configuration. It was concluded that an effective deployment notation requires mapping rules between agent instances and threads.

Further research and evolution of the current deployment notations is needed. The increasing power of code generation tools implicates a need for a powerful notation for controlling them.

References

1. Specification and Description Language (SDL), ITU-T Standard Z.100. International Telecommunication Union. November 1999.
2. Floch, J. *Using UML for Architectural Design of SDL Systems*, SINTEF Report No. STF40 A00009. February 2000.
3. Douglass, B.P., *Doing Hard Time Developing Real-Time Systems with UML, Objects, Frameworks, and Patterns*, Addison-Wesley, Massachusetts. 1999.
4. Message Sequence Charts (MSCs), ITU-T Standard Z.120. International Telecommunication Union. 1996.
5. SDL Combined with UML, ITU-T Standard Z.109. International Telecommunication Union. November 1999.
6. Rumbaugh, J. Jacobson, I. Booch, G., *The Unified Modeling Language Reference Manual*, Addison-Wesley, Massachusetts. December 1998.
7. OMG Unified Modeling Language Specification, version 1.4 Beta R1. Object Management Group. November 2000.

ETSI Testing Activities and the Use of TTCN-3

Anthony Wiles

ETSI PTCC, Sophia Antipolis, France
anthony.wiles@etsi.fr

Abstract. This paper provides an introductory overview of ETSI testing activities related to interoperability and conformance testing. It explains why testing is important and why languages such as TTCN-3 are considered to be key components in the development of ETSI testing specifications. This paper highlights the relationship and harmonization between TTCN-3, MSC, SDL and ASN.1.

1 About ETSI

ETSI, the European Telecommunications Standards Institute, develops a wide range of standards and other technical documents as Europe's contribution to world-wide standardization of telecommunications and associated domains. Based in the Sophia Antipolis science park in southern France, ETSI brings together well-over 800 companies and some 5000 technical experts from around the world.

2 ETSI and Testing

ETSI members have long recognized the important role that validation and testing play in the development of ETSI standards, and of the products based on those standards. Even in a climate where time-to-market is paramount, manufacturers are prepared to put time, effort and valuable expertise into testing in order to ensure interoperability.

ETSI has two permanent entities involved in testing activities. The PTCC (Protocol and Testing Competence Centre) and the Bake-off Service.

2.1 The PTCC

The PTCC provides support and services to ETSI Technical Bodies (TBs) on the application of modern techniques for specifying protocols and test specifications. The PTCC is also responsible for the technical management of the ETSI Specialist Task Forces (STFs) which develop conformance test specifications for ETSI standards. In the past 10 years test suites have been produced for many leading ETSI technologies, including 3G UMTS, GSM, DECT, TETRA, Hiperlan/2, TIPHON, INAP, B-ISDN etc.

R. Reed and J. Reed (Eds.): SDL 2001, LNCS 2078, pp. 123–128, 2001.

While certain areas quite rightly require regulatory testing, the policy today is to keep this to a minimum. As a consequence, there has been a subtle but fundamental change in the development and application of ETSI test suites over the last few years. ETSI test specifications now concentrate on maximizing the chances of interoperability, for example by increased focus on critical functionality or error recovery behaviour. The test suites are not part of a bureaucratic testing program but are increasingly requested for by fora and manufacturers as an integral step in product development processes.

Note that the PTCC does not perform the actual testing but does have close contact with the companies and organizations that do. Neither is ETSI involved in certification.

2.2 The Bake-Off Service

The ETSI Bake-off service organizes bake-offs or interoperability events for ETSI members and non-members alike. It provides the logistical and organizational support (often at ETSI premises) for such events. Recent events include Synch-Fest, IPv6 InterOp, IMTC SuperOp and Bluetooth UnplugFest. The intention with bake-offs is to validate (debug) both the standards and early products or prototypes of those standards as they are developed. This activity is considered to be one of validation, rather than testing.

2.3 Testing for Interoperability

The fact that ETSI develops conformance test suites and provides bake-off services to its members shows that both approaches are considered valuable by the ETSI membership. In some cases only conformance testing is considered adequate, in others the reverse is true. It has been clearly demonstrated in GSM and 3G, for example, that conformance testing of mobile terminals provides a very high degree of confidence that handsets from different manufacturers will interoperate. So much so that with the reduction in regulatory testing for GSM in Europe, operators and manufacturers are co-operating to put in place a voluntary testing scheme for GSM terminals.

In the IP world, where the culture of bake-offs is well-established, it is obvious that this is an excellent way to develop and validate the base standards.

The trend, however, indicates that the industry realizes that it is often not a case of either-or but rather a combination of the two. A focused set of conformance tests can provide an excellent complement to bake-offs and/or interoperability testing. This is a view that is shared by other, non-ETSI bodies such as Bluetooth.

3   Conformance Testing Methodology

Generally, ETSI follows the testing methodology defined in ISO/IEC9646. The methodology defines precise test methods, architectures etc. The test specifications themselves comprise two parts: Test Purposes written in prose and detailed tests, or Test Cases, written in the test language TTCN.

All ETSI conformance test specifications are developed by groups of experts, better known as a Specialist Task Force or STF, recruited from the ETSI membership. Experts from the relevant Technical Bodies, manufacturers, test system developers and other interested parties are also closely involved in the development of test specifications.

In many cases, this activity gives valuable feedback to the base standards. At ETSI the test specifiers make a special effort to ensure that their experiences are fed-back to the relevant technical body. However, if this input is to be effective it is essential that the development of the base standard and the test specifications is done in a co-ordinated and timely manner.

An important spin-off from conformance testing, and one that is widely used even in cases where no test specifications are developed, is the ICS (Implementation Conformance Statement). This document, which is essentially a checklist of the capabilities supported by a given system, provides an overview of the features and options that are actually implemented in any given product. A comparison of the ICS for two (or more) different products can give an early indication of potential interoperability problems.

4   Use of TTCN-3

Nearly all ETSI conformance test specifications are written in the standardized test language TTCN. The latest version of this language, known as TTCN-3 [1], has been developed by ETSI Technical Committee MTS (Methods for Testing and Specification).

Typical areas of application of TTCN-3 are protocol testing including mobile and Internet protocols (including text-based protocols), supplementary service testing, module testing, testing of CORBA based platforms, testing of APIs etc. TTCN-3 is no longer restricted to conformance testing. It can be used for many types of test specification including interoperability, robustness (torture), regression, system and integration testing.

Use of a standardized test language, such as TTCN benefits both ETSI(who produce the tests) and ETSI members (who use the tests). For example:

- education and training costs are rationalized and reduced;
- maintenance of test suites (and products) is easier;
- off the shelf tools and TTCN-based test systems are readily available;
- universally understood syntax and operational semantics;
- tests concentrate on the purpose of the test (not the test system);
- allows application of a common methodology and style.
- constant maintenance and development of the language.

4.1   Main Capabilities of TTCN-3

Most importantly, the language incorporates well-proven testing-specific capabilities which as a whole are not present in any other programming language. For example

- dynamic concurrent testing configurations;
- synchronous and asynchronous communication mechanisms;
- encoding information and other attributes (including user extensibility);
- data and signature templates with powerful matching mechanisms;
- optional presentation formats (e.g., tabular format, MSC-like format);
- assignment and handling of test verdicts;
- test suite parameterization and test case selection mechanisms;
- combined use of TTCN-3 with ASN.1 (and other languages such as IDL);
- well-defined syntax, interchange format and static semantics;
- precise execution algorithm (operational semantics).
- type and value parameterization.

The top-level unit of TTCN-3 is a module. A module cannot be structured into sub-modules. A module can import definitions from other modules. Modules can have parameter lists to give a form of test suite parameterization similar to the PICS and PIXIT parameterization mechanisms of TTCN-2.

A module consists of a definitions part and a control part. The definitions part of a module defines test components, communication ports, data types, constants, test data templates, functions, signatures for procedure calls at ports, test cases etc.

The control part of a module calls the test cases and controls their execution. The control part may also declare (local) variables etc. Program statements (such as **if-else** and **do-while**) can be used to specify the selection and execution order of individual test cases. The concept of global variables is not supported in TTCN-3.

TTCN-3 has a number of pre-defined basic data types as well as structured types such as records, sets, unions, enumerated types and arrays. As an option, ASN.1 types and values may be used with TTCN-3 by importation.

A special kind of data value called a template provides parameterization and matching mechanisms for specifying test data to be sent or received over the test ports. The operations on these ports provide both asynchronous and synchronous communication capabilities. Procedure calls may be used for testing implementations which are not message based.

Dynamic test behaviour is expressed as test cases. TTCN-3 program statements include powerful behaviour description mechanisms such as alternative reception of communication and timer events, interleaving and default behaviour. Test verdict assignment and logging mechanisms are also supported.

Finally, most TTCN-3 language elements may be assigned attributes such as encoding information and display attributes. It is also possible to specify (non-standardized) user-defined attributes. The display attributes have been used to define two standardized (but optional) presentation formats of the core language.

The tabular format [2] is a simplified form of the traditional TTCN tables. The graphical format [3] presents TTCN-3 in the form of sequence charts.

4.2   Differences between TTCN-2 and TTCN-3

TTCN-3 is not backward compatible with TTCN-2. The fundamental paradigm shift of making the core language into a modern programming language means that the TTCN syntax has been completely redefined. There is, however, an initial mapping document and tools which successfully translate TTCN-2 test suites to TTCN-3. The Tabular Presentation Format has been considerably simplified, with the tables being more generic. This means that the number of different tables has been reduced to about one-third. Naming conventions may be used to distinguish between objects which are syntactically similar but play different roles in a test suite (for example, ASPs and PDUs). The designers of TTCN-3 have tried to ensure that the basic operational semantics of the language are unchanged. A TTCN-2 test suite should execute in the same way that its counterpart in TTCN-3 would.

Major additional functionality includes the dynamic creation of test components, synchronous communication mechanisms, the introduction of new types, a generic template mechanism, cleaner harmonization with ASN.1 and the Graphical Presentation Format.

There is a significant investment in TTCN-2 test suites both in industry and in standardization bodies such as the ITU-T and ETSI. The purpose of TTCN-3 is to expand the use of the language into application areas where it is not currently used. It is ETSI's policy that the two versions of the language will exist for some time to come. It is probable that technologies that have made extensive use of TTCN-2 in the past (such as ISDN, GSM, Hiperlan, IN and DECT) will continue to do use TTCN-2 in order to preserve that investment. However, it is expected that as commercial tools become available new ETSI test specifications, especially for technologies that make use of protocols such as the Session Initiation Protocol (SIP) as specified in RFC 2543 or the XML-based Open Settlement Protocol (OSP) will be written in TTCN-3. It is planned that some 3G test specifications will start using TTCN-3 in early 2002. There are no plans for a wholesale upgrade of existing TTCN-2 specifications to TTCN-3, though this may be done on an as needed basis for some individual test suites

4.3   Harmonization of TTCN-3 with Other Languages

The testing process is never isolated. A good test specification goes hand-in-hand with a good requirements specification. ETSI encourages the use of MSC and SDL in its base specifications to ensure technical quality. The MSCs not only clarify the behaviour of a protocol but can also form the basis for an initial set of test purposes and, if a suitable SDL model exists, test case generation tools can be used. For example, both MSC (including HMSC) and SDL have been used very effectively in the specification of the ETSI Hiperlan/2 RLC (Radio Link Control) protocol. The developers of TTCN-3 have taken care to ensure the harmonization of the language with ASN.1, MSC and SDL. ASN.1 type specifications can be cleanly imported and used in a TTCN-3 test suite. The semantics of TTCN-3 (especially the communication mechanisms) are compatible with the underlying semantics of MSC and SDL. This ensures that the languages can be used together in all stages of the development process, for example using SDL for the base specification, MSC for the test purposes and TTCN-3 for the test specification.

5 Conclusions

ETSI has long been committed to improving the quality of its standards and related products through validation and testing. To this end ETSI actively contributes to the development and standardization of modern testing techniques (such as TTCN-3). The use of techniques such as TTCN and SDL is well-accepted by many ETSI Technical Bodies.

There are already indications that TTCN-3 is fulfilling its promise of opening the use of the technique to new areas of testing. Now that TTCN-3 and some early tools are available TIPHON has started the production of conformance test specifications for the Session Initiation Protocol (SIP) as specified in RFC 2543. Work will begin shortly on the test specifications for the TIPHON XML-based Open Settlement Protocol (OSP). This work is expected to continue throughout 2001, the goal being "Conformance testing for interoperability".

The near future is expected to see an increased use by ETSI Members of the combined services of the ETSI Bake-off Service and the PTCC.

References

1. ETSI ES 201 873-1, Methods for Testing and Specification; The Tree and Tabular Combined Notation version 3; Part 1: TTCN-3 Core Language.
2. ETSI ES 201 873-2, Methods for Testing and Specification; The Tree and Tabular Combined Notation version 3; Part 2: TTCN-3 Tabular Presentation Format.
3. ETSI TR 101 873-3, Methods for Testing and Specification; The Tree and Tabular Combined Notation version 3; Part 3: TTCN-3 Graphical Presentation Format.

HyperMSCs with Connectors for Advanced Visual System Modelling and Testing

Jens Grabowski[1], Peter Graubmann[2], and Ekkart Rudolph[3]

[1] Universität zu Lübeck, Institut für Telematik
Ratzeburger Allee, D-23538 Lübeck, Germany
grabowsk@itm.mu-luebeck.de
[2] Siemens AG, Corporate Technology, Software and Engineering
Otto-Hahn-Ring 6, D-81739 München, Germany
peter.graubmann@mchp.siemens.de
[3] Technische Universität München, Institut für Informatik
Arcisstraße 21, D-80290 München, Germany
rudolphe@informatik.tu-muenchen.de

Abstract. Experiences with the use of the MSC language for complex system specifications have shown that certain extensions are necessary in order to arrive at sufficiently transparent and manageable descriptions. Extended HMSCs, where MSC reference symbols may either be presented by hypertext-like descriptions or, in an expanded form, as detailed MSCs, appear to be especially suitable for a compact and transparent MSC representation. For an effective usage of such advanced MSC constructs, a corresponding tool support seems to be mandatory where interactively the event structures of special paths can explicitly be expanded while others remain hidden as MSC references that contain solely textual descriptions. The name '*HyperMSCs*' is proposed for such extended HMSCs. Beyond that, the communication between MSC references, operator expressions or HMSCs demands a generalisation of the gate concept. For that purpose, the introduction of MSC *connectors* denoting logical connections is suggested. MSC connectors may be expanded similar to MSC references. HyperMSCs enhanced by MSC connectors also provide a means for a selected visualisation of large MSCs in an interactive manner where, depending on the current selection, some parts are exhibited in full detail whereas other parts are presented in an abbreviated form. The same concepts may be applied for system modelling based on stepwise refinement starting with HyperMSCs, decomposed instances and MSC connector communication and for system testing.

1 Introduction

While Message Sequence Charts (MSCs) without any doubt are amongst the most popular and successful description techniques now, their real potential has not yet been exploited. Although the MSC language [8,15,22] contains very powerful composition mechanisms and other structural concepts, the language is still

R. Reed and J. Reed (Eds.): SDL 2001, LNCS 2078, pp. 129–147, 2001.
© Springer-Verlag Berlin Heidelberg 2001

used essentially to define a set of sample behaviours. However, with the increasing popularity of the MSC language, a more comprehensive application is demanded by some user communities. Recently, MSCs have been applied for a graphical presentation format for TTCN-3 test cases [16,17,19] and appears more intuitive than the tabular presentation format of TTCN-3. Experience with a comprehensive MSC based specification has shown that the MSC language needs certain extensions to facilitate reading and understanding of MSC diagrams, and to support and ease the handling of MSC documents. These extensions are strongly related to a corresponding advanced tool support. Hypertext-like mechanisms have been suggested for an appropriate handling of large MSC documents [6,16,17]. Because of this hypertext analogy, the term 'HyperMSC' has been introduced.

More generally, HyperMSCs can be viewed as a means for a selective detailed visualisation of certain parts of MSC, and consequently a hiding of other parts. With suitable tool support, such a selective visualisation may be quite flexible:

- MSC references may be folded and unfolded presenting their contents in a highly interactive manner; or
- decomposed instances can be used as a special kind of MSC reference to hide the message interaction description between selected instances if these messages actually are not in the focus of interest.

In addition to MSC references interpreted in a hypertext-like manner, the introduction of the *MSC connector concept* into the MSC language may be valuable for the same purpose. For complex MSC operator expressions MSC connectors have been proposed as a generalisation of the gate concept in [12,13]. They may be used quite generally for the hiding of the detailed description of the message flow between MSC references or decomposed instances. Furthermore, MSC connectors may be used to predefine communication patterns that could be re-used in different places thus providing for the compositionality of MSC specifications.

In the next section, a brief description is given of the MSC presentation format developed for TTCN-3 test cases at ETSI. This provides a motivation for the concepts of HyperMSCs and MSC connectors. Within Sect. 3, the concept of HyperMSCs is elaborated and demonstrated using an example of the CCBS service specification. Sect. 4 is dedicated to the introduction and elaboration of the MSC connector concept and in particular to the inclusion of MSC connectors into the HyperMSC concept. In Sect. 5, a summary and outlook is provided.

2   Motivation: A HyperMSC-Based Presentation Format for TTCN-3 Test Suites

An MSC-based presentation format for TTCN-3 [19] has been developed as part of the ETSI Specialist Task Force on "Specification of a Message Sequence Chart/UML format, including validation for TTCN-3" – [16,17, and ETSI STF 156]. Experiments with different variants of MSC representations have shown that extensions of the MSC language are required in order to obtain a sufficiently transparent and readable MSC test description.

The most obvious and straightforward way to represent TTCN test cases by MSC diagrams would be to use inline operator expressions for alternatives, iterations, etc. Practice has shown that apart from simple cases, such a naive translation does not lead to diagrams that are easy to read and understand. In particular, inline operator expressions obscure the message flow for important cases: paths leading to a PASS verdict (PASS cases) are mixed with INCONC cases (paths leading to an INCONCLUSVE verdict) and FAIL cases (paths leading to a FAIL verdict).

The obvious drawback of this representation is the fact that PASS, INCONC, and FAIL cases are all presented in the same way. Certainly, it would be advantageous to have a means to emphasise the PASS cases in the test representation. The MSC language already contains several structuring mechanisms, and for this special purpose, the MSC reference mechanism seems to be suited. To highlight the PASS cases, all other cases may be represented as MSC references. Yet this notation has an immediate drawback: without explicitly looking into the definitions of the referenced MSCs, there is no immediate information about the hidden cases. This means that, the representation of the complete test case is not very intuitive except for the presented message exchange.

A much more satisfactory representation is obtained if a comment together with the test verdict is included in the MSC reference symbols instead of reference names. However, in the case of many alternatives the resulting representation is still not sufficiently transparent. As a general rule, inline expressions should be used only in a very limited manner and remain restricted to one or two alternatives or loops. In more complex situations, HMSCs are much more transparent since they abstract from details and focus on the compositional structure. However, if the inline representation is translated into an HMSC, we are faced with the problem to represent the expanded parts because (according to the standard) an HMSC contains only non-expanded MSC reference symbols. In order to overcome this deficiency, HyperMSCs are introduced which admit expanded MSC references within the HMSC formalism [10]. By shifting the branching points to the borderline, the PASS case is shown in expanded form in a coherent manner whereas the INCONC and FAIL cases are indicated as non-expanded MSC references. With advanced tool support the roles of the PASS and the INCONC cases may be changed interactively, and the HyperMSC document becomes a powerful tool for reading, presenting and analysing test cases.

A significant difference between TTCN and MSC shows itself in the behaviour description of several test components. TTCN-3 defines the behaviour of components of the same test case in separate functions. In corresponding MSCs this partitioning results in test component specifications that are merged by join operations. For an appropriate specification of such a join operation, an MSC connector has been proposed which allows description of the exchange of co-ordination messages among test components on a suitable level of abstraction.

HyperMSCs and MSC connectors are general concepts that allow an intuitive presentation of comprehensive communication patterns and behaviour. Their usage will not be restricted to the testing community only and thus, both concepts, we propose, should find their way into the MSC language.

3   From HMSCs to HyperMSCs

In the case of a great number of alternatives, the inline branching constructs for MSCs are not very transparent. In particular, they obscure the presentation of the complete message flow. In UML Sequence Diagrams, there is a tendency to indicate branching and iterations at the borderline of the diagram either using special graphical constructs, such as **for** loops, or even using a program-like form. In this way, the diagrams may be focussing on the representation of the pure message flow. In practice, however, such a notation again soon becomes quite intricate and clumsy, particularly for nested alternatives or loops. In the following, we re-formulate High Level MSCs (HMSCs) in such a way that they serve for a similar purpose. The obtained notation turns out to be very intuitive and simple, even in more complicated situations.

HMSCs describe the composition of MSCs as a graph with MSC references and conditions as nodes [10,15]. This way, they abstract from instances and messages which are not shown in the diagram. Each MSC reference has a name. Its meaning is given by a corresponding MSC with the same name defined in another place in the same MSC document. HMSC diagrams with many references that refer to many fairly small MSC definitions, soon become again quite complex and difficult to maintain and handle.

In order to arrive at user-friendlier handling, HMSCs may be re-interpreted in a way that has an analogy in hypertext-like specifications. MSC references may be shown optionally also in an expanded manner within the HMSC and non-expanded MSC references may contain hypertext-like descriptions instead of pure reference names [6,16]. This implies a real extension of the MSC standard, but essentially concerns the handling and the graphical layout only: it has no effect on the semantics. Thereby, we assume a corresponding tool support where the MSC references can be interactively expanded within the HMSC in which they are contained or are possibly in a separate window (for example double clicking into an MSC reference symbol to fold or unfold it, etc.). Obviously, in this context tools should have appropriate mechanisms and adequate means to define MSC references inline.

Drawing expanded MSC references in HMSC presentations has been already suggested in [10], but without the additional dynamic mechanisms of the HyperMSCs, this only inflates the diagrams and makes them even harder to understand. In addition, gates were introduced in [10] which proves to be not completely appropriate. Our proposed solution for this problem is the connector concept which will be presented in Sect. 4.

In Fig. 1, the concept of HyperMSCs is indicated schematically. Fig. 1(a) shows an HMSC with connection points between the MSC references R1 and

$R2/R4$ and between the MSC references $R2$ and $R3/R5$. In Fig. 1(b) the MSC references $R1$, $R2$, and $R3$ appear in expanded form within the HMSC diagram.

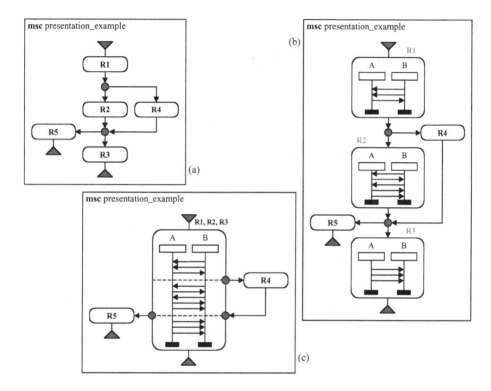

Fig. 1. Various representations of one HyperMSC
 (a) as HMSC,
 (b) with expanded MSC references,
 (c) with expanded MSC references as coherent "path of interest".

The HyperMSC that results if we only admit expanded MSC references still has the drawback that the main event flow usually is split into many separate parts. In the case of many alternatives describing exceptional or erroneous behaviour, this splitting is very disturbing since it is not possible to show the main flow in a coherent manner in expanded form. One would like to have a coherent expanded representation of a whole path not only in a separate window, but also in inline-form within the HyperMSC itself. Therefore, a further extension of HMSCs has been suggested which somehow may be viewed also as a unification of High Level MSCs and plain MSCs. Several expanded MSC references, which are interrupted by branching points, may be combined to one coherent MSC reference. As a consequence, the branching (or connection) points have to be shifted to the borderline of the resulting MSC reference. In Fig. 1(c), the MSC references $R1$, $R2$ and $R3$ are combined to one coherent message flow while the connector points are shifted to the borderline.

Such a coherent representation is possible also in the case of cyclic or compound HMSCs, and thus may cope with much more complex situations than just few alternatives. For nested alternatives the hierarchical structuring of HMSCs is a major advantage. HMSCs are hierarchical in the sense that a reference in an HMSC may again refer to an HMSC.

Particular attention has to be paid to keep the different appearances of the MSC references identified. Within HyperMSCs, MSC references can be displayed in several ways:

1. as MSC reference symbol with the proper name of the reference inscribed (see Fig. 1(a));
2. as MSC reference symbol with an explanatory text, the inscription (see Fig. 3(b));
3. as expansion of the MSC reference (see Fig. 1(b) and (c)).

One may imagine other representations as well: program code fragments representing the behaviour of the MSC, TTCN-3 test definitions, etc. However, it is important, that in all but the case (1), the MSC reference name is attached additionally to the presentation, to unambiguously identify expanded elements. There are several possibilities to attach the MSC reference name (see Fig. 1(b) and (c) where the names of the expanded elements are placed on top of the expansion, or Fig. 6 where the reference to the expanded MSC references is to be found after the dashed line which separates the expansion from the previous one).

It has to be pointed out, that the expansion of MSC references within the HyperMSC concept is not purely a tool issue, but means a real graphical extension of the MSC language because MSC reference symbols containing detailed MSCs are not allowed in the MSC standard language. This extension is in our view most important since the dynamics behind the HyperMSCs (the expanding and folding of MSC references, etc.) is an essential means to increase the understandability of MSC diagrams, in particular if self-contained parts of a system have to be described completely.

The concept of HyperMSCs can also be carried over to MSC operator expressions. For that, we want to recall the relation between MSC operator and inline expressions. As can be seen in Fig. 2, inline expressions just describe unfolded MSC reference operator expressions. For MSC operator expressions the unfolding is well established (except for the operator **seq**).

3.1   Integration of HyperMSC into the MSC Language

HyperMSCs provide the concepts for an MSC presentation that allow an interactive folding or unfolding of parts of the MSC diagrams. Of course, the full benefits of these concepts can be only realised with adequate tool support. In general, it has to be decided manually what is to be folded and what is to be displayed in full detail. However, sophisticated tools may analyse HMSCs and MSC

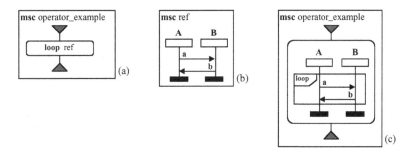

Fig. 2. Expansion of an MSC operator expression
(a) with its MSC reference (b) by means of an inline expression which results in (c).

diagrams and display them in an optimal form (according to built-in strategies or user predefinitions). For tool interaction and the exchange of diagrams, the MSC/PR language plays an important role, thus, the HyperMSC concepts have to be reflected in the textual presentation form, too. We propose the integration of the HyperMSC concepts into the MSC/PR language by augmenting of the MSC/PR with XML tags.

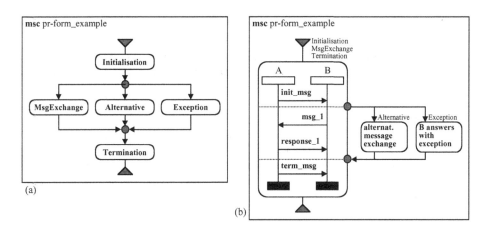

Fig. 3. HyperMSC presentation of an MSC
(a) completely folded; (b) partially expanded

A small example may provide a first idea of what an HyperMSC consistent extension of the MSC/PR may look like. Fig. 3 shows two variants of the graphical representation of a HyperMSC. Fig. 3(a) is the classical HMSC presentation, Fig. 3(b) is one of its graphical presentation variants with three expanded MSC references and two folded ones, showing the explanatory 'inscription' in lieu of the reference name.

The corresponding PR form is presented in Fig. 4. In addition to the basically unchanged structural description of the HyperMSC, it contains the XML tag

```
msc pr-form_example;
expr L1;
L1: <hyperref
       representation=folded&name
       inscription=no-inscription
       expansion='C:\>Initialisation.msc'>
    (Initialisation) <\hyperref> seq (L2);
L2: connect seq (L3 alt L4 alt L5);
L3: <hyperref
       representation=folded&name
       inscription='expected message exchange'
       expansion='C:\>MsgExchange.msc'>
    (MsgExchange) <\hyperref> seq (L6);
L4: <hyperref
       representation=folded&name
       inscription='alternat. message exchange'
       expansion='C:\>Alternative.msc'>
    (Alternative) <\hyperref> seq (L6);
L5: <hyperref
       representation=folded&name
       inscription='B answers with exception'
       expansion='C:\>Exception.msc'>
    (Exception) <\hyperref> seq (L6);
L6: connect seq (L7);
L7: <hyperref
       representation=folded&name
       inscription=no-inscription
       expansion='C:\>Termination.msc'>
    (Termination) <\hyperref> seq (L8);
L8: end;
endmsc;
```
(a)

```
msc pr-form_example;
expr L1;
L1: <hyperref
       representation=expanded&init
       inscription=no-inscription
       expansion='C:\>Initialisation.msc'>
    (Initialisation) <\hyperref> seq (L2);
L2: connect seq (L3 alt L4 alt L5);
L3: <hyperref
       representation=expanded&continued
       inscription='expected message exchange'
       expansion='C:\>MsgExchange.msc'>
    (MsgExchange) <\hyperref> seq (L6);
L4: <hyperref
       representation=folded&inscription
       inscription='alternat. message exchange'
       expansion='C:\>Alternative.msc'>
    (Alternative) <\hyperref> seq (L6);
L5: <hyperref
       representation=folded&inscription
       inscription='B answers with exception'
       expansion='C:\>Exception.msc'>
    (Exception) <\hyperref> seq (L6);
L6: connect seq (L7);
L7: <hyperref
       representation=expanded&continued
       inscription=no-inscription
       expansion='C:\>Termination.msc'>
    (Termination) <\hyperref> seq (L8);
L8: end;
endmsc;
```
(b)

Fig. 4. Textual representation (PR form) of the HyperMSC 'pr-form_example' in Fig. 3

(a) textual representation of the completely folded HTML representation;
(b) textual representation of the partial unfolded variant of Fig. 3(b).

hyperref which controls the GR form presentation of the MSC references. The attributes of the hyperref tag are **representation** (indicating whether the MSC reference is to be presented folded or unfolded, showing its proper name or with an explanatory description attached, etc.), **inscription** (containing the text of the explanatory description), **expansion** (referencing the location of the MSC reference expansion), etc.

The instrumentation may not be restricted to special MSC language constructs. It should be possible to identify arbitrary parts of MSC and HMSC diagrams that can be folded to HyperMSC references and when required can be re-expanded. The HyperMSC mechanism adds the possibility to structure information on the presentation level to the MSC language.

3.2   Usage of HyperMSC

HyperMSC is a means to emphasise selected behaviour and to abstract behaviour alternatives that are currently less relevant. Hypertext-like inscriptions within MSC reference symbols provide a natural interface to different MSC user communities such as system designers, system developers and test engineers. For system designers the HyperMSC concept provides a tool that allows specifying

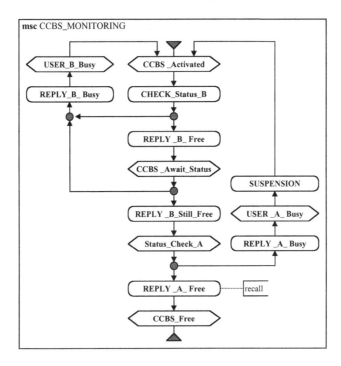

Fig. 5. CCBS example: Monitoring

system behaviour in the form of purely textual descriptions that later on are refined into the form of concrete MSCs. For other user communities, Hyper-MSCs may be used to guide the understanding and manipulation of MSCs by emphasising behaviour relevant for the momentary analysis and investigation and providing additional support in the form of explanatory descriptions placed into MSC reference symbols instead of the MSC reference name.

As has been outlined in Sect. 3, the HyperMSC concept was stimulated by the development of a graphical test format. However, HyperMSCs may in fact have a much larger area of application. In Fig. 5 and Fig. 6, the usage of the HyperMSC concept is demonstrated by means of an extract of the CCBS (Completion of Calls to Busy Subscriber) service specification (ISDN) [15]. The HMSC in Fig. 5 which describes only a fairly small and simplified part of the CCBS service contains already quite a considerable number of MSC references. Since in standard MSC these MSC references are defined by separated MSC diagrams within the MSC document, so that the representation consists of many small pieces rather than providing a coherent view. As a consequence, such a specification would fail the main purpose of the MSC language. Obviously, a much more satisfactory and convincing representation is provided by means of the corresponding Hyper-MSC in Fig. 6, in which the main path is shown in an expanded and coherent manner. The side cases contained in the non-expanded MSC references may be expanded individually in a hypertext-like manner. The chosen CCBS example demonstrates that within HyperMSC alternatives as well as cyclic behaviour may be represented in a convincing and transparent form, even in more complex cases.

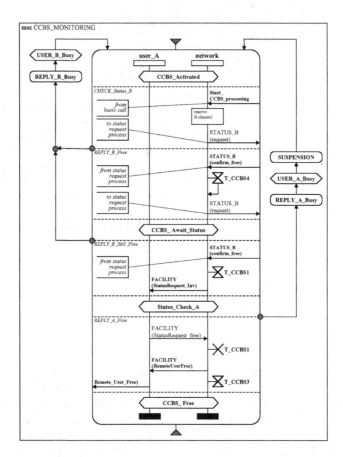

Fig. 6. CCBS example: Monitoring partially expanded

4   MSC Connectors

For a long time the MSC language only offered the communication pattern of basic message exchange: a non-blocking asynchronous dispatch of information from a sender to a receiver. The means to express a blocking synchronous communication has only recently been introduced (see also [6,14]). However, there are not yet language constructs that provide for further abstractions of the information interchange – in contrast to the well known MSC references and operator expressions which serve as behaviour abstractions. The need for communication abstractions, however, is obvious. There are three main reasons for their introduction:

1. *MSC language completeness:* Behaviour abstractions indeed need the accompanying abstractions of the information exchange, which in this case is just the well-defined folding of the messages coming in and going out of MSC operator expressions (find more details below).

2. *Practicability of MSCs for describing complex systems:* To exploit MSCs as a description technique for practical system design (beyond the mere specification of a few exemplary behaviour traces), abstraction is still the key issue in handling the system complexity. Compositionality and re-use are in the centre of modern description techniques, because today's system development is component oriented and deals with system families and product lines. In this context, the communication between the parts of a system (components, objects, etc.) will follow interaction patterns which are expressed in the component oriented world as software connectors [11]. They are defined once then implemented and then re-used over and over again (see system families where interaction patterns are essential). MSC connectors mirror these facts and allow separate definition of recurring interaction patterns and joint application with the other MSC abstractions. The same issue is stressed in Sect. 2 where the adequate abstractions for the test descriptions are presented.

3. *Satisfactory HyperMSC presentation techniques:* MSC connectors appear as a by-product of the HyperMSC concept. Folding and unfolding of parts of an MSC diagram is essential to present an MSC in a form where its interesting parts are shown in full detail but all others are folded away. To provide a correct and comprehensive MSC representation in spite of the folding, a well-defined abstraction mechanism has to be in place (see Fig. 13). The consequent elaboration of the HyperMSC concept leads to mechanisms that allow either folding or unfolding (parts of) instances to MSC references and to subsume groups of messages into MSC connectors.

In this paper, we focus our discussion on the issue (1) and the issue (3) above, where MSC connectors can be employed in a natural, somewhat simplified and default-based manner. Issue (2), which shows the great potential of the MSC connector concept, is bound to utilise more general mechanisms to reach its full effectiveness. This issue will be detailed in a further publication (see also [7]).

The introduction of MSC connectors appears to be inevitable in order to clearly define the message communication between general MSC operator expressions [10,12,13]. The current MSC standard is arguably not precise enough in this point by using (or perhaps abusing) normal messages as sort of MSC connectors. Actually the standard MSC language uses message gates to define the connection points for messages with respect to the interior and exterior of MSC references and inline expressions. The new MSC connector concept is supposed to elegantly subsume and generalise the gate construct.

Within Fig. 7(a), an example is provided showing two inline loop expressions connected by one message which in fact has the meaning of three connecting messages (as indicated in Fig. 7(b)). Yet, the message m in Fig. 7(a) between the two loop expressions in fact denotes an MSC connector and thus should be graphically distinguished (see Fig. 7(c)).

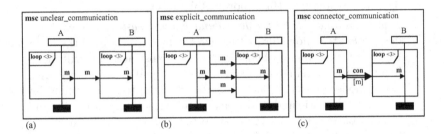

(a)   (b)   (c)

Fig. 7. Message communication between loop expressions
(a) rather unclear description according to the current MSC standard;
(b) explicit description; (c) description using an MSC connector.

From this example, we see that a message specification in an MSC defines an actual occurrence of a message (the sending and receiving event) whereas an MSC connector alone only denotes the *possibility* of message occurrences. Actually, the MSC connector needs a definition that determines what kind of communication pattern it represents. In the above example, the connector *con* may be defined as accepting an arbitrary number of messages m (in Sect. 4.1, the explicit definition of MSC connectors will be briefly discussed).

MSC connectors between MSC operators and references bundle the messages that are crossing the environment of the references (see Fig. 8). Within an MSC reference, the *connector pointer* indicates to which connector a message is sent (`<message_name>` → `<connector name>`) or from which connector it is received (`<connector_name>` → `<message_name>`). The connector itself is presented graphically as a double lined arrow (possibly with two arrow heads, if the connector represents a bi-directional communication). As well as its name the connector is associated with the message list which indicates the names of the messages that are admissible to the connector (see Fig. 8(a) with the message list [a, b] attached to the connector *con*).

Experiments with MSC connectors in large applications will determine how defaults for connector pointers, message lists, etc., will be optimally chosen. There could be rules such as "For an MSC reference, all messages from and to the environment that are listed in the message list of an attached connector are passed through this connector".

4.1   MSC Connector Specifications

MSC connectors do not only bundle a bunch of messages between MSC references, they denote a particular behaviour. Depending on the application, various communication mechanisms may be assigned to MSC connectors. Therefore it becomes necessary to explicitly specify the behaviour of each MSC connector

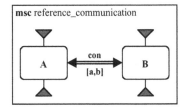 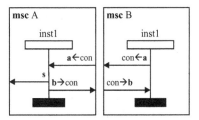

Fig. 8. Communication between MSC references via an MSC connector

prior to its usage or to use pre-defined default connector types (as we did in the examples given here, assuming the consistency as given[1]). In the following, we discuss a few MSC connector types that may be useful in general and we briefly indicate which mechanisms can be used to define the connector behaviour:

The unrestricted connector: In its most general form an MSC connector transmits messages in arbitrary order if they are named in its message list. Whether these messages cross each other is not determined. Therefore the order relation between sending and consumption of messages is exclusively specified by the respective event definitions given by the connected components.

The FIFO connector: The FIFO connector is probably the connector type which describes the most common situation. It is capable of accepting messages in an arbitrary order, as long as they are named in its message list. However, in contrast to the unrestricted connector, messages from the connected components come out of the connector in the same order in which they are put onto it: messages must not overtake (FIFO property). A FIFO connector can transport a message only if it has been put onto the MSC connector first. As a consequence, alternative or loop operands which are initiated by an input message coming out of an MSC connector are enabled only if this message has been put onto the connector first – and (of course) if any guards that are present evaluate to true.

Similarly, a LIFO connector may be defined which conveys messages to a connected component in inverse input order.

The unrestricted message list: Message lists determine the names of the messages that are admitted into an MSC connector (an example of the usage of message lists is to be found in Fig. 8). It proves convenient to define an unrestricted message list, which allows all messages to be sent and retrieved from a connector. This unrestricted message list is denoted as [∗], but may be omitted.

Regular expressions for the description of more complex communication patterns: Because MSC connectors are assumed to support the specification of systems or components, a certain behaviour is associated with them. Both connectors described above represent connectors with a very unrestricted behaviour. They can be conveniently applied in many cases. In other cases, however, where systems are composed out of existing or separately defined components, it is

---

[1] We also suppressed the necessary typing of the MSC connectors in the examples to clearly present the concepts without going into mere technicalities.

appropriate to associate connection semantics with connectors and to use them with a more restricted behaviour. To allow for such a behaviour definition, regular expressions over message names may be assigned to an MSC connector (see the example in Fig. 9) or separately to the connector endpoints (this allows an explicit specification of crossing messages, though some consistency requirements apply). The regular expression is attached to the connector and enclosed in guillemet brackets (for example «{ab}∗»). An omitted regular expression means the behaviour of the connector unrestricted.

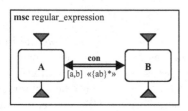 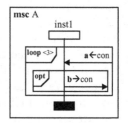 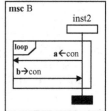

Fig. 9. Communication of MSC references via an MSC connector with explicit behaviour specification

The MSC reference A is able to produce the message sequences «{{$a|ab$}^{3}}», B is able to produce the message sequences «{ab}∗». Due to the connector definition, the composed MSC produces the message sequence «{ab}^{3}».

Any behaviour of the connected components that does not match with the connector definition is excluded (see Fig. 9 for an informal explanation).

MSC representation of connector behaviour: It is easy to imagine, that the most convenient way to define the behaviour of an MSC connector is by providing it as another MSC. This also will provide an appropriate framework to present the formal semantics of the connector concept based upon the synchronisation of event traces. However, a few technicalities are involved in presenting this coherently, therefore a detailed presentation is postponed to a forthcoming paper (meanwhile, see [7] for details).

It may also be of interest to point out the differences of MSC connectors and SDL channels. In SDL, channels are part of a static architectural specification: they connect outputs and input queues of processes. With respect to the dynamic specification, an output corresponds to the send event and the input queue to the reception event of messages. In MSC, however, the message events normally are interpreted as sending and processing events. To specify the reception event, the inclusion of additional instances representing the behaviour of the input queue would be necessary (which could be done transparently with an appropriately defined MSC connector). Beyond that, the MSC connector construct is introduced on a highly abstract level denoting a purely logical communication construct whereas in SDL, a channel is predominantly static and usually closer to the description of a realisation mechanism.

4.2 Combining MSC Connectors with the HyperMSC Concept

The inclusion of MSC connectors into the MSC language also leads to a generalisation of the HyperMSC concept described in Sect. 3. Where simple MSCs without branching or loops define MSC references that are joined by a connector, this generalisation appears to be quite straightforward: the MSC references may be expanded as usual. In addition, the MSC connector may be expanded thus exhibiting the detailed message communication between the MSC references. An example is provided in Fig. 10. Figure 10(a) shows the HMSC *reference_communication* of Fig. 8 with expanded MSC references. In Fig. 10(b) also the connector *con* is expanded, thus connecting the messages a and b of MSC reference A and MSC reference B to one coherent message flow.

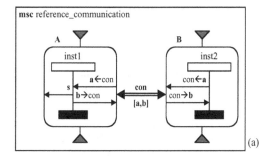 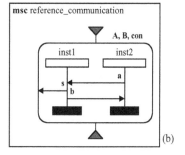

(a)

(b)

Fig. 10. Unfolding of MSC references and MSC connectors (see Fig. 8)

The HyperMSC presentation with folded MSC references A and B): (a) Unfolding of the MSC references and (b) jointly unfolding of the MSC references and the MSC connector.

The situation is more complicated if the connected MSC references are defined by means of MSCs containing alternatives as in Fig. 11. The MSC references A and B in MSC *alternative_communication* are defined in form of HMSC A and HMSC B which contain as alternative branches the MSC references $A1/A2$ and the MSC references $B1/B2$, respectively. Obviously, MSC $A1$ only can be matched with MSC $B1$ and MSC $A2$ only with MSC $B2$. As a rule, it should therefore be only allowed to expand corresponding alternatives which fit together with respect to the connector communication. For example, in Fig. 12(a) the corresponding alternatives MSC reference A1 and MSC reference $B1$ are presented in expanded form. Fig. 12(b) provides a representation showing MSC reference $A1$ and $B1$ together with the connector *con* in completely expanded form.

The previous examples demonstrate how HyperMSC diagrams present themselves if MSC references and MSC connectors are folded and unfolded. There is a large number of possibilities to select, group, and present the information contained in the MSC diagrams in a way that is optimal and shows exactly what is considered relevant for a particular analysis step, a structured walk-through,

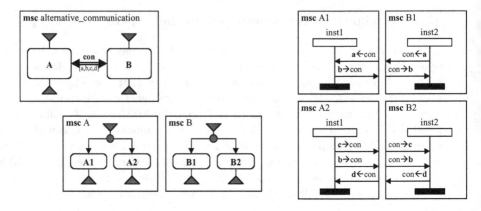

Fig. 11. Definition of the MSC alternative_communication

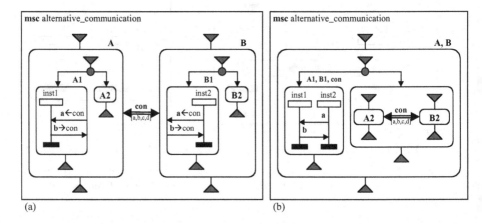

Fig. 12. Two variant HyperMSC presentations of the MSC defined in Fig. 11

etc. The folding and unfolding mechanisms of HyperMSCs therefore provides a means to cope with the complexity of a system design and its large amount of information. However, the examples given so far started with already defined MSC references and MSC connectors. This is a natural way, since system design and modelling just produces all these abstractions that are consequently represented by MSC references and connectors. Yet, there is a further potential in the flexibility of the presentation means proffered by HyperMSCs: the ad-hoc folding/unfolding of diagrams and their parts. Such presentation does not impact on the structure of the model, but focuses on the display of those parts of the MSC diagram that are relevant at the very moment of viewing it. Therefore ad-hoc folding may be used to highlight a statement or an idea during a discussion, to concentrate on a particular trait during analysis, etc. Altogether, folding supports the designers' actual work with the diagrams. However, it is essential that the context of a selected diagram part does not completely vanish, but remains visible even if only in a condensed form.

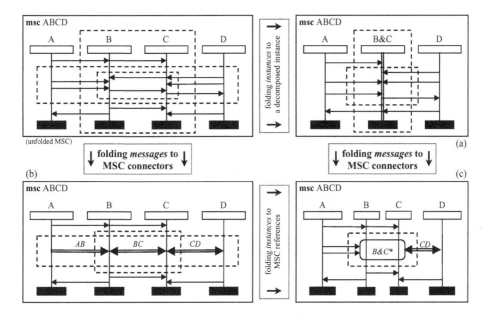

Fig. 13. Ad-hoc folding of a HyperMSC

(a) Folding of instances; (b) folding of a set of messages; (c) folding of a selected part of an MSC.

Figure 13 tries to give an impression about what ad-hoc folding (and unfolding) may look like. Folding away instances (Fig. 13(a)) produces decomposed instances (the instance $B\&C$) which are, tentatively, drawn with a double line, just in analogy to the double lined arrow of the connector symbol. Folding away groups of messages (Fig. 13(b)) produces ad-hoc defined connectors which take over the characteristics of the messages that are subsumed within it (cf. the one-directional connector AB, containing two messages directed from instance A to instance B). The most interesting folding operation is to fold away a part of the diagram that covers both, instances and messages (see Fig. 13(c)). Interactions between the part of the instances that are going to be hidden result in MSC references (see the reference $B\&C*$), messages can be grouped into connectors (as is done with the generation of the MSC connector CD), or may be displayed in the classical way as messages just entering the MSC reference. The wide variety of possibilities and the high flexibility to switch between different presentations allows to quickly come up with a presentation that is optimally adjusted to the actual demands.

5 Summary and Outlook

Within this paper, the HyperMSC concept which has been introduced recently [6,16,17] has been further elaborated and enhanced by MSC connectors representing communication patterns on a highly abstract level. HyperMSCs including MSC connectors are ideally suited to present MSCs on different levels of detail and, what is most important, to easily switch between different views. Assuming an advanced tool support, such enhanced HyperMSCs may be also used for a visual preparation of large MSCs which is highly interactive. By defining MSC references, decomposed instances and MSC connectors in an interactive manner, certain details of a large MSC may be hidden in order to exhibit the momentarily relevant parts. HyperMSCs have already been successfully applied to the specification of test cases based on TTCN-3 and Use Case modelling within UML. At present, the ETSI test format based on HyperMSCs with MSC connectors is under preparation within OMG for inclusion as a profile to UML. The MSC connector concept may be further generalised, eventually representing a complex message interface between system components.

References

1. F. Belina, D. Hogrefe, A. Sarma: *SDL with Applications from Protocol Specification*. Prentice Hall, 1991.
2. G. Booch, J. Rumbaugh, I. Jacobson: *The Unified Modelling Language User Guide*. Addison-Wesley, 1999, 3rd edition.
3. D.F. D'Souza, A. C. Wills: *Objects, Components and Frameworks with UML. The Catalysis Approach*. Addison-Wesley, 1999.
4. A. Egyed, N. Metha, N. Medvidovi: *Software Connectors and Refinement in Family Architectures*. In: Proceedings of the 3rd International Workshop on Software Architectures for Product Families, Las Palmas de Gran Canaria, Spain, March 15-17, 2000.
5. J. Grabowski, A. Wiles, C. Willcock, D. Hogrefe. *On the Design of the new Testing Language TTCN-3*. In: Testing of Communicating Systems - Tools and Techniques (H. Ural, R.L. Probert, G. von Bochmann, editors), Kluwer Academic Publishers, August 2000.
6. P. Graubmann, E. Rudolph: *HyperMSCs and Sequence Diagrams for Use Case Modelling and Testing*. In: UML2000, 3rd International Conference on The Unified Modeling Language (A. Evans, S. Kent, B. Selic, editors), 02-06 October, 2000, York, UK, Springer 2000.
7. P. Graubmann, R. Wasgint: *Methods for Interface Annotations and Component Selection*. SAG-WP2-0106-16, ESAPS internal report, 2001.
8. I. Krüger: *Distributed System Design with Message Sequence Charts*, PhD Thesis, Technische Universität München, 2000.
9. S. Loidl, E. Rudolph, U. Hinkel: *MSC'96 and Beyond-a Critical Look*. In SDL'97 Time for Testing-SDL, MSC and Trends, Proceedings of the 8th SDL Forum in Evry, France (A. Cavalli and A. Sarma editors), North Holland, September 1997.
10. S. Mauw, M. A. Reniers: *High Level Message Sequence Charts*. In: SDL'97 - Time for Testing-SDL, MSC and Trends, Proceedings of the 8th SDL Forum in Evry, France (A. Cavalli ,A. Sarma, editors), North Holland, September 1997.

11. N. Mehta, N. Medvidovic, S. Phadke: *Towards a Taxonomy of Software Connectors*. University of Southern California, Center of Software Engineering, Technical Report 99-529, 1999.
12. E. Rudolph: *Putting Extended MSC-2000 to Practice*, Contribution to the ITU-SG 10 Meeting, Geneva, November 1999.
13. E. Rudolph: *Advanced MSC- A Unifying Modeling Language for the Next Millennium*, Contribution to the ITU-SG 10 Meeting, Geneva, November 1999.
14. E. Rudolph, J. Grabowski, P. Graubmann*: Towards a Harmonization of UML-Sequence Diagrams and MSC*. In: SDL'99 - The Next Millennium, Proceedings of the 9th SDL Forum in Montréal, Québec, Canada (R. Dssouli, G.V. Bochmann, Y. Lahav, editors), Elsevier Science B.V., Amsterdam, 1999.
15. E. Rudolph, J. Grabowski, P. Graubmann: *Tutorial on Message Sequence Charts (MSC-96)*. Forte/PSTV'96. Kaiserslautern, Germany, October 1996.
16. E. Rudolph, I. Schieferdecker, J. Grabowski: *Development of an Message Sequence Chart/ UML Test Format*. In: Proceedings of FBT'2000 - Formale Beschreibungstechniken für verteilte Systeme, Lübeck, Germany (J. Grabowski, S. Heymer, editors). Shaker-Verlag, Aachen, 2000.
17. E. Rudolph, I. Schieferdecker, J. Grabowski: *HyperMSC - A Graphical Representation of TTCN*. Proceedings of the 2nd Workshop of the SDL Forum Society on SDL and MSC (SAM'2000), Grenoble, France, June, 26 - 28, 2000.
18. ETSI TC MTS: *TTCN-3 — Core Language*. European Norm EN00063-1 (provisional)[2], 2000.
19. ETSI TC MTS: *TTCN-3 — Graphical Presentation Format*. European Norm EN00063-3 (provisional), 2000.
20. ETSI TC MTS: *TTCN-3 — Tabular Presentation Format*. European Norm EN00063-2 (provisional), 2000.
21. ITU-T Rec. Z.120 (MSC-96): *Message Sequence Chart (MSC).*, Geneva, 1996.
22. ITU-T Rec. Z.120 (MSC-2000): *Message Sequence Chart (MSC).*, Geneva, 1999.

---

[2] The EN-00063 numbers are only provisional ETSI Work Item numbers. The actual EN numbers will not be the same.

Graphical Test Specification –
The Graphical Format of TTCN-3

Paul Baker[1], Ekkart Rudolph[2], and Ina Schieferdecker[3]

[1] Motorola Labs, Basingstoke, Hampshire, UK
`paul.baker@motorola.com`
[2] Technische Universität München, Institut für Informatik
Arcisstraße 21, D-80290 München, Germany
`rudolphe@informatik.tu-muenchen.de`
[3] GMD FOKUS, Berlin, Germany
`schieferdecker@fokus.gmd.de`

Abstract. Recently, the European Telecommunications Standards Institute (ETSI) approved the third edition of the Tree and Tabular Combined Notation (TTCN-3) as a requirement to modernise and widen its application beyond pure OSI conformance testing. As part of this evolution, TTCN is embracing Message Sequence Charts (MSCs) as a natural notation for specifying and visualising test suites. This paper defines the role of MSCs during test development, and more specifically introduces an MSC profile called the Graphical Format for TTCN (GFT) that facilitates the effective specification of TTCN-3 test suites.

1 Motivation

TTCN is a language used for test specification. However, experience has shown that the second edition [1, TTCN-2], a semi-graphical representation by means of a tabular format, has turned out not to be very intuitive for behaviour description, even if tools are used. For example, within TTCN-2 tables are used for the graphical representation of test cases, where statements are written on successive lines with either successively incremented indentations to indicate subsequent statements, or with equal indentations to indicate alternatives. In the case of highly nested alternatives, such a notation becomes very user-unfriendly. Consequently, with the third edition [2,3, TTCN-3] a textual language was developed that now looks more like a common programming language e.g., C or C++ or Java. Even though TTCN-3 makes the description of complex distributed test behaviour much easier there is still a requirement from the TTCN user committee to provide a visualisation means.

Message Sequence Charts (MSC) appeared to be a particularly attractive candidate as a graphical means for visualising TTCN. Therefore, in addition to the pure textual core language, the definition of other presentation formats has been admitted within TTCN-3. At present, two presentation formats are defined: a tabular conformance format [4] that resembles the tabular format of TTCN-2, and an MSC presentation format denoted as the Graphical Format for TTCN-3

R. Reed and J. Reed (Eds.): SDL 2001, LNCS 2078, pp. 148–167, 2001.

(GFT) [5]. GFT supports the presentation and development of TTCN-3 test descriptions on an MSC level. Thereby, the TTCN-3 core language may be used independently of GFT, but GFT cannot be used without the core language. Use and implementation of GFT shall be done on the basis of the core language. In the following, the TTCN-3 core representation is denoted briefly as TTCN-3.

GFT is based on the ITU Recommendation Z.120 for Message Sequence Charts [6] using a subset of MSC with test specific extensions, as well as extensions of a general nature. A main advantage of the MSC language is its clear graphical layout, which immediately gives an intuitive understanding of the described behaviour. Within the area of conformance testing, MSC is already well established for the specification of test purposes, and as such for the automatic generation of TTCN test cases [10]. Beyond that, MSCs have been proposed for a selected visualisation of TTCN descriptions by means of simulation techniques [7]. Although MSC has been used for limited test specification in the past, the latest version of the language now contains constructs that make the comprehensive MSC specification of test suites feasible. Such language constructs include MSC composition, object oriented modelling, as well as data. However, it should be pointed out that GFT is not intended as a standalone language, but as a basis for the generation of TTCN-3 descriptions. It may be possible that hybrid representations may turn out to be most effective, where only the main parts of the test description are visualised by means of MSCs, whilst the remaining parts are provided in the form of TTCN descriptions. Such a hybrid representation appears to be ideally tailored for a smooth transition from an MSC test specification to TTCN test case descriptions.

The possibility to clearly discriminate between different parts of a test description, and between different language constructs is one of the main points that are strongly in favour of using GFT.

The second advantage of GFT in comparison with TTCN-3 refers to the description of the communication behaviours between test components and their ports, and between test components via connected ports, and between test components and the system under test via mapped ports. Note that within GFT all ports may be represented by different port instances. Consequently the test events belonging to different ports are clearly separated visually, in contrast to TTCN-3 where all events appear in a mixed form. Within TTCN-3, the communication between test components via connected ports is provided in a fairly indirect manner. In contrast to that, GFT has the possibility to show the communication via connected ports in a more explicit manner either by means of 'MSC connectors' (which are introduced as an extension of the MSC language), or even by means of the explicit message flow for selected cases. The combined use of GFT and TTCN-3 can be compared in this respect with the combined use of MSC and SDL. Both SDL and TTCN-3 are component oriented, whereas MSC is communication oriented.

Using MSC as a presentation format for TTCN-3 may considerably improve the readability of test cases and make them more understandable. At the same time, MSC in the form of Sequence Diagrams forms a central constituent of UML and is employed for the formalisation of Use Cases. Since there is at the time of writing no accepted test notation in UML, this is an ideal opportunity to bring TTCN-3 in form of GFT to the attention of the UML world [8]. In this context, a graphical format is of particular importance since UML is exclusively based on graphical modelling techniques. Therefore, a purely textual test language would not have any chance of acceptance. At present, GFT is already under preparation within OMG for inclusion as a test profile to UML.

The paper is structured as follows: in two subsequent sections the main ingredients for GFT are shortly presented: TTCN-3 and MSC. Afterwards, the use of MSC in the test development process is discussed. The main concepts of graphical test specification are presented in Sect. 5. Section 6 introduces GFT and discusses the extensions to MSC in order to enable its use for test case specifications. A short example shows the application of GFT. The concluding section gives a summary and describes the next steps.

2   Overview on TTCN-3

The Tree and Tabular Combined Notation (TTCN) [1] is a language for writing test specifications. TTCN was first published as an ISO standard in 1992, where OSI conformance testing is understood as functional black-box testing i.e. an Implementation Under Test (IUT) is given as a black-box and its functional behaviour is defined in terms of inputs and outputs from the IUT. Since TTCN was standardised the use of the language has steadily grown within the telecommunications industry. TTCN has been used to specify tests for technologies such as GSM, DECT, INAP, N-ISDN, Q-Sig, TETRA, VB-5, Hiperlan, 3G (terminals) and VoIP.

ETSI (European Telecommunication Standards Institute) recently funded a team to evolve TTCN into a language whose look and feel is of a modern programming language with test specific extensions, called the TTCN third edition (TTCN-3) [2]. These extensions consist of: test verdicts, matching mechanisms to compare the reactions of the IUT with the expected range of values, timer handling, distributed test processes, ability to specify encoding information, synchronous and asynchronous communication, and monitoring. In providing these extensions it is expected that test specifiers and engineers will find this general-purpose language, more flexible and user-friendly and easier to use than its predecessor. As illustrated in Fig. 1 TTCN-3 also encompasses different presentation formats: a tabular format [4] and a graphical like format [5] based upon Message Sequence Chart (MSC).

TTCN-3 is intended for the following application areas: protocol testing (such as mobile and internet protocols), supplementary service testing, module testing, testing of CORBA based platforms, testing of API's etc.

TTCN-3 is on a syntactical and methodological level a drastic change to TTCN-2, however, the main concepts of TTCN-2 have been retained and improved and new concepts have been included, so that the expressive power and applicability of TTCN-3 are increased.

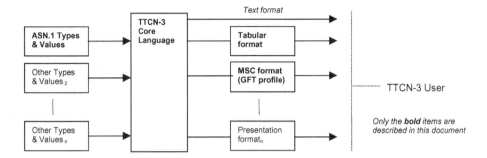

Fig. 1. Users view of TTCN-3 and the various presentation

New TTCN-3 concepts are:

- test execution control to describe relations between test cases such as sequences;
- repetitions and dependencies on test outcomes;
- dynamic concurrent test configurations;
- test behaviour in asynchronous and synchronous communication environments.

Improved concepts in TTCN-3 are:

- integration of ASN.1 [9];
- the module and grouping concepts to improve the test suite structure;
- the test component concepts to describe concurrent and dynamic test setups.

The top-level unit of a TTCN-3 test suite is the module, which can import definitions from other modules. A module consists of a definitions part and a control part. The definitions part of a module covers definitions, which for test components are: their communication interfaces (so called ports), type definitions, test data templates, functions, and test cases. The control part of a module calls the test cases and describes the test campaign. For this part, control statements similar to statements in other programming languages (e.g. if-then-else and while loops) are supported. They can be used to specify the selection and execution order of individual test cases.

Test cases describe the probes during the test campaign: they specify the test behaviour. One can express a variety of test relevant behaviour within a test case such as the alternative reception of communication events, their interleaving and default behaviour to cover unexpected reactions from the tested systems, for example. In addition to the automatic test verdict assignment, more powerful logging mechanisms such as detailed tracing are provided.

3   Overview on MSC

Message Sequence Chart (MSC) is a language to describe the interactions between a number of independent components of a system. The basic model of interaction is that of asynchronous communication by means of passing messages between the components, which are called instances. An MSC describes the order in which interactions and other events take place. MSC diagrams are used to graphically present the pattern of interaction using different constructs.

Core constructs are instance, message, timer, coregion, conditions, and inline expressions. Instances of an MSC represent interacting components of a system. Horizontal arrows between the instances present the message flow and their interaction. The head of the message arrow denotes the reception of the message and the opposite end the sending of the message. The sending and receiving of messages are also called communication events. The message name is assigned to the arrow.

Along each instance axis a total ordering of the described communication events is assumed. Events of different instances are in general unordered. The only order for events of different instances is implied via the interaction with messages: a message must be sent before it is consumed. The total ordering of events along each instance may not always be appropriate for describing interaction patterns in general. Therefore, a coregion is used for the specification of unordered events on an instance. A coregion may contain an arbitrary mixture of communication events. The general ordering relation is used to imply an order for unrelated events.

Conditions denote special states for the set of instances they refer to. A condition is either global by referring to all instances contained in the MSC or local by referring to a subset of instances only. A setting condition is a state-like condition and is used to associate a state with the covered instances. Guarding conditions contain Boolean expression and enable the subsequent interaction pattern only if it evaluates to true.

Timer handling in MSC encloses the setting of a timer and a subsequent timeout (timer expiration) or the setting of a timer and a subsequent timer reset (time supervision).

Composition of event structures may be defined inside an MSC by means of inline expressions. The operators of inline expressions can refer to alternative (**alt**), iteration (**loop**), and optional (**opt**) regions. The **alt** operator defines alternative executions of MSC sections, i.e. only one of them will be executed. The **loop** construct for iteration can have several forms. The most basic form is "**loop** $<n, m>$" where n and m are natural numbers. This means that the operand may be executed at least n times and at most m times. The **opt** operator indicates optional MSC sections and is the same as an alternative where the second operand is the empty MSC.

More elaborate language features for structuring are MSC references and High-Level MSC (HMSC). With the MSC-2000 version data and time concepts and object-oriented aspects have been introduced.

4 Using MSCs during the Test Development Process

MSC is used throughout the engineering process of test development: for the specification of test purposes to define the specific objective of a test, via the specification of test cases to define the concrete test behaviour, up to the visualization of test executions.

According to the OSI conformance testing methodology [11], testing normally starts with the development of a test purpose, defined as follows:

A prose description of a well-defined objective of testing, focusing on a single conformance requirement or a set of related conformance requirements as specified in the appropriate OSI specification.

Having developed a test purpose, an abstract test suite is produced that consists of one or more abstract test cases. An abstract test case defines the actions necessary to achieve part (or all) of the test purpose. Applying these terms to Message Sequence Charts (MSCs) we can define three categories for their usage:

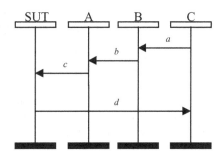

Fig. 2. MSC illustrating how the *SUT* interacts with its environment

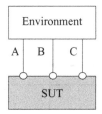

Fig. 3. Illustration of the architecture that is normally represented by a test purpose

1. Test Purposes – Typically, an MSC specification that is developed as a use-case or as part of a system specification. For example, Fig. 2 illustrates a simple MSC describing the interaction between instances representing the System Under Test (*SUT*) and its environment (represented by instances *A*,

B, and C). Such MSC specifications can represent many different behaviours. Where the complete behaviours cannot be represented by a single test case, more abstract test cases may be required to ensure that the *SUT* conforms to the specification. Note that the inclusion of *SUT* instances is optional, and that both the *SUT* and Environment can be defined using more than one instance. Figure 3 illustrates the typical configuration used during the development of test purposes, where A, B, and C represent the potential connections to the environment.

2. *Abstract Test Cases (or Test Suite)* – Typically, an MSC written solely for the purpose of describing the behaviour of a test case. For example, Fig. 4 illustrates a simple MSC defining the interactions between different elements of the test configuration. Instances can represent test components (*MTC*, *TC_A*, *TC_BC*), ports mapped to the SUT and ports connected to other test components (A, B, C, $C1$, $C2$). Figure 5 illustrates the test configuration used by the MSC specification.

3. *Traces* – Typically MSCs derived from simulation or test logs.

In identifying these three categories of MSC usage we can define three distinct areas of work for the development of a graphical test specification format:

1. The development of test specific extensions for Message Sequence Charts involving both syntactical and semantic extensions that are needed for test purpose and test case specification. In particular, syntax extensions are needed to simplify the development of test specifications, whereas semantic extensions are needed when the necessary behaviours cannot be modelled using the current MSC framework. For example, in TTCN-3 the test case verdict is updated when a test component terminates. In this case, representing such behaviour using MSC would lead to inappropriate specifications. Therefore

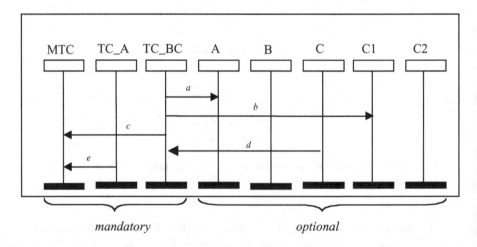

Fig. 4. An MSC illustrating both the behaviour and configuration of a test case

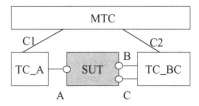

Fig. 5. Configuration used by the above MSC

we introduce new syntax to simplify its description, where the new syntax can be considered as a macro that expands into a more complex, yet valid MSC representation.

2. The development of various semantic models for representing MSC test purposes and their mapping to test cases. For example, test purposes can be written assuming full MSC semantics (that is with both send and receive events interleaved) as supported by the ptk tool [10] from Motorola Labs, or using a restricted subset of MSC semantics (where only those events received from the SUT are interleaved) as supported by the MSC2TTCN3 from Lübeck University.

3. The mapping of graphical test cases onto the underlying textual test specification language TTCN-3 for further processing.

This paper is only concerned with the graphical presentation of TTCN-3 test specifications.

5   Concepts of Graphical Test Specifications

Graphical specification techniques are gaining more acceptance and wide spread use as they are easier to develop, understand, and maintain. Prominent examples are MSC [6], SDL [12], and UML [13]. With the work on graphical test specifications the attempt was made to use graphics for the definition and description of test behaviour as well.

As such, a test system is like a distributed system consisting of various components (the so called test components), which interact with each other (for synchronization and control during the test execution), and which interact with the system under test (for performing the test behaviour in order to assess the correctness of the system under test). Beyond the typical characteristics of distributed systems the following specific characteristics of test systems exist:

– The differentiation of communication links between test components and the system under test (so called points of control and observation, which denote the tested system interface and via which the interaction to the system under test is realized) and between test components only (so called coordination points). The difference lies in the ability to control all end points of coordination points by the test system, while for points of control and observation only one side is controllable.

- The differentiation of master and parallel test components in order to identify one test component, which has the authority to assign the overall test verdict, and to which all other test components have to report.
- The need to classify incoming responses from the system under test in order to decide whether the response belongs to the expected ones or not. Specifically for testing, the expressions to denote possible matches have to be quite powerful. Although this characteristic exists for any "monitoring" and/or "metering" component, in testing the specific need is to match efficiently in order to respect the speed of the system under test: to be as fast as needed without classifying an incoming response wrongly.
- The existence of so called default behaviours, which enable a test component to react also on unexpected events and which make it robust. Default behaviours are comparable to exception handlers of object-oriented languages but differ in their handling: they can be dynamically activated and deactivated during test execution and define low-priority alternatives to expected responses from the system under test.
- The existence of a predefined verdict handling in order to track detected malfunctioning of the system under test and to prevent corruption of test results. This is also known as the *only-get-worse* rule of testing.

Not every detail of a test case is graphically represented, as it may not be practical or feasible. For example, for data declarations and their use within the test case it is more practical to use normal text. However, aspects such as the behaviour, structure, and configuration of a test case should be graphically represented.

The core behaviour of a test case is the interaction between test system and system under test and, on a lower level of detail, between the test components. MSC instances can be used to represent the constituents and MSC messages to represent their interaction. In addition, it would be advantageous to support a graphical differentiation between (1) test components, and (2) communication links to the system under test. This would visualize better the exchange of messages to the system under test and within the test system.

As well as supporting plain test behaviour specifications only (as is done in TTCN-3 with functions and function calls), to support the structuring of the test behaviour, a graphical way to structure graphical test case specifications is needed. MSC structuring concepts like MSC references and High-Level MSC with slight extensions are well suited for this purpose.

A critical point is the graphical representation of test configurations consisting of dynamically created and terminated components and the connections to the system under test and to other test components. Test configurations allow not only point-to-point but also point-to-multipoint connections. In addition, they are dynamic. This would require the graphical representation of types of test components (and not the individual test components) but that destroys the ability to represent the connections, which are always between concrete test components. In MSC, there is no such kind of diagram or graphical symbol. In addition, the MSC gate concept of MSC turned out to be inadequate as it is based on the concept of gating singular events to the environment and does not enable the accumulation of gate events in terms of connectors, for example. Due to the problems in representing dynamically created test components and their communication links (this is problematic even with static configurations), this aspect is not yet addressed graphically within GFT. The textual TTCN-3 operations to set-up configurations are used within action boxes instead. However, further work is planned in this area.

6   The GFT Extensions to MSC

Since the appearance of the first MSC recommendation in 1992 it has become common practice to use MSC for validation purposes. In this case, the system (for example described in SDL) is checked to see if it can execute the sequence of events described by an MSC. However, only with the powerful language concepts contained in MSC-2000 [14], together with some test specific extensions listed below, can MSC be used as a real test language. Beyond the basic MSC language (instances, message events, actions, timer, conditions, instance creation and termination), constructs for MSC composition and object oriented modelling as well as data concepts are necessary for this purpose. MSC composition constructs comprise of: MSC references, High-level MSCs (HMSCs), and in-line expressions. In addition, coregions can be used to specify interleaving behaviour in a limited manner. MSC-2000 has adopted concepts for object oriented modelling from UML, offering the possibility to define synchronous communication and control flow comparable to UML sequence diagrams. Another important addition to MSC is its generalised data concept, where no specific MSC data language is defined. Instead, MSC can be used together with a data language of the user's choice by means of a suitable interface. Since an MSC specification in general consists of a whole collection of MSCs, an MSC document is defined on the top level of the MSC language. Within the MSC document the contained MSCs are assigned either to a defining part or to a utility part. Furthermore data definitions are given within the MSC document header.

MSC contains some further structural concepts, which at present do not play a role within GFT:

1. Instance decomposition allowing instance refinement;
2. Generalised ordering that serves for the modelling of general event structures;
3. Gates that can be used to define connection points for messages and order relations with respect to the interior and exterior of MSC references and in-line expressions.

Real time concepts, which are also part of MSC-2000, have entered GFT only partially. On the other hand extensions to MSC have become necessary in order to make MSC applicable for test case specifications.

This section contains the list of extensions introduced into GFT, thereby presenting an overview about the necessary enhancement of MSC language concepts. These enhancements can be considered as part of a test specific profile for

MSC, denoted by the term GFT.[1] Such enhancements may include the introduction of new syntax in order to improve the readability of test specifications.
In general, most of the test specific extensions introduced within GFT imply
some semantic restriction on the current MSC framework. However, in other
cases the semantic framework has to be extended to accommodate the needed
behaviours. For all graphical extensions either substitutes within standard MSC
are provided or these extensions are optional. This allows MSC tools with some
modifications to the textual syntax to be used for GFT.

6.1   MSC Documents

A GFT document (see Fig. 6) contains new sections for control, test cases, functions, and named defaults. These are denoted using new keywords and additional
separator lines. These can been seen as a syntactic extension to MSC-2000 in
which MSCs contained within the control and test case parts are mapped onto
defining MSCs, and MSCs contained within the functions and named default
parts are mapped onto utility MSCs. GFT allows parameterization of MSC documents. As a substitute, comments could be used instead.

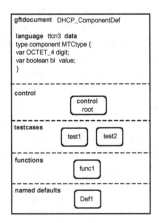

Fig. 6. Example of a GFT Document

6.2   MSC Headings

An MSC heading provides the MSC name together with its formal parameter list.
In GFT, the MSC head has been extended to include the **system** and **return**
keywords. These keywords allow the user to define the test system interface type,
and the value to be returned by the defining GFT respectively.

---

[1] MSC does not currently provide a means for defining application specific profiles
such as GFT.

6.3   Instances

In GFT, instances either represent test components or ports. In order to differen-
tiate the two kinds of instances, and their semantics, different graphical symbols
may be used. Ports may be represented explicitly as particular 'environmental'
instances using a new graphical symbol in the form of a dashed instance. Such
port instances mean that there is no special event structure defined apart from
FIFO ordering on connecting ports. Whereas, for test components the standard
instance event ordering is assumed. As an alternative representation, a standard
MSC instance may be used to denote ports using the keyword **port** in front of
the port type. For a compact notation, ports may also be implicitly represented
by means of a special textual syntax notation on messages sent to the environment. Within this implicit port representation, the message name is pre-fixed
for sending ports by the port identifier and post-fixed for receiving ports by the
port identifier. As a consequence, the implicit port representation can just be
reflected as a textual extension.

From a GFT semantic perspective the events placed upon a port instance
are not considered when mapping to TTCN-3. Therefore, we can think of GFT
diagrams (containing port instances) as a subset or restriction of MSC, indicated
by the use of the keyword **gft**. However, some static restrictions are imposed
on what can be drawn to avoid the production of invalid test specifications
(deadlocks). For example, we wish to restrict the fact that messages passed
between component instances are not allowed to overtake each other; again this
is just a restriction on current MSC semantics. And, even though users may
draw the messages entering and leaving a port instance in different orders this
does not violate MSC semantics. Note that this also represents a deadlock test
specification, but this is also possible within TTCN-3.

6.4   Messages and Special Test Operations

Within GFT, messages describe the asynchronous communication in test cases:
sending and receipt of events at components and ports. Thereby, the standard
graphical MSC message symbol is used, but the textual message inscription
differs slightly. On top of the message arrow the message type can be given,
and below the message arrow the message constraint can be specified either by
referring to a named TTCN-3 template or by giving an in-line template definition
(in parenthesis). As an option, either the message type or the message template
may be omitted. Where implicit ports are used, the message type (if present)
is pre-fixed for sending ports by the port identifier and post-fixed for receiving
ports by the port identifier. In the case where the message type is omitted, the
same notation is used with an empty message type string – see Fig. 7.

In GFT several special test operations, which are represented by the form of
special messages, are introduced in addition to the normal test events:

- The *trigger* operation filters messages with a certain matching criteria from
  a stream of received messages on a given incoming port. In GFT, trigger is

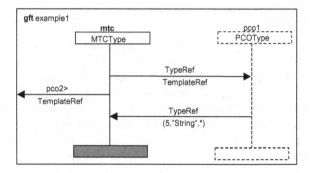

Fig. 7. Example illustrating different usage of messages

defined as a special message with the keyword *trigger*. Because MSC has no notion of queuing, the trigger operation is seen as a message event.

- The *check* operation allows access to the top element of message–based and procedure-based incoming port queues without removing the top element from the queue. In GFT, check is defined as a special message with keyword *check*. Again, because MSC has no notion of queuing, the check operation is seen as a message event.
- By means of the *start* operation for test components, the execution of a component's behaviour is started. Graphically, a dashed arrow represents the start operation. This implies a new use of dashed line messages. As a substitute, standard message arrows can be used instead of dashed line arrows. The GFT start construct uses the textual TTCN-3 syntax.
- The *return* operation terminates execution of a function. Graphically, a dashed arrow represents the return operation. As in the case of the start operation, this implies a new use of dashed line messages. Likewise, standard message arrows can be used instead of the dashed line arrow as a substitute. The GFT return construct uses the textual TTCN-3 syntax. However, because return represents the termination of a function there are some restrictions on when it can be used. For example, it is not possible to place events after a return symbol, therefore it must always be the last event placed upon an instance axis.
- The port operations clear, start and stop are interpreted as special messages that control a port. Where, *Clear* removes the contents of an incoming port queue, *Start* activates listening and gives access to a port, and *Stop* halts listening and disallows sending operations at a port. In GFT, dashed line arrows together with a new textual syntax represent these graphically. As a substitute, standard message arrows can be used instead of dashed line arrows. Semantically these events are treated as message events.

6.5   Timers

For the use of timers in GFT, slight graphical and textual extensions are introduced. Unnamed timers are used to supervise call operations. In this case a timer started without a timer identifier is directly attached to the beginning of a suspension region, and a corresponding timeout is directly attached to the end of a suspension region. A possible substitute would be to place the start/timeout timer near the begin/end of a suspension region. Named timers represent the setting, resetting and timeout of timers within TTCN-3.

6.6   Control Flow

Control flow consists of procedure based synchronous communication mechanisms defined by means of calls and replies. GFT uses special keywords for the call messages: **call**, **getcall**; and for reply messages: **getreply**, **catch**, **reply**, **raise**. For the reply messages solid message arrows are used instead of dashed arrows since they are already distinguished from call messages by keywords. As a substitute, dashed message arrows may be used. Special binding to variables for parameters and return values of a call is employed.

6.7   Verdicts

In GFT, conditions with the keywords **none**, **pass**, **inconc**, and **fail** are employed to denote the setting of a local test verdict. Test verdicts assign a special meaning and handling to conditions. However, as any identifier is allowed within MSC conditions, this does not really impact the syntax of MSC. Semantically the notion of using a condition with a label is different to 'setting' conditions within MSC. Hence, these conditions can be seen as a macro for setting the local test verdict variable, which is implicitly declared for each component instance. The updating of the global test verdict takes place when the test component is terminated.

6.8   Actions

Action boxes in GFT can contain TTCN-3 statements. One extension is the ability to invoke functions in addition to assignments. The other extension relates to the fact that in MSC the expressions placed within an action box are evaluated concurrently. Therefore, within the GFT profile we restrict the MSC semantics to only allow the sequencing of expressions placed within an action box.

6.9   Create

In GFT, the create operation is used by test components to dynamically create other test components, the exception being the Main Test Component (MTC) which is implicitly created when a test case is executed. The GFT create construct uses the TTCN-3 inscription instead of the MSC inscription.

6.10   In-Line Expressions

In GFT, in-line expressions are used to define alternative, cyclic and interleaved behaviour. GFT differs from MSC-2000 with respect to the treatment of the environment by adopting the rule from MSC-96 that messages may be propagated to the next higher environment. Outgoing and incoming messages respectively come from the in-line expression frame only.

In GFT, a special kind of TTCN-3 interleaving behaviour is introduced which differs slightly from full interleaving used in MSC. Those parts of an event sequence that are initiated by a reception statement (receive, trigger, getcall, getreply, catch or check) are treated as non-interruptible: these parts are treated as atomic events with respect to interleaving. To distinguish this kind of restricted interleaving from full interleaving, a new interleaving operator is introduced indicated by a keyword **int** in the left top corner of the in-line expression frame. The non-interruptible parts are defined by the TTCN-3 semantics.

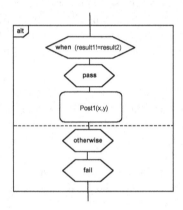

Fig. 8. Example of a guarded alternative expression in GFT

In TTCN-3 the arguments of an alternative expression are evaluated top-down, whereas, in MSC all arguments are evaluated. In general this does not cause a problem, however: if guards are used for different alternatives then these must be transformed to imply a top-down ordering – see Fig. 8. For example, if we have a GFT containing an alt expression with three arguments each guarded by a Boolean expression: A, B, and C, it can be mapped onto an equivalent MSC by modifying the guards to reflect the top-down ordering.

In GFT, the termination of a test component is allowed also within an argument of an in-line expression.

6.11 Default Behavior

In GFT, the control of default behaviours is represented by a new graphical symbol together with the keywords **activate** and **deactivate**. Where, default behaviour acts as an extension to an alt statement, or a single receive operation – see Fig. 9. Semantically, default behaviour fits with the current MSC framework, and therefore can be seen as syntactic simplification for a set of recursive transformation rules that can be used to map GFT onto MSC. A substitute for the new default symbol is an action box.

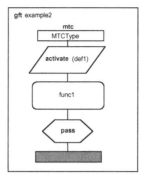

Fig. 9. Example illustrating how default behaviours are activated within GFT

6.12   MSC References / Hybrid MSCs

In GFT, some of the MSC references may point to MSC reference definitions, which are provided in the form of TTCN-3 descriptions instead of MSC diagrams. MSCs employing such textual reference definitions are denoted as *hybrid*. Since GFT is intended as the basis for creating TTCN-3 diagrams, in practice such a hybrid representation is often easier to handle than a complete graphical description. Probably a hybrid representation of test cases is most important where the main path (pass case) is described in form of MSC diagrams and the side cases (inconclusive/fail cases) by means of MSC references defined in form of TTCN-3 descriptions.

6.13   High-Level MSCs

Within GFT, the module control part is described by means of HMSCs. For this purpose several graphical and textual extensions are necessary. Variable declarations are allowed within the HMSC header. Action boxes are included in HMSCs. As a possible substitute, a reference to an MSC including the action box may be used. Furthermore, value returning MSC references are introduced. The extension of HMSCs in the form of HyperMSCs is described in the next paragraph.

6.14   HyperMSCs

Practice has shown that apart from simple cases, the representation of test cases by means of standard in-line expressions and HMSCs leads to diagrams that make the visualisation of pass and fail branches of test script sometimes less obvious. Therefore, GFT provides an expanded form of MSC references, where the referenced MSC can be visualised within a reference symbol. This concept is referred to as HyperMSCs [15]. The HyperMSC concept is tailored particularly for HMSCs. Within HMSCs, several expanded MSC references may be combined to one coherent expanded MSC reference with the connection points being shifted to the borderline of the MSC reference. This way, a convincingly transparent representation is obtained even in case of many alternatives and loops.

6.15   Data Concepts

MSC-2000 introduced the novel approach for the inclusion of data. Instead of defining a data language, it provides an interface by which the data syntax can be checked and the MSC semantics can be evaluated. This allows a user to adopt the data language of their choice.

GFT represents the parameterisation of MSC with TTCN-3 data types and values, together with some extensions and modifications. For example, the declaration of variables and timers are removed from the document header. Instead the declaration of component instance variables is given within the TTCN-3 data definition string. Other additions include:

- The concept of using message names and instance types as references to TTCN-3 types;
- Implicit and explicit typing for component instances, where either component types are generated from the GFT specification, or the events placed on a component instance must adhere to the component type respectively;
- HMSCs can contain local variable declarations;
- Local and global test verdicts;
- Value returning MSC references;
- Formal parameter lists for MSC documents.

However, because TTCN-3 provides its own constraint/pattern matching facility MSC wildcards are not needed for TTCN-3 data parameterisation.

7   An Example: A Complete Test Case

A complete GFT test case example is given in this section. The MSC document contains the MTC type definition, has no control function defined, and contains one test case definition for AA_1.

The MSC for test case AA_1 contains one test component: the MTC. The MTC activates at first the default behavior $Default\_1$. Afterwards, the preamble Pre_1 is performed. This is represented as an MSC reference to the MSC containing the definition for the preamble Pre_1. Within an action box, the variable $cltid1$ is initialised. A $DISCOVER$ message at port $LTPCO1$ is sent and the timer $Tshort$ is started. Afterwards, the response - an $OFFER$ message from port $LTPCO1$ - is awaited. The information elements of the received message are stored in variables – again within action boxes. If the response is not received within the timer period of $Tshort$, the timer will timeout. Its timeout will be treated by the default behavior $Default\_1$.   If the response is received in time, the timer will be cancelled. Subsequent to that, an alternative is defined for the comparison of the received data values with the expected ones. If

Fig. 10. Sample GFT Document

the received data values are as expected, a **pass** is assigned and the postamble *Post_1* is invoked. Again, an MSC reference to the MSC defining the behaviour of the postamble *Post_1* is used. In all other cases (that is, whenever the received data is not as expected) are represented with **otherwise**, and a **fail** verdict is assigned. Finally, the test case terminates.

For comparison, the corresponding TTCN-3 textual representation is given in addition. The close relationship of the graphical format of TTCN-3 and its

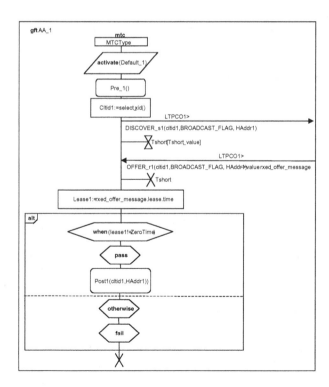

Fig. 11. Example GFT test case

core language can be seen here. The module *DHCP_Srv* [16] is constituted by the declarations of the GFT document and by the definitions of all individual MSCs defined in the context of the GFT document. In this example, the MSC representing the test case *AA_1* is reflected in the test case definition for *AA_1* in the core language module for the test suite. Also the other GFT constructs have a straightforward representation in the core language. For example, the activation of the default *Default_1* and the invocations of the preamble *Pre_1* are reflected with **activate**(*Default_1*); and *Pre_1*(); in TTCN-3 core.

```
module DHCP_Srv_ComponentDef {
    /* --- Data definition part --- */
import DHCP_declarations;
type component MTCtype {
      var OCTET_4 cltid1;
      var OCTET_4 lease1;
      var OFFER_rxed offer_message;
      timer Tshort := Tshort_value;
   }
   /* --- Test case definition part --- */
testcase AA_1 () runs on MTCType {
   activate (Default_1);
   Pre_1 ();
   cltid1 := select_xid ( );
   LTPCO1.send ( DICOVER_s1 (cltid1, BROADCAST_Flag, HAddr1) );
   Tshort.start;
   LTPCO1.receive ( OFFER_r1 (cltid1, BROADCAST_Flag, Haddr1) )
                     ->value rxed_offer_message;
   lease1 := rxed_offer_message.opts.lease.time;
   Tshort.stop;
   If (lease1!=ZeroTimer) {
        verdict.set (pass);
        Post_1 (cltid1, HAddr1);
   } else {
        verdict.set (fail);
   }
   stop;
   } /* testcase AA_1 */
} /* module DHCP_Srv */
```

8   Summary and Outlook

This paper presents a definition for the different uses of MSCs within the test development process. In doing so, a number of tasks are identified that need to be addressed when MSC is used for the different aspects of graphical test specification: test purposes and test specifications. An overview of the Graphical Format for TTCN (GFT) is presented as a test specific profile for MSCs. In particular GFT is tailored towards the development of MSC test specifications, which are mapped onto TTCN-3 test scripts. In presenting GFT the paper starts to address both the syntactical and semantic extensions that are needed to MSC to facilitate the adequate visualisation of TTCN-3 test suites.

However, the development of GFT is still not complete. Further work is needed to complete both the language definition and mapping to TTCN-3. For that, the issue of MSC convergence also needs to be addressed: demonstrating how GFT constructs are really just a restriction on the current MSC semantic framework. In those cases where this is not possible then new proposals to the MSC standard are needed. Finally, with the interest in MSC within OMG, as a potential candidate for UML v2.0, there are further possibilities to enhance the GFT as a UML test specification profile.

References

1. ISO/IEC 9646-3 (1998): "Information technology - Open systems interconnection - Conformance testing methodology and framework - Part 3: The Tree and Tabular combined Notation (TTCN)"
2. ETSI DES-00063-1 TTCN-3: Core Language.
3. J. Grabowski, A. Wiles, C. Willcock, D. Hogrefe: On the Design of the new Testing Language TTCN-3. '13th IFIP International Workshop on Testing Communicating Systems' (Testcom 2000), Ottawa, 29.8.2000-1.9.2000, Kluwer Academic Publishers, August 2000.
4. ETSI DES-00063-2 TTCN-3: Tabular Presentation Format.
5. ETSI DES-00063-3 TTCN-3: Graphical Presentation Format.
6. ITU-T Recommendation Z.120 (11/99): "Message Sequence Charts (MSC)" – to be published.
7. J. Grabowski, T. Walter. Visualisation of TTCN Test Cases by MSCs, SAM98, Proceedings of the 1st Workshop of the SDL Forum Society on SDL and MSC , Humbold Universität Berln, 1998.
8. E. Rudolph, I. Schieferdecker, J. Grabowski: Development of an MSC/UML Test Format. FBT'2000 - Formale Beschreibungstechniken für verteilte Systeme (Editors: J. Grabowski, S. Heymer), Shaker Verlag, Aachen, June 2000.
9. ITU-T Recommendation X.680 (1997): "Information Technology – Abstract Syntax Notation One (ASN.1): Specification of basic notation"
10. P. Baker, C. Jervis, D. King, *An Industrial use of FP: A Tool for Generating Test Scripts from System Specifications*, Scottish Functional Programming Workshop/Trends in Functional Programming, 2000.
11. ITU-T Recommendation X.290 (1995): "OSI Conformance Testing Methodology and Framework – General Framework"
12. ITU-T Recommendation Z.100 (11/99): "System Specification and Description Language (SDL)".
13. OMG: Unified Modelling Language v1.0 (UML).
14. E. Rudolph, P. Graubmann, J. Grabowski: Tutorial on Message Sequence Charts (MSC-96).-Forte/PSTV'96, Kaiserslautern, October 1996.
15. E. Rudolph, I. Schieferdecker, J. Grabowski: HyperMSC - a Graphical Representation of TTCN. Proceedings of the 2nd Workshop of the SDL Forum, Society on SDL and MSC (SAM'2000), Grenoble (France), June, 26 - 28, 2000.
16. IETF Network Working Group, RFC 2131: "Dynamic Host Configuration Protocol", March 1997.

Some Implications of MSC, SDL and TTCN Time Extensions for Computer-Aided Test Generation

Dieter Hogrefe, Beat Koch, and Helmut Neukirchen

Institute for Telematics, University of Lübeck
Ratzeburger Allee 160, D-23538 Lübeck, Germany*
{hogrefe,bkoch,neukirchen}@itm.mu-luebeck.de

Abstract. The purpose of this paper is to describe how computer-aided test generation methods can benefit from the time features and extensions to MSC, SDL and TTCN which are either already available or currently under study in the EC Interval project. The implications for currently available test generation tools are shown and proposals for their improvement are made. The transformation of MSC-2000 time concepts into TTCN-3 code is described in detail.

1 Introduction

Computer-aided test generation (CATG) from system specifications has been an active field of research for many years [1,5,10,24]. This research has resulted in the development of a number of test generation tools [2,8,9]. Today, two industrial-strength, commercially available CATG applications exist [6,18]. These tools take formal system specifications with the 1992 edition of the *Specification and Description Language (SDL-92)* and test purpose descriptions with the 1996 version of *Message Sequence Charts (MSC-96)* as input and produce test suites based on the second edition of the *Tree and Tabular Combined Notation (TTCN-2)* [14].

Meanwhile, the standards of both SDL and MSC have been updated (SDL-2000 [16], MSC-2000 [15]) and a thoroughly new version of TTCN has been standardized (TTCN-3 [7]). In addition, the European Commission has set up the *Interval* project [21] to prototype an SDL, MSC and TTCN-based tool chain for the development and testing of systems with real-time constraints. During the first project stage, the Interval consortium identifies constructs which are suitable for capturing, specifying, modelling and testing real-time requirements. Based on these constructs, the consortium proposes time extensions to the formal languages as ITU-T recommendations. In the second project stage, tools will be developed which include the new time constructs.

Taking existing test generation tools as reference implementations, this paper evaluates the implications of existing and proposed time extensions to CATG.

---

* Part of this work has been sponsored by the European Commission under contract IST-1999-11557.

R. Reed and J. Reed (Eds.): SDL 2001, LNCS 2078, pp. 168–181, 2001.
© Springer-Verlag Berlin Heidelberg 2001

It is structured as follows: Section 2 shows when and why time constructs are needed during testing. Section 3 contains an overview of the timer concepts in MSC, SDL and TTCN. In Sect. 4, timer support of the test generation tools TestComposer and Autolink is discussed. Section 5 is the main part of this paper. It examines first how the test generation process may be improved through the use of SDL-2000 together with the proposed extensions. Second, the benefits of using MSC-2000 for test purpose description are shown and a concrete mapping of MSC-2000 time concepts to TTCN-3 is presented. Section 6 concludes this paper.

2 Timer in Test Purpose Descriptions

Timers in test sequences have one of the following purposes:

- assuring that test cases end even if they are blocked due to unexpected behavior of the system under test (SUT);
- checking constraints on the response time of the SUT;
- delaying the sending of messages to the SUT in order to
  - allow the SUT to get into a state where it can receive the next signal (if the tester is too fast);
  - check the reaction of the SUT if a signal is delayed too long (invalid behavior specification);
  - check that the SUT does not send any signal for a given amount of time.

To guarantee the conclusion of a test case, one or more *global timers* are used. In the case of a single-tester test architecture, one timer is started at the beginning of test case execution. Its duration is chosen to be longer than the expected execution time of the test case. At the end of each possible test sequence, the timer is reset. In the exception handling section of the test case, the timeout of the global timer is caught and handled. If a distributed test system is used, a global timer is started within each test component participating in the test execution. In the case of a timeout in any test component, the other test components have to be notified in order to let them conclude the test execution gracefully.

Time constraints are checked through the use of one or a pair of *guarding timers*. Guarding timers are started when a signal is sent to the SUT. If a lower bound is specified in the time constraint, one timer has to expire before the response signal from the SUT is received. The second timer — which checks the upper bound of the time constraint — is reset immediately upon reception of the response signal. Premature reception of the response signal or the expiration of the second timer is caught in the exception handling section of the test case and result in a FAIL verdict.

A *delaying timer* is specified by inserting a timer start operation immediately followed by a timeout event into the test sequence.

3   Timer in Formal Languages

The formal languages MSC, SDL and TTCN all contain timer support. In this Section, an overview of the timer concepts of these languages is given.

3.1   MSC-96

Timer support in MSC-96 [13] is very basic: there exist events to *set* and *reset* a timer, and a *timeout* event. Timer events are identified by a mandatory timer name and an optional timer instance name. The specification of a timer duration is optional; if it is specified, it has no semantics. Pairs of timer set and reset/timeout events must be specified on the same MSC instance.

3.2   MSC-2000

MSC-2000 [15] supports the same basic timer events as MSC-96, with some changes and refinements. First of all, the *set* event has been renamed to *starttimer* and *reset* is now called *stoptimer*. If a duration is specified, then it must be done in the form of an interval with an optional lower bound (default value: zero) and an optional upper bound (default value: infinite). This means that the timer can expire within the specified period.

In addition to the basic timer concepts, MSC-2000 also provides a timed semantics for constraining and measuring the time of events (though a formal semantics for MSC-2000 is still missing). Using the external data language approach introduced in MSC-2000, variables of type *Time* may be declared. There are two operators to measure time and store it in time variables: one to determine the absolute time at the moment of the execution of a given event, and one to determine the amount of time which passes between two events. It is also possible to specify time constraints: the lower and upper bound of a time interval between a pair of events may be defined in order to specify the allowed delay between those events. For a single event, the absolute time of occurrence can be constrained, too.

An extension to the MSC-2000 standard has been proposed by the Interval consortium to ITU-T Study Group 10 in [20]. A new symbol is proposed to express periodic occurrence of repetitive events which are folded into a loop.

3.3   SDL-2000

In SDL-2000 [16], it is mandatory to declare timers to use them. As part of the declaration, a constant default duration may be defined. With the *set* statement, a timer is activated. With the *reset* statement, a timer is put back to the inactive state. If an active timer expires, a signal with the same name as the timer is put into the input queue of the process which contains the timer. This corresponds to the *timeout* event in the MSC language. Whereas timers in SDL-2000 and MSC-2000 are basically equivalent, the new time constraint concept of MSC-2000 has no equivalent in standard SDL-2000.

Timer handling has been a weak point of SDL since its first introduction and there has been no improvement with the publication of SDL-2000. Therefore, several timer and time semantics related extensions to the SDL standard have been proposed by research groups [19] and Interval consortium members [3,4,11]. The latter proposals include:

- the addition of cyclic timers which are automatically restarted after expiration;
- mechanisms to read a timer value;
- the introduction of interruptive signals and timeouts.

Furthermore, a real-time semantics is introduced. This semantics allows to assign urgencies to transitions and to model time progression caused by actions which are annotated with a corresponding assumption on time consumption.

3.4 TTCN-2

In TTCN-2 [12], there are three timer operations: the common *START* and *CANCEL* operations to activate and deactivate a timer, as well as the *READ-TIMER* operation which returns the amount of time which has passed since a timer has been activated. Timer expiration is caught with the *TIMEOUT* event.

There are several problems with the implementation of timers in TTCN-2. First, timers have to be declared at test suite level. According to the standard, a full set of timers must be allocated for each test component, potentially wasting scarce hardware resources. The second problem concerns the applicability of TTCN-2 to the testing of real-time time constraints: timeout events are stored in a list until they match an alternative in the test sequence. As a consequence, timeout events may remain unnoticed for some time. This in turn may lead to incorrect test execution and verdict.

Moreover, due to the snapshot semantics of TTCN, it has to be noted that when using the existing timer concepts, a coherent and valid test verdict for real-time tests can only be found if the test equipment is reasonably fast. Since the snapshot semantics may summarize time-critical events arriving at different queues in one snapshot, important timing or ordering information might get lost. In this case, it is not decidable whether a violation of real-time constraints has occurred or not. The test verdict will rather depend on the question of how the triggering events of an alternative are ordered in the TTCN dynamic behavior description.

To solve this problem, a refinement of the standard snapshot semantics is proposed in [25]. Instead of testing time constraints using standard timers, additional columns for earliest and latest execution times of TTCN events are proposed. Since this way of specifying time constraints is orthogonal to the evaluation of alternatives, the test verdict does not depend on the speed of the tester or the ordering of alternatives.

Nevertheless, even with [25] the test system has to be fast enough in order to avoid the overflow of input queues. Therefore, sufficient processing capabilities of the tester are in any case a necessary prerequisite of real-time testing.

3.5 TTCN-3

TTCN-3 [7] renames some of the timer operations of TTCN-2: the keyword to deactivate a timer is now *stop* and the elapsed time of an active timer can be queried with the *read* operation. In addition, the *running* operation returns *true* if a given timer is running, *false* otherwise. The *start* operation and *timeout* event remain unchanged. Timer functionality is also included in the synchronous *call* operation. A timeout value may be provided as an optional parameter to this operation. If a timeout occurs, it may be handled as an exception with the *catch* operation.

No concrete proposals have been published so far regarding the extension of time concepts in TTCN-3. However, since TTCN-3 uses the same snapshot semantics as TTCN-2, the weakness of this semantics concerning real-time testing still holds for TTCN-3. A forthcoming proposal to overcome this problem is currently under study by the Interval consortium. Rather than using the standard timers to test real-time requirements, it is intended to separate the description of functional requirements (such as signal reception and "functional" timeouts) and non-functional (that is real-time) constraints. Since these extensions are currently under study, the test cases given in this paper are written using standard TTCN-3 notation.

4   Timer in Current Test Generation Tools

At the time of writing this paper, there are two major test generation tools on the market which take SDL-92 and MSC-96 specifications as input and produce TTCN-2 as output: TestComposer [18] and Autolink [6]. In this Section, the current status of these tools with respect to timer support is presented.

4.1   TestComposer

TestComposer automatically generates four types of timers during the computation of test cases:

- a timer *TAC* is set whenever a test component waits for a response from the SUT. A fail verdict is assigned in case of a timeout. With respect to timer purposes introduced in Sect. 2, *TAC* corresponds to a guarding timer;
- the timer *TWAIT* is another guarding timer: it checks that time to execute an implicit send does not exceed a predefined amount of time;
- the timer *TNOAC* is set to check that the SUT does not send a message to the tester for a specific amount of time. *TNOAC* is a delaying timer;
- *TEMPTY* is a delaying timer which is used to force a timeout in the SUT.

4.2   Autolink

Autolink generates the declaration of a global timer *T_Global* automatically. Depending on the test architecture, timer statements for *T_Global* are added to the test case behavior description and the top-level test steps of all parallel test components.

Guarding and delaying timers can be specified by the user in test purpose MSCs with timer set, reset and timeout events; these events are translated into corresponding TTCN-2 statements during test generation.

4.3   Discussion

With the methods available in the current test generation tools, the common cases for using timers in test cases can be handled fairly well. However, both tools do not offer optimal timer support. On the one hand, unnecessary timer events may be generated with the fully automatic method in TestComposer. These events have to be removed from the test suite manually. On the other hand, while Autolink offers complete flexibility regarding timers, the manual specification with MSC-96 may be laborious. This is especially true if an SUT response has to fall within a time interval: with the MSC-96 notation used by Autolink, two timers must be drawn, which increases the effort to specify the test purpose MSC and reduce its readability (see Fig. 1). Neither tool supports the reading of timer values.

5   Improving the Test Generation Process

Both TestComposer and Autolink are test-purpose-based test generation tools. This means that they need a formal description of the test purpose which they

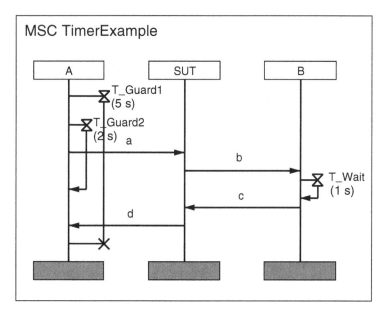

Fig. 1. Timer specification for Autolink with MSC-96

can transform into a TTCN test case. The transformation is done either by direct translation from MSC to TTCN or by performing a state space exploration of an SDL specification. Test purpose descriptions may be provided in the form of MSC-96 diagrams for both tools.

CATG tools may benefit from the use of formal languages with time extensions in a number of ways:

- reduction of the state space during exploration-based test generation with timed SDL;
- automatic generation of time requirements for test equipment with timed SDL;
- improvement of the capabilities to efficiently describe timing constraints in test purpose descriptions by using MSC-2000.

5.1 Test Case Generation with Timed SDL

The extensions proposed by the Interval consortium for SDL are mainly intended for verifying and validating a specification with respect to time properties. Nevertheless, automatic test generation benefits for two reasons from such time annotations.

First, the state space of a timed SDL model can be reduced in comparison to an untimed specification. The reason is that an untimed specification allows a lot of unrealistic scenarios which cannot occur in practice, because it contains paths where time does not progress at all. By using a real-time semantics and a timed SDL model, the state space can be reduced to the realistic scenarios. As CATG is mainly based on representing observable events of paths allowed by an SDL model, unrealistic test cases can be avoided.

Second, if additional timing information is given for all symbols contained in an SDL transition, the exact moment when observable events are allowed to take place can be determined. Test cases which take this information into account can be derived automatically. However, this topic is subject to further study. If additional timing information is not available for a whole transition, it is still possible to specify real-time requirements using MSC-2000. The usage of MSC-2000 for test description is shown in Sect. 5.3.

5.2 Generation of Time Requirements for Test Equipment

TTCN assumes that the test equipment is always fast enough to test the IUT. While this assumption is legitimate if only time non-critical functional behavior is tested, it may not hold for real-time applications. The processing speed of the tester may not be fast enough to keep track with the test events that happen at the PCOs. As an example, during the development of the GSM test suite at ETSI, there were various occasions where the possible lack of sufficient speed of the test devices had to be taken into account. In some cases, this problem was resolved by letting the tester respond to a signal from the SUT before it even receives the signal, just by assuming that the signal will arrive eventually. If the tester had to wait for the reception of the signal, it would be not fast enough to respond to it. While such workarounds are possible, they are problematic, because the order of test events has to be changed. As a consequence, the prose test purpose description no longer corresponds to the formal description.

In general, it seems more feasible to require some speed of the tester and treat these requirements as part of the test suite. If this approach is taken, time constraints for the tester have to be defined somehow. This means that the tester is required to execute test events within a certain time interval in order to test the SUT accurately and successfully.

A very detailed idea about the timing behavior of the SUT is required to determine the time intervals between test events. The test designer or the test generation tool needs to know at which points in time the SUT may be stimulated or which events from the SUT may be observed. Traditionally, this timing information has not been part of the SDL specification. However, if such timing information is added to the SDL specification, the minimal time interval between test events can be derived which the test equipment must be able to process. Based on this information, it is possible to generate benchmarks for the tester. An example for a tester benchmark is given below using the TTCN-3 notation. This benchmark checks whether the test equipment is fast enough to send two consecutive messages within a duration specified by *required_time*:

```
1:  timer T;
2:  T.start(required_time);
3:  A.send(a);
4:  A.send(b);
5:  if (T.running)
6:  {
7:    T.stop;
8:    verdict.set(pass);
9:  }
10: else
11: {
12:   verdict.set(fail);
13:   MyComponent.stop;
14: }
```

5.3   Using MSC-2000 for Test Purpose Description

Figure 2 shows the test description of Fig. 1 in MSC-2000 notation. The use of the time interval notation instead of four separate timer symbols (two set events, one reset and one timeout) to specify two guarding timers makes the diagram much more readable. Given a time interval where the start event is a send to the SUT and the end event is a receive from the SUT, the test generation tool has to perform the following actions:

1. Check the time interval to establish the number of timers which are needed to represent the interval with TTCN. If just one boundary value is specified,

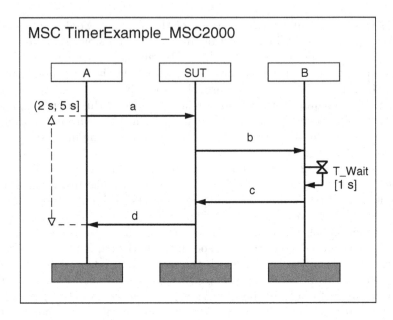

Fig. 2. Time constraint specification with MSC-2000

only one timer is needed. If both a minimal and a maximal time point are specified, two TTCN timers are needed. Since no timer name is specified with the time interval notation, the tool has to select the timer names by itself. Preferably, the user should be able to define timer name templates such as *T_Guard_Min* and *T_Guard_Max*. The tool then must check if timers with these name are in use already. If they are, new timers (such as *T_Guard_Min_2* and *T_Guard_Max_2*) must be declared. In order to minimize the number of timers which have to be declared, the test configuration has to be taken into account in this step.

2. If the time interval contains expressions with measurements (see Sect. 5.3), replace the time patterns with the corresponding variable identifier. If necessary, convert time values to seconds.

3. Create the appropriate TTCN timer statements. This step depends on the number of time points specified in the MSC. In the examples below, TTCN-3 is produced from the MSC in Fig. 2, assuming that there is a test component handling just PCO *A*. A similar transformation can be done for TTCN-2.

If only the lower boundary value is specified, create the following statements:

```
1:   timer T_Guard_Min;
2:   A.send(a); T_Guard_Min.start(2);
3:   alt {
4:   []   T_Guard_Min.timeout;
5:   []   A.receive(d)
```

```
 6:         {  verdict.set(fail);
 7:             MyComponent.stop;
 8:         }
 9:  }
10:  A.receive(d);
```

The case where d is received prematurely by PCO A (lines 5 to 8) may as well be handled in a default. If only the upper boundary value is specified, create the following statements:

```
 1:  timer T_Guard_Max;
 2:  A.send(a); T_Guard_Max.start(5);
 3:  alt {
 4:  []  A.receive(d)
 5:         {  T_Guard_Max.stop;  }
 6:  []  T_Guard_Max.timeout
 7:         {  verdict.set(fail);
 8:             MyComponent.stop;
 9:         }
10:  }
```

If the lower and the upper boundary values are specified, create the following statements:

```
 1:  timer T_Guard_Min;
 2:  timer T_Guard_Max;
 3:  A.send(a); T_Guard_Min.start(2); T_Guard_Max.start(5);
 4:  alt {
 5:  []  T_Guard_Min.timeout;
 6:  []  A.receive(d)
 7:         {  verdict.set(fail);
 8:             MyComponent.stop;
 9:         }
10:  }
11:  alt {
12:  []  A.receive(d)
13:         {  T_Guard_Max.stop;  }
14:  []  T_Guard_Max.timeout
15:         {  verdict.set(fail);
16:             MyComponent.stop;
17:         }
18:  }
```

The case where d is received prematurely by PCO A (lines 6 to 9) may as well be handled in a default. TTCN-3 has a special notation to put a timeout value on a procedure call. If the start event of a timer interval in an MSC is a method call (cf. Fig. 3), then the following code should be generated in order to guard the call by an upper time bound of 3 seconds (for example):

```
1:   A.call(x, 3);
2:   {
3:       [] A.getreply(x);
4:       [] A.catch(timeout);
5:           { verdict.set(fail);
6:               MyComponent.stop;
7:           }
8:   }
```

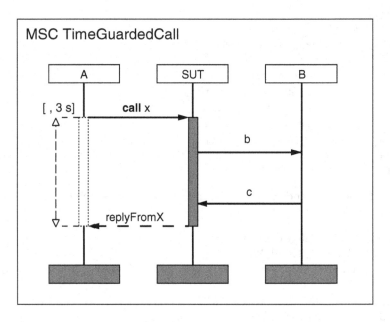

Fig. 3. Time constraint for a method call in MSC-2000

Time Measurement. In MSC-2000, time can be measured and stored in variables. These measurements can be reused to specify time intervals. Figure 4 shows the two kinds of time measurements provided by language: $\&t1$ is a relative measurement; the time which passes between the sending of a and the reception of d is stored in a variable $t1$ of type *Time*. $@t2$ is an absolute measurement, which means that the value of an existing global clock is stored in a variable $t2$ of type *Time*. The global clock is started when the first event in the MSC is executed.

Transforming a relative time measurement into TTCN is straight-forward. The test generation tool needs to declare a special timer used for the measurement. This timer is activated after the first event in the MSC has occurred. After the second event, the timer is read and deactivated. The only problem is the fact that to start a timer in TTCN, a duration has to be defined, which in this case is not known in advance. As a solution, the test designer has to provide a maximum value for time measurement. Most likely, this value is available anyway

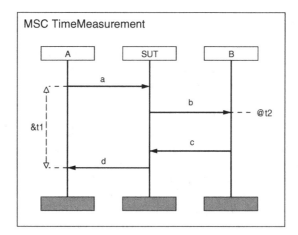

Fig. 4. Time measurements in MSC-2000

because a timeout period has to be defined for a global test case timer. From the MSC in Fig. 4, the tool should generate the following TTCN-3 statements:

```
1:   timer T_Measure := 10000;
2:   var float t1;
3:   A.send(a); T_Measure.start;
4:   A.receive(d); t1 := T_Measure.read;
5:   T_Measure.stop;
```

To measure an absolute time value, a global timer has to be started at the beginning of the test case. The measurement can then be done by using the *read* operation on the global timer. Below is the TTCN-3 code generated for the measurement of *t2*:

```
1:   timer T_Global = 10000;
2:   var float t2;
3:   T_Global.start;
4:   B.receive(b); t2 := T_Global.read;
5:   B.send(c);
```

MSC-2000 also allows to measure or to constrain the amount of time which passes between a pair of events on different instances. This kind of time interval cannot be represented with standard TTCN-3 timers, since the start and read or timeout operations of the same timer cannot be distributed between different parallel test components. If there is more than one test component, coordination messages might be used to synchronize the parallel test components concerning the two relevant events. However in this case, the time needed to transmit these coordination messages has to be taken into account.

6   Conclusion

In this paper, the current state of SDL, MSC, TTCN and test generation tools with regard to timer support has been presented. Commercially available test generation tools already allow to generate TTCN-2 test suites which reflect time requirements expressed by standard MSC-96 timers. However, due to the simple timer concepts of MSC-96, the specification of time constraints in test purpose descriptions may be quite laborious. It has been shown that the use of the MSC-2000 time interval notation can facilitate the specification of time constraints for events on the same instance. A translation of MSC-2000 time constructs into TTCN-3 code has also been presented. The mapping of MSC-96 timers to MSC-2000 time constructs and the transformation of TTCN-2 to TTCN-3 is straightforward. Therefore it will be possible to automatically generate TTCN-3 test cases which test the conformance to such MSC-2000 time constraints.

Nevertheless, many challenges remain. The testing of time constraints in a distributed test architecture has not been solved yet. Currently, it is neither possible to derive test cases in an automated way nor to test real-time requirements if several parallel test components observing time critical events are involved. Care has to be taken to synchronize these parallel test components not only in a functional manner but also with regard to their local clocks.

It has also not been shown yet that it is possible to generate real-time tests from time extended SDL models. The accuracy of automatically derived test cases depends on how exhaustively an SDL model is enriched with time annotations. Research by the Interval consortium will show whether this is feasible. Moreover, as an underlying basis, existing test theory has to be extended in the area of real-time testing.

Due to the problems introduced by the snapshot semantics of TTCN, standard timers should not be used to test real-time constraints where a high resolution of timing information is required. Therefore in the testing domain, the next step which will be done in the Interval project is to present real-time extensions for TTCN-3 which allows deterministic real-time testing.

References

1. C. Bourhfir, R. Dssouli, E. Aboulhamid, and N. Rico. Automatic executable test case generation for extended finite state machine protocols protocols. In IWTCS'97 [17], pages 75–90.
2. C. Bourhfir, R. Dssouli, E. Aboulhamid, and N. Rico. A test case generation tool for conformance testing of SDL systems. In SDL'99 [23], pages 405–419.
3. M. Bozga, S. Graf, A. Kerbrat, L. Mounier, I. Ober, and D. Vincent. SDL for real-time: what is missing? In *SAM 2000 – 2nd Workshop on SDL and MSC*, pages 108–122, Grenoble, France, June 2000.
4. M. Bozga, S. Graf, L. Mounier, I. Ober, J.-L. Roux, and D. Vincent. Timed extensions for SDL. In *SDL Forum 2001*, Copenhagen, Denmark, June 2001.
5. M. Clatin, R. Groz, M. Phalippou, and R. Thummel. Two approaches linking a test generation tool with verification techniques. In *Proceedings of IWPTS '95 (8th*

Int. Workshop on Protocol Test Systems, pages 151–166, Evry, France, September 1995.

6. A. Ek, J. Grabowski, D. Hogrefe, R. Jerome, B. Koch, and M. Schmitt. Towards the industrial use of validation techniques and automatic test generation methods for SDL specifications. In SDL'97 [22], pages 245–259.

7. ETSI, Sophia Antipolis, France. *Methods for Testing and Specification (MTS); The Tree and Tabular Combined Notation version 3; TTCN-3: Core Language*, v1.0.10 edition, November 2000. DES/MTS-00063-1.

8. J.-C. Fernandez, C. Jard, T. Jéron, and C. Viho. An experiment in automatic generation of test suites for protocols with verification technology. *Science of Computer Programming*, 29, 1997.

9. J. Grabowski. *Test Case Generation and Test Case Specification with Message Sequence Charts*. PhD thesis, University of Bern, Bern, Switzerland, February 1994.

10. J. Grabowski, D. Hogrefe, and R. Nahm. Test case generation with test purpose specification by MSCs. In *SDL'93: Using Objects*, pages 253–265, Darmstadt, Germany, October 1993. Elsevier Science Publishers B.V.

11. S. Graf. Timed extensions for SDL, November 2000. Delayed Contribution No. 13 to ITU-T Study Group 10, Questions 6&7.

12. ISO/IEC. *Information technology – Open Systems Interconnection – Conformance testing methodology and framework*, 1994. International ISO/IEC multipart standard No. 9646.

13. ITU-T, Geneva, Switzerland. *Message Sequence Charts*, 1996. ITU-T Recommendation Z.120.

14. ITU-T, Geneva, Switzerland. *Information technology – Open Systems Interconnection – Conformance testing methodology and framework – Part 3: The Tree and Tabular Combined Notation*, 1997. ITU-T Recommendation X.293-ISO/IEC 9646-3.

15. ITU-T, Geneva, Switzerland. *Message Sequence Charts*, November 1999. ITU-T Recommendation Z.120.

16. ITU-T, Geneva, Switzerland. *Specification and Description Language (SDL)*, 1999. ITU-T Recommendation Z.100.

17. *Testing of Communicating Systems*, Cheju Island, Korea, September 1997. Chapman & Hall.

18. A. Kerbrat, T. Jéron, and R. Groz. Automated test generation from SDL specifications. In SDL'99 [23], pages 135–151.

19. A. Mitschele-Thiel. *Systems Engineering with SDL – Developing Performance-Critical Communication Systems*. Wiley, Chichester, England, 2001.

20. H. Neukirchen. Corrections and extensions to Z.120, November 2000. Delayed Contribution No. 9 to ITU-T Study Group 10, Question 9.

21. Interval Consortium Web Page. http://www-interval.imag.fr/, 2000.

22. *SDL'97 – Time for Testing*, Evry, France, September 1997. Elsevier.

23. *SDL'99 – The Next Millennium*, Montréal, Québec, Canada, June 1999. Elsevier.

24. G. v. Bochmann, A. Petrenko, O. Bellal, and S. Maguiraga. Automating the process of test derivation from SDL specifications. In SDL'97 [22], pages 261–276.

25. Th. Walter and J. Grabowski. Real-time TTCN for testing real-time and multimedia systems. In IWTCS'97 [17].

Verification of Quantitative Temporal Properties of SDL Specifications[*]

Iulian Ober[1,2] and Alain Kerbrat[2]

[1] Institut National Polytechnique/IRIT
2 Rue Camichel, 31000 Toulouse, France.
iulian.ober@enseeiht.fr
[2] Telelogic, 150 Rue N. Vauquelin, 31106 Toulouse, France.
{iulian.ober,alain.kerbrat}@telelogic.com

Abstract. We describe an approach for the verification of quantitative temporal properties of SDL specifications, which adapts techniques developed for timed automata [2]. With respect to other verification approaches applied to SDL, our approach broadens the class of analyzable specifications and improves the handling of non-deterministic systems, such as open systems communicating with an unspecified environment. Compared to the initial framework of timed automata, the application of these verification techniques to SDL raises two interesting issues, discussed in the paper. They are: expressing the semantics of time in SDL in terms of timed automata concepts, and employing a user friendly automata-based property specification language (GOAL [1]) to express and verify temporal properties. The paper also presents a verification tool prototype for SDL which implements these ideas.

1 Introduction

Automatic verification of behavioral properties is an important feature which can be offered by SDL tools, and which distinguishes SDL from other modeling languages used for specification and design of complex systems. Automatic verification is made possible by the existence of a formal semantics and syntax for the language [13], which can be used as a basis for mathematical reasoning about SDL specifications. Verification is an important task in the development of safety critical systems, which constitute an important share of the systems built with SDL.

Automatic verification capabilities implemented in industrial SDL tools (e.g. [19,17]), employ techniques derived from model checking. Properties which can be verified range from simple safety properties (e.g. absence of deadlocks, invariance of a logical condition) to more complex safety or liveness properties

---

[*] Work supported by the European project INTERVAL IST-1999-11557, *Formal Design, Validation and Testing of Real-Time Telecommunications Systems* (http://www.cordis.lu/ist/projects/99-11557.htm) and by the French RNRT Project PROUST (http://www-verimag.imag.fr/PROUST/).

R. Reed and J. Reed (Eds.): SDL 2001, LNCS 2078, pp. 182–202, 2001.

(e.g. linear properties specified by (Büchi) automata). The value added by SDL compared to other automatic verification approaches is twofold:

1. model checking is done on the analysis or design model (a functional model), without the need for building more abstract models in another formalism;
2. model checking is brought to the level of non specialists, by using simple and intuitive property specification languages (such as GOAL [1]) and tools.

As SDL is increasingly used for designing and implementing real-time and embedded systems, the verification methods and tools for SDL must be extended to account for quantitative temporal properties. This paper presents the results of a work aiming to adapt analysis techniques initially developed for timed automata [2] to SDL. This work also points out a series of deficiencies of SDL, which diminish its expressivity and the power of the analysis tools based on it. For the identified problems, we suggest some ways to improve the language.

The rest of the paper is structured as follows: Sect. 2 presents the state of the art in the specification and verification of temporal behavior using timed automata. We complete this section with a discussion of the limitations of timed automata, and their impact on our SDL approach. In §3 we show how the SDL execution model relates to the timed automata model. We also discuss possible improvements of SDL suggested by this relation. In Sect. 4 we examine how GOAL [1] can be used to describe temporal properties of SDL systems. In Sect. 5 we present the temporal verification facilities implemented in an extended version of the *Object*GEODE tool [19]. Finally, we draw the conclusions of this work and examine further advancement possibilities.

1.1 Related Work

The idea of applying timed automata techniques in higher level formalisms was intensively studied during the past few years. SDL is not the only language to benefit from this trend: proposals for the improvement of both formal description techniques (LOTOS, ESTELLE) and more informal modeling languages (UML) using timed automata are being studied.

For relating SDL and timed automata, [5] proposes an intermediate formalism, called IF, in which one can describe a system of asynchronously communicating processes much in the same way as in SDL. The authors propose a methodology in which SDL specifications are translated to IF, and IF models are analyzed using timed automata methods and tools.

[8] presents an approach in which a variant of timed automata, called Timed Finite State Machines (TFMS), are used as an intermediary form for generating SDL specifications and timed test cases from timed scenario descriptions. Timed automata techniques are not used for verification of properties, but the authors point out some of the same deficiencies of SDL that we reveal in this paper.

The general problems with expressing time-related behavior, discovered during this work and the work of our project partners are systematized in [6].

2 Specification and Verification of Temporal Behavior

2.1 Reasoning about Timed Systems

SDL contains constructs for describing time-driven behavior: the designer can use timers or enabling conditions involving the time variable **now** in order to describe such behavior. Thus, the execution of an action may be triggered or conditioned by time.

While these constructs can model infinitely complex behavior, SDL has two major drawbacks when one wants to *verify* temporal properties of timed systems:

1. The formal semantics of the language [12] is loose about time progress: indefinite amounts of time may pass while a process is in a state even if it has a valid input signal waiting in the queue, and actions take indefinite times to execute. The only system component which behaves strictly with respect to time is the underlying agent responsible for keeping track of timers and sending timer expiration signals. With such loose assumptions about the performance of the underlying execution machine, it is difficult to guarantee almost any time-related property about system behavior. This problem was examined in more detail in [6].

2. Timers and conditions on **now** can describe infinitely complex behavior, for which it is difficult to conceive analysis methods and algorithms.

Several formalisms have been proposed for modeling time-conditioned behavior. Among them, we look into timed automata because they are extended versions of finite automata, and thus semantically related to SDL. Timed automata cope with both problems mentioned above:

1. They provide stronger requirements on time progress, which can be constrained by the state of the automaton. Thus, one can specify actions that occur at a specific moment or within a bounded time, unlike in SDL.

2. Time conditions can only have simple forms. In timed automata, the only mechanism to measure time is the clock. An automaton may use several clocks at a time, all of which progress at the same rate and can be initialized and tested separately. Time conditions are represented by conditions on clocks, which can only have the following forms: $x \sim c$ and $x - y \sim c$, where x, y are clocks, $c \in \mathbb{Z}_+$ is a constant, and $\sim$ is one of $\leq, <, =, >, \geq$.

These restrictions make it possible to solve analytically a series of problems on timed automata, such as the reachability problem ("given a configuration of the timed automaton, i.e. a state and a set of values of the clocks, is there an execution which leads from the initial state to that configuration"), or various model checking problems.

For timed automata techniques to be applied to SDL, an SDL specification has to conform, in a way, to the restrictions mentioned above. We will discuss the implications of this in Sect. 3.

2.2   Timed Automata

There are several, more or less restrictive ways of defining timed automata. We consider the following definition which includes the notion of transition urgency[1]:

Definition 1 (timed automaton).
A timed automaton is a tuple $A = (\Sigma, \mathcal{X}, Q, q_0, E, \mathbf{inv})$ where:

1. *Σ is a finite set of transition labels.*
2. *$\mathcal{X}$ is a finite set of clocks.*
3. *Q is a finite set of discrete states.*
4. *q_0 is a distinguished state of Q called* initial state.
5. *E is a set of transition edges between the states from Q, each edge $e \in E$ having the following components: $e = (q, \zeta, u, a, X, q')$. $q, q' \in Q$ are the source and destination states. ζ, also denoted $guard(e)$ is the guard of the transition and it is a conjunction of atomic conditions on the clocks $\mathcal{X}$ of A. An atomic condition has one of the following two forms: $x \sim \mathbf{c}$ or $x - y \sim \mathbf{c}$ where $x, y \in \mathcal{X}$, $\sim \in \{<, \leq, >, \geq\}$ and $\mathbf{c} \in \mathbb{Z}_+$ is a constant. We will denote $\mathcal{CP}(\mathcal{X})$ the set of conjunctions of atomic conditions over the clocks of $\mathcal{X}$. $u \in \{eager, lazy, delayable\}$ is an attribute called the* urgency *of the transition. $a \in \Sigma$ is the label of the transition edge e. $X \subseteq \mathcal{X}$ is the set of clocks reset during the transition e, also denoted $reset(e)$.*
6. *$\mathbf{inv} : Q \longrightarrow \mathcal{CP}(\mathcal{X})$ is a function that associates to each state q of A a conjunction of atomic conditions on the clocks from $\mathcal{X}$ called the* invariant *of the state q.*

Timed automata as labeled transition systems. A semantics is given to timed automata by associating an (infinite) labeled transition system (LTS) G_A to each timed automaton A. This infinite LTS, called the *semantic graph* of the timed automaton, is defined by the following:

1. The nodes of G_A are called *configurations* or *dynamic states* of A. They are pairs $(q, \mathbf{v})$ where $q \in Q$ is a discrete state and $\mathbf{v}$ is a *valuation* of the clocks of the automaton, $\mathbf{v} : \mathcal{X} \to \mathbb{R}_+$. In order for a configuration to be consistent, the $\mathbf{v}$ must satisfy the invariant $\mathbf{inv}(q)$.
2. The edges of G_A correspond to transitions of A from one configuration to another. There are two kinds of transitions allowed in a state $(q, \mathbf{v})$:
   - **Discrete transitions** occur when a transition edge $e = (q, \zeta, u, a, X, q')$ is taken. e is *enabled* in $(q, \mathbf{v})$ iff $\mathbf{v}$ satisfies the condition ζ. When the transition e is taken, the system moves to state $(q', \mathbf{v}')$ where $\mathbf{v}'(x) = \mathbf{v}(x), \forall x \in \mathcal{X} \setminus X$ and $\mathbf{v}'(x) = 0, \forall x \in X$. The transition is denoted by $(q, \mathbf{v}) \overset{e}{\longrightarrow} (q', \mathbf{v}')$.
   - **Time transitions** happen when an amount $\delta \in \mathbb{R}$, $\delta > 0$ of time elapses without any discrete transition being fired in the meantime. A time transition moves the system from state $(q, \mathbf{v})$ to state $(q, \mathbf{v}+\delta)$ where $\mathbf{v} + \delta$ denotes the valuation $\mathbf{v}'$ such that $\mathbf{v}'(x) = \mathbf{v}(x) + \delta, \forall x \in \mathcal{X}$. The

---

[1] Urgencies for timed automata transitions were proposed in [3,4]

transition is denoted $(q, \mathbf{v}) \xrightarrow{\delta} (q, \mathbf{v} + \delta)$. Time is considered to progress only if the automaton is prepared for that, i.e. the time transition is *enabled* iff the following conditions hold:

(a) $\forall \delta' \in (0, \delta]$, $\mathbf{v} + \delta' \in \mathbf{inv}(q)$

(b) $\forall \delta' \in [0, \delta)$, there is no *eager* transition $e = (q, \zeta, u, a, X, q')$ enabled in $(q, \mathbf{v} + \delta')$ (i.e. $u = eager$ and $\mathbf{v} + \delta' \in \zeta$).

(c) $\forall \delta'$, δ'' such that $0 \leq \delta' < \delta'' \leq \delta$, there is no *delayable* transition $e = (q, \zeta, u, a, X, q')$ enabled in $(q, \mathbf{v} + \delta')$ and disabled in $(q, \mathbf{v} + \delta'')$ (i.e. $u = delayable$ and $\mathbf{v} + \delta' \in \zeta$ and $\mathbf{v} + \delta'' \notin \zeta$).

It is considered that G_A contains only those vertices which are reachable from the initial configuration of the system, which is $(q_0, \mathbf{v})$ with $\mathbf{v}(x) = 0$, $\forall x \in \mathcal{X}$.

Decidable problems for timed automata. An interesting problem concerning an automaton A is whether a particular state $(q, \mathbf{v})$ is reachable from the initial state $(q_0, 0)$. This is called the reachability problem of timed automata. It is important in that many properties of a timed automaton (e.g. invariance properties) may be solved if reachability is solved.

[2] proves the decidability of the reachability problem for timed automata[2], by constructing a finite abstraction (the *region graph*) from the potentially infinite semantic graph G_A, and showing that it preserves reachability. The same abstraction can be used to solve other problems, such as satisfaction of certain temporal logic formulas.

The simulation graph. For verification problems involving only reachability or linear properties (see Sect. 2.3), there are more efficient analysis methods than the region graph mentioned above. The *simulation graph* is one of them, which we use for verifying temporal properties of SDL specifications.

First, note that in a configuration $(q, \mathbf{v})$ of the automaton A, $\mathbf{v}$ can be regarded as a point in the space $\mathbb{R}^{|\mathcal{X}|}$. Then, the set of valuations $\mathbf{v}$ which satisfy a condition $\zeta \in \mathcal{CP}(\mathcal{X})$ form a convex polyhedron in $\mathbb{R}^{|\mathcal{X}|}$ (which can be identified with the condition ζ and denoted identically). Finally, let $\mathcal{NCP}(\mathcal{X})$ denote the set of non-convex polyhedra in $\mathbb{R}^{|\mathcal{X}|}$, i.e. finite unions of convex polyhedra from $\mathcal{CP}(\mathcal{X})$. We will call *zone* an element from $\mathcal{NCP}(\mathcal{X})$.

The *simulation graph* of A has couples (q, S) as vertices, where $q \in Q$ is a discrete state and $S \in \mathcal{NCP}(\mathcal{X})$ is a zone. The following operations are defined on (q, S) pairs:

$$\text{time-succ}((q, S)) = (q, \{\mathbf{v}' \mid \exists \mathbf{v} \in S, \delta \in \mathbb{R}. (q, \mathbf{v}) \xrightarrow{\delta} (q, \mathbf{v}')\})$$
$$\text{disc-succ}(e, (q, S)) = (q', \{\mathbf{v}' \mid \exists \mathbf{v} \in S. (q, \mathbf{v}) \xrightarrow{e} (q', \mathbf{v}')\})$$

where e is an edge between q and q'.

---

[2] [2] refers to timed automata without urgencies. The proof can be extended to timed automata with urgencies, by showing that time restrictions imposed by urgencies preserve the form of the regions.

It can be proved that if S is a zone, time-succ$((q, S))$ and disc-succ$(e, (q, S))$ also yield zones[3]. The simulation graph of the automaton A is the smallest graph $SG(A)$ such that:

1. time-succ$((q_0, 0))$ is a node of $SG(A)$
2. for every node (q, S) of $SG(A)$ and every discrete transition edge e from q to q', if $(q', S') = $ time-succ$($disc-succ$(e, (q, S)))$ and $S' \neq \emptyset$ then (q', S') is also a node of $SG(A)$ and $(q, S) \overset{e}{\longrightarrow} (q', S')$ is an edge of $SG(A)$.

As a zone S is a finite reunion of regions (from the *region graph* [2]), it can be easily proved that the simulation graph is always finite. Moreover it contains all and only the reachable states of A, and every run of A is contained in a path from $SG(A)$.

2.3   Temporal Properties

As for modeling systems, several formalisms have been proposed for expressing temporal properties of systems (and in particular quantitative temporal properties). Complexity limits apply on these formalisms: there are properties which cannot, in general, be algorithmically verified on a timed automaton.

We present here some classes of properties which have been studied for timed automata, and can be solved using the *simulation graph*. The effective method for solving them is discussed later on in the case of SDL.

Absence of deadlocks. A deadlock in a timed automaton is a situation in which the automaton, arrived at a state $(q, \mathbf{v})$, cannot take any more discrete transitions (there is no delay δ, edge e, and state $(q', \mathbf{v}')$, such that $(q, \mathbf{v}) \overset{\delta}{\longrightarrow} \overset{e}{\longrightarrow} (q', \mathbf{v}')$.

Non-zenoness of runs. An infinite run of a timed automaton is called *zeno* if all along the run, time remains below a fixed finite limit. A *zeno* run corresponds to the situation where the automaton makes an infinite number of actions (transitions) within a bounded amount of time.

Invariance properties. An invariance property is a condition on the state and clocks of an automaton, which holds along all possible executions of the automaton. Only particular forms of invariants can be verified using the simulation graph: propositional logic formulas having as atoms either conditions on the discrete state q of the automaton, or simple atomic conditions on the clock values (conditions from $\mathcal{CP}(\mathcal{X})$).

Timed linear properties. More complex properties than the ones mentioned above are sometimes useful to express and verify. Take for example the following property: the action a of the system is always followed by b within at most d time units.

This property belongs to a class called *linear properties*, which are, informally defined, assertions about the set of possible executions of a system. Such properties generally express things like: there is an execution of the system which has a certain form, or all executions have a certain form.

---

[3] for timed automata *without urgencies*, convexity is also preserved, and therefore all zones of the simulation graph are from $\mathcal{CP}(\mathcal{X})$. See [18].

Linear properties may be expressed in various (equivalent) ways: as linear logic formulas, as finite automata – if they make assertions about finite executions of a system, as Büchi automata – if they make assertions about infinite executions of a system. We are interested in properties expressed through (finite or Büchi) automata, since we use GOAL [1], an automata-based property specification language, in connection with SDL.

The verification of properties that are expressed through timed Büchi automata (TBA), a simple extension of timed automata with Büchi acceptance conditions, has been studied in [18]. This verification problem is decidable using the *simulation graph*. We use a decision method similar to the one described in [18] to verify GOAL properties on SDL.

2.4   Decidability Limits of Timed Models

Other models, more general than timed automata, have been studied. From a practical modeling point of view, for example, it is interesting to generalize the variation laws for the clocks, such that:

- clocks may vary with different relative speeds (*multirate automata*),
- a clock may vary with different speeds in different states (*integrator automata*), etc.

However, such models are usually decidable only under strong restrictions. For example, *multirate automata* are decidable only if clocks are not compared with one another, and *integrator automata* are decidable only under strong structural restrictions (for a synthesis of decidability results, see [10] and the work cited therein). Thus, timed automata give in a way a *complexity limit* up to which "general" timed models are decidable.

Semi-decision procedures can sometimes be developed for undecidable models such as those mentioned above. Such procedures are important from a practical point of view, in situations where timed automata are not expressive enough and only a generalized model can capture the behavior of a system. Nevertheless, these semi-decision procedures are usually complex and difficult to apply to large models. Therefore, in our work we have restricted to basic timed automata and equivalent SDL models.

3   A Timed Automata Model for SDL

In order to apply analysis techniques developed for timed automata to SDL specifications, we select the SDL constructs that can be mapped to timed automata constructs (clocks, operations and tests on clocks).

3.1   The Semantics of SDL

The execution model for SDL is given by the formal semantics of the language in [12]. The model of an SDL system is a multi-agent abstract state machine

(ASM). A multi-agent ASM, which is formed of a (dynamic) set of agents and a (dynamic) set of programs, has in turn its behavior defined in terms of *states* and *transitions* (modulo partial ordering, see [9]). Thus the semantic model of an SDL system is a labeled transition system (LTS) formed with the aforementioned *states* and *transitions*.

The reachability problem for SDL is undecidable, since the language does not impose any restrictions on variables and message queues (and therefore two-counter machines – which are undecidable in general [11] – may be simulated by an SDL system). However, this difficulty is overpassed in verification tools by imposing additional restrictions on the range of variables and message queue length.

However, free-formed expression containing **now** are allowed in enabling conditions, continuous signals, decision discriminants, and generally in any expression appearing in an SDL statement. This can also lead to non-decidability of the reachability problem[4].

This latter problem is more abruptly avoided by SDL verification tools. In most SDL verification tools (including *Object*GEODE and Telelogic TAU), in verification mode **now** is always considered to be 0, and bookkeeping is correctly done only for timers with relative expiration delays (i.e. set with a deadline **now** + d, where d is an integer not depending on **now**).

This method rules out from the verification process all SDL systems in which **now** is used for other things than setting relative timers.

By defining a relation between the SDL execution model and timed automata, we aim to broaden the class of systems on which behavior properties can be verified, to include systems in which **now** is used not only for setting relative timers. The class of verifiable SDL systems will still be limited by the complexity limits of timed automata, which are however looser than the one mentioned above.

3.2   Mapping SDL Constructs to Timed Automata Constructs

Among the SDL constructs which can be used to specify time-related behavior, there are two categories which can be represented through timed automata constructs:

Timers. Timers in SDL can be set to a deadline which can be computed in an arbitrary way. Absolute timers (set with a constant deadline d) and relative timers (set with a deadline of **now** + d) are the most common. However, timers may be set with deadlines computed by arbitrarily complex expressions, such as 2 ∗ **now** or **now** ∗ **now** (these expressions are type-incorrect, but equivalent expressions involving type casts may be written).

Absolute and relative timers can be mapped to timed automata constructs, while timers set with general expressions cannot. The mapping of absolute and relative timers is based on the observation that each timer can be represented

---

[4] This is proved by the fact that an *integrator automaton* (Sect. 2.4) can be simulated by an SDL system.

by a clock, which is tested for equality with the deadline in all states following the arming of the timer (see the mapping details in Sect. 3.4).

Relative delays measured with now. Free-formed expressions involving the SDL global clock **now** cannot be mapped, in general, to timed automata constructs. There is however a case in which they can: if y is a variable, and the assignment $y :=$ **now** appears on a transition, from that point on, the value of the difference **now** $- y$ can be simulated by a TA clock (x_y).

The assignment $y :=$ **now** equates to resetting x_y to 0. An enabling condition, continuous signal or decision discriminant which compares **now** $- x$ with an integer constant is equivalent to a TA guard testing the clock x_y (see the mapping details in Sect. 3.4).

3.3   Expressivity Considerations

Although we impose the above mentioned restrictions on the statements of SDL specifications, they are looser than the ones imposed by current verification tools. In this section we argue, based on an example, that from a temporal point of view our restricted SDL is strictly *more expressive* than the restricted SDL analyzed by current tools.

By allowing inequalities involving **now** in enabling conditions, continuous signals and decision discriminants, the SDL modeler has a means to specify time-non-deterministic behavior. Such behavior cannot otherwise be specified using only timers. There are at least two cases in which time non-determinism must be modeled in the SDL specification:

1. open systems – in which the behavior of the environment is not completely specified but may conform to some conditions (e.g. a certain signal is sent within given time bounds, etc.).
2. incompletely specified systems – in which the timing of the system depends on an unspecified component (such as the underlying hardware), for which some timing information is available (e.g. minimum/maximum execution times, etc.).

One more construct is necessary for the accurate expression of non-deterministic behavior: a notion of *urgency*, similar to the one defined in Sect. 2.2 for timed automata. The *urgency* specifies the degree of non-determinism of a transition: *eager* transitions are deterministically triggered as soon as they are enabled (i.e. they block the progress of time) while *lazy* and *delayable* transitions are non-deterministically triggered after they are enabled, but they do not block the progress of time (*delayable* transitions still block time progress if they reach the upper limit of their enabling condition, beyond which they are disabled).

In the example below (Fig.2-3), urgency is attached to SDL transitions as a formal comment.

Example. We consider the SDL specification of the exchange level of the SpaceWire protocol stack. SpaceWire [16] is a protocol for high-speed data links, used by the European Space Agency for handling payload data on-board

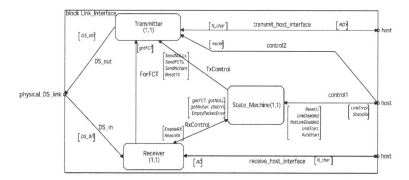

Fig. 1. The structure of a SpaceWire Link Interface

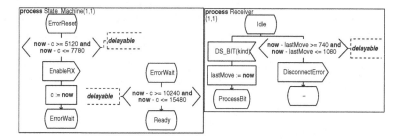

Fig. 2. Non-deterministic behavior of the SpaceWire Link Interface

a spacecraft. The SpaceWire standard covers all the protocol levels from the physical links up to the transport layer. The Exchange Level referred to in the sequel corresponds to the Data Link layer in the OSI stack.

The Exchange Level is built on top of an *unreliable serial full-duplex point-to-point* link, and offers functionality such as:

- Error detection through parity checking. No re-transmission functionality is provided.
- Disconnection detection based on timeouts.
- Flow control using credit counters for avoiding buffer overflow.

The SDL specification of the exchange level of SpaceWire consists in an SDL block describing a SpaceWire Link Interface (see Fig. 1). The block contains three processes: a Transmitter, a Receiver and a State Machine, which correspond to the components described in the informal version of the SpaceWire standard [16].

The SpaceWire standard stipulates the timing requirements a link interface must satisfy during functioning. Examples of requirements taken from the standard are:

- the disconnect timeout of 850 *ns* nominal shall be between 740 *ns* and 1080 *ns*.

- after reset, the link interface remains in the *ErrorReset* state for a period of 5.12μs to 7.78μs.
- during the reset cycle, the link interface remains in the *ErrorWait* state for 10.24μs to 15.48μs.

Figure 2 contains excerpts from the behavior specification of the SDL processes composing the Link Interface. It shows how the three timing requirements mentioned above are modeled using *delayable* transitions triggered by continuous signals involving inequalities on **now**.

Such non-deterministic behavior cannot be specified in SDL using timers, which have a deterministic deadline. Moreover, existing industrial SDL verification tools work in discrete time and make simplified assumptions on the progress of time such as: time progresses only when there is no transition that can be triggered, and it always progresses up to the deadline of the next timer which is to expire [19]. In consequence, many time lines corresponding to realizable system executions are not explored by "classical" SDL verification tools.

The non-deterministic behavior can be modeled using inequalities on time as shown in Fig. 2. As we will see later, the analysis techniques employed allow exploration of all realizable time lines described by such specifications.

The behavior of the environment of most open SDL systems is also non-deterministic. For example, a component of the environment of a SpaceWire Link Interface is the host system. The host system may fail and reset the link at random moments, and it transmits characters over the link also at random. For verification purposes, the above properties of the host system may be specified in SDL. Fig. 3 shows how this is done using *lazy* transitions (which may let time pass with any amount before they are triggered). Modeling the host system like this leads to a more complete exploration of the realizable behaviors of the Link Interface.

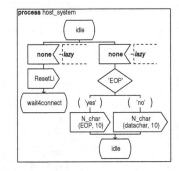

Fig. 3. Non-deterministic behavior of the host systems

3.4   Formalizing the Mapping of SDL to Timed Automata

We take here only a brief look at how the relation between SDL systems (with the structural constraints described before) and timed automata can be defined formally. This provides an insight into the internal workings of the verification tool described in Sect. 5. Since the point is to define a mapping that is usable by this tool, which is an extended version of the *Object*GEODE verification tool, we start from the semantics of SDL implemented in the original tool [19] rather than from the standard semantics [12].

The *Object*GEODE discrete-time semantics of SDL. For a given SDL system S, the *Object*GEODE verification tool builds a labeled transition system $\mathcal{G}_S$ corresponding to the state space of $\mathcal{S}$. $\mathcal{G}_S$ is used to verify behavioral properties.

The nodes of $\mathcal{G}_S$ are global states of the SDL system, comprising the discrete state of each process instance, procedure or service, the values of all variables, the content of all queues, as well as the relative delay until expiry for each active timer in the system. The predefined variable **now** is not part of the system state, since this would cause $\mathcal{G}_S$ to be infinite. (A consequence is that certain expressions involving **now** are wrongly interpreted in $\mathcal{G}_S$.) $\mathcal{G}_S$ may contain the following kinds of transitions between states:

1. **internal discrete transition.** $q_1 \xrightarrow{t} q_2$ iff there is a discrete transition identified by t enabled in the state q_1 and which takes the system into the state q_2. A discrete transition is caused by an input, a priority input, a continuous signal, a signal save/discard, etc.

2. **feed discrete transition.** $q_1 \xrightarrow{t} q_2$ iff t is an input transition, and the signal that causes it may be sent by the (unspecified) environment of the system.

3. **time transition.** $q_1 \xrightarrow{time(c)} q_2$ iff there is no internal discrete transition and no timeout transition enabled in q_1, and the next timer to expire has c time units until expiring. q_2 is equal to q_1 except for the delays of active timers, which are decreased by c.

4. **timeout transition.** $q_1 \xrightarrow{timeout\ t} q_2$ iff the timer t is active in q_1 and its relative delay until expiry is 0. q_2 is equal to q_1 except for the active status of t and the queue to which the signal t is appended.

Continuous time and loose time passage conditions. As mentioned in Sect. 3.3, the main advantage of using timed automata techniques in SDL tools is that time can be given a less constrained semantics, so that more of the realistic time lines of SDL system execution can be explored by a verification tool.

The time lines included in the $\mathcal{G}_S$ graph built by *Object*GEODE are restricted by the conditions in which *time* transitions may be taken. We can define a continuous-time semantics by relaxing time passage conditions, thus obtaining a larger LTS, $\mathcal{G}_S^\tau$, called the *timed semantic graph* of the system $\mathcal{S}$.

$\mathcal{G}_S^\tau$ is defined similarly to $\mathcal{G}_S$, except that the value of **now** is made part of the system state (i.e. the vertices of $\mathcal{G}_S^\tau$). We denote by $q.$**now** the value of **now** in the state q. For timers, what is kept in the states of $\mathcal{G}_S^\tau$ is not the relative delay until expiry (as in $\mathcal{G}_S$) but the absolute deadline. Transitions are redefined in $\mathcal{G}_S^\tau$ in the following way:

1. **time transition.** $q_1 \xrightarrow{time(c)} q_2$ iff $c \in \mathbb{R}$, $c > 0$, and:
   (a) All components of q_2 are equal to the corresponding components of q_1, except for the value of **now**, for which $q_2.$**now** $= q_1.$**now** $+ c$. In this case we denote q_2 by $q_1 + c$.
   (b) $\forall \delta', \delta''$. $0 \le \delta' < \delta'' \le c$, there is no discrete *eager* transition and no *timeout* transition enabled in $q_1 + \delta'$, and there is no *delayable* transition enabled in $q_1 + \delta'$ and disabled in $q_1 + \delta''$.

2. **timeout transition.** $q_1 \overset{timeout\ t}{\longrightarrow} q_2$ iff the timer t is active in q_1 and its deadline is equal to $q_1.\textbf{now}$. q_2 is equal to q_1 except for the active status of t and the queue to which the signal t is appended.

Note that the *urgency* attributes of the transitions enabled in the source state q_1 are used in this definition to constrain time passage, as in timed automata (see Sect. 2.2). Therefore, all transitions need to have a (default) urgency. Internal discrete transitions and *timer* transitions are *eager* by default, *feed* transitions are *lazy* by default (since it is logical to put only loose conditions on the behavior of the unspecified environment which sends *feeds*).

With the new conditions on time progress, it can be easily seen that $\mathcal{G}_S \subseteq \mathcal{G}_S^\tau$ (the inclusion refers to both vertices – modulo the **now** part – and edges) .

Relation to timed automata. The mapping to timed automata is given by defining a timed automaton $\mathcal{A}_S$ which has the property that its *semantic graph* (see Sect. 2.2) is strongly equivalent to the *timed semantic graph* $\mathcal{G}_S^\tau$ of S. We will not go into all the details of this construction. A similar construction is described in [14] for an extended version of SDL, and the strong equivalence property is formally proved.

The idea is that the states/transitions of $\mathcal{A}_S$ are defined similarly to those of $\mathcal{G}_S$ (the "untimed" semantic graph of S), with the following differences:

1. **Time transitions** from $\mathcal{G}_S$ are deleted in $\mathcal{A}_S$. Time passage in $\mathcal{A}_S$ is represented by the modification of the clocks, which are defined below.
   As a consequence, the delay until expiry of each active timer (kept in the states of $\mathcal{G}_S$ and updated by *time* transitions) will never be updated in the states of $\mathcal{A}_S$ (i.e. will remain equal to the initial delay of the timer, until a timer transition is triggered, see below).
2. A **timer transition** $q_1 \overset{timer\ t}{\longrightarrow} q_2$ leaves each state q_1 of $\mathcal{A}_S$ in which t is active, regardless of the delay of t.
3. A **discrete transition** $q_1 \overset{t}{\longrightarrow} q_2$ leaves each state q_1 of $\mathcal{A}_S$ in which t could be triggered provided a condition on **now** (having the form defined in §3.2) is satisfied.

$\mathcal{A}_S$ further has a set of clocks $\mathcal{X}$, which contains:

1. A clock $x_{t,p}$ for each timer t of a process instance p which can be activated during an execution of the system. Defined like this, the set $\mathcal{X}$ is potentially infinite. However, our implementation described in Sect. 5 handles dynamic clock creation and deletion, and therefore can partially analyze systems in which the set of timers poses problems.
2. A clock $x_{y,p}$ for each variable y of a process instance p which is used in assignments $y := \textbf{now}$ and in tests over $\textbf{now} - y$. Moreover, these variables y are not kept in the states of $\mathcal{A}_S$, and assignments such as $y := \textbf{now}$ are represented in $\mathcal{A}_S$ by the reset of $x_{y,p}$ on the corresponding transition.

To simulate the temporal behaviors described by $\mathcal{G}_S^\tau$, the transitions of $\mathcal{A}_S$ have correspondingly defined guards:

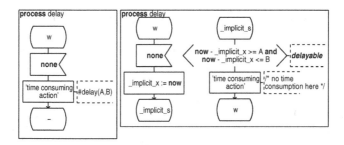

Fig. 4. Implicit transformation of time consuming actions

1. A **timer transition** $q_1 \xrightarrow{\text{timer } t} q_2$ has the guard $\zeta = [x_t = v_t(q_1)]$, where x_t is the clock in $\mathcal{X}$ corresponding to t and $v_t(q_1)$ is the relative delay until expiry of t, which is part of the state q_1.

2. A **discrete transition** $q_1 \xrightarrow{t} q_2$ for which the transition t is guarded with a condition of the form $\textbf{now} - y \sim \textbf{c}$ in the original SDL system $\mathcal{S}$, has in $\mathcal{A}_{\mathcal{S}}$ the guard $\zeta = (x_y \sim \textbf{c})$. SDL transitions guarded with conjunctions of conditions over **now** translate similarly.

With this definition, it can be proved that the *semantic graph of the automaton* $\mathcal{A}_{\mathcal{S}}$ *is strongly equivalent to the timed semantic graph* $\mathcal{G}_{\mathcal{S}}$ *of* $\mathcal{S}$.

In the verification tool presented in Sect. 5, we use the timed automaton defined here as a semantic model for an SDL system. However, the automaton $\mathcal{A}_{\mathcal{S}}$ is built *on the fly*, at the same time with its *simulation graph*, which is actually used for verification.

3.5 Time Consuming SDL Tasks

The methods described in the previous sections can be used to analyze SDL systems in which transition execution times for specific statements are given by the designer. Previously, several extensions have been proposed for the modeling time consuming tasks in SDL [15,7].

We consider the extensions proposed in [15], in which a delay can be associated to each SDL action statement. The delay is either a precise value A or an interval $[A, B]$. The semantics described in [15] is such that when the system reaches the time-consuming action, it stays blocked for a time $\delta = A$ (or $\delta \in [A, B]$) after which it executes the action in zero-time.

Time consuming tasks can be translated in an equivalent form in our model, by using an implicit state, an additional clock and a *delayable* transition exiting the implicit state to model the waiting conditions (given by A or $[A, B]$). The translation of time consuming tasks into delayable transitions is shown in Fig. 4. This translation is done implicitly by our verification tool (Sect. 5).

4   Specification of Temporal Properties in GOAL

There are several classes of properties which can be verified for SDL systems. While simple properties, such as absence of deadlocks or invariance of a logical condition do not need a complex formalism to be expressed in (propositional logic with an appropriate set of atomic propositions suffices), there are more complex properties which cannot be specified by these means.

Linear properties mentioned in Sect. 2.3 are from this class. The discrete-time verification tool of *Object*GEODE employs an automata-like language, GOAL, for the specification and verification of such properties. An introduction to GOAL can be found in [1].

Briefly, a GOAL observer has states, which are classified as **normal, error**, or **success** states, and transitions which are triggered by events occurring in the SDL system. The observer is executed synchronously with an SDL system. There are two verification modes possible:

1. **safety** verification, in which the GOAL observer is considered a finite automaton, whose accepting states are the **error** and **success** states. The verification tool looks for particular executions which lead to these states.
2. **liveness** verification, in which the GOAL observer is considered a Büchi automaton whose accepting states are the **success** states. The verification tool looks for *infinite* executions of the SDL system which are not accepted by this Büchi automaton (i.e. do not pass infinitely often through a **success** state), and which correspond to lack of progress ad infinitum.

The GOAL language may be used for verifying timing properties of the SDL system, by using the value of **now** in the same way it is used in SDL (Sect. 3). The difference between time driven transitions in GOAL and SDL is that GOAL transitions can only have *lazy* urgency. Allowing them to be *eager* or *delayable* would cut out potentially realizable system behaviors from the simulation graph, by blocking time passage.

Fig.5 shows how a GOAL observer may be used to verify that a SpaceWire link is always re-established in at most $30\mu s$ after a fault (safety property). Furthermore, Fig.6 shows an observer which can be used in liveness mode to verify that an SDL system has *no zeno runs* (on each infinite zeno run, the observer will eventually remain in the state ko forever).

5   The Temporal Property Verification Tool

In this section we describe a tool for verifying temporal properties of SDL systems, based on building the timed *simulation graph*. This tool is derived from the *Object*GEODE SDL Simulator [19], and therefore supports all the features of SDL–96 in addition to the extensions presented in the previous sections.

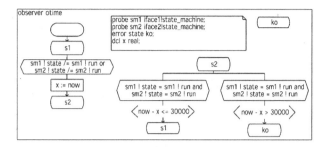

Fig. 5. GOAL observer for checking a timed safety property

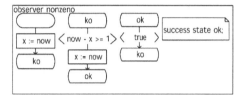

Fig. 6. GOAL observer for checking non-zenoness in *liveness* mode

5.1   The Simulation Graph

The tool handles SDL specifications according to the timed automata model for SDL described in Sect. 3. Verification is based on building the simulation graph of the timed automaton $\mathcal{A}_S$ corresponding to the SDL system S.

The timed automaton $\mathcal{A}_S$ itself is not explicitly built by the verification tool. The rules for building $\mathcal{A}_S$ described in Sect. 3.4 are instead used for constructing the simulation graph of $\mathcal{A}_S$ directly.

Simulator states. The dynamic states handled by the verification tool are pairs (q, S). q is the discrete state of the SDL system (containing the components described in the definition of $\mathcal{A}_S$ (Sect. 3.4), i.e. no value kept for **now**, initial relative deadline kept for timers). S is a convex polyhedron over the clocks of $\mathcal{A}_S$, which represents the clock values reachable via the scenario executed so far. The tool eliminates non-active clocks from S, such as clocks corresponding to inactive timers, so that S keeps only constraints on clocks which are relevant in the state q.

The discrete state and the clock constraints of the GOAL observers executed in parallel are part of q and S. Thus $\mathcal{A}_S$ is in fact the synchronous product (actually the product synchronized only on observable events) of the timed automaton corresponding to the system S and the timed automata corresponding to observers.

Simulator transitions. When the simulator executes a transition t in a state (q, S), it performs several steps corresponding to the computation of

time-succ(disc-succ($t, (q, S)$))

from the definition of the simulation graph (Sect. 2.2). The steps are detailed below.

We note one difference with the functioning of the standard *Object*GEODE Simulator: in case of an SDL model executed in parallel with a GOAL observer, there may be more than one successor of the state (q, S) by the transition t. The cause is that, by executing t on the SDL model first, the simulator reaches another state (q', S') (see steps 1 and 2 detailed below), on which *more than one* GOAL *transition of a same observer may be fireable*. A simple example is the observer in Fig.5: consider the two transitions exiting state s2. If the SpaceWire link is restored in a state (q', S') which has non-void intersection with both guards **now** - x <= 30000 and **now** - x >= 30000, there will be two successors of the initial simulator state on the same SDL transition, but on two different GOAL transitions.

The steps of the execution of the transition t are:

1. The zone S of the state (q, S) is intersected with the guard of t.
2. The actions of transition t are executed on the contents of the discrete state q. If there are clocks reset by t – i.e. assignments of the form $x := $ **now** – the zone resulted from step 1 is modified accordingly. The result (q', S') of these two steps corresponds to disc-succ$(t, (q, S))$.
3. For each GOAL observer, the fireable transitions are determined. There may be several fireable transitions $g_1, ..., g_k$. However, the parts of the simulator state (q', S') on which these transitions are fireable must not overlap – i.e. $guard(g_j) \cap S'$ must be disjoint for $j = \overline{1, k}$. This condition ensures that the behavior of the observer is deterministic for every actual system state $(q', \mathbf{v}) \in (q', S')$. The determinism condition is essential for GOAL observers, see [1].
4. For each GOAL transition g_j, there will be a successor state (q'_j, S'_j) computed in two steps similar to steps 1 and 2 above:
   (a) q'_j is computed from q' by applying the transformations described by the transition g_j.
   (b) S'_j is obtained from $S' \cap guard(g_j)$, by applying the clock resets described in g_j, if there are such resets.
   Additionally, if $S' \setminus \bigcup_{j=1}^{k} guard(g_j) \neq \emptyset$, no GOAL transition may be taken on this part of the state (called residual state). Therefore there will be a successor $(q', S' \setminus \bigcup_{j=1}^{k} guard(g_j))$.
5. If there are additional GOAL observers executing in parallel, they further split each state (q'_j, S'_j) (and the residual state) in substates.
6. Lastly, on each state (q', S') among the states resulted from the previous steps, the possible temporal successors $(q'', S'') = $ time-succ(q', S') are computed. The states (q'', S'') are the possible successors of (q, S) by t.

5.2   Verification of Properties

The simulation graph built by our verification tool as shown above is a finite symbolic representation for the *reachable states* of the SDL system (and associated GOAL observers). It can therefore be used for the verification of linear properties, such as those described in Sect. 2.3.

Absence of deadlocks in an SDL system is verified with no attached GOAL observer. Thus for each state and each fireable SDL transition there is only one successor. A state (q, S) contains a deadlock if one of the following conditions holds:

1. (q, S) has no successor in the simulation graph, or
2. (q, S) has the successors $(q'_1, S'_1), ..., (q'_k, S'_k)$ by the transitions $t_1, ..., t_k$, but $S \setminus \bigcup_{j=1}^{k} guard(t_j) \neq \emptyset$ (meaning that there is an actual SDL system state $(q, \mathbf{v}) \in (q, S)$ on which no transition among $t_1, ..., t_k$ is fireable).

Invariance properties are given by a propositional logic formula that must hold in each state of the system. The formula is formed of atoms α which can test:

1. The *discrete state* of the SDL system, such as the value of a variable, the length of a queue, etc. In this case, the satisfaction of α by (q, S) is decided based on the discrete part q.
2. *Time*. Time may only be tested by means of atomic conditions of the form $\mathbf{now} - y \sim c$, where y is an SDL system variable interpreted as a clock. In this case, (q, S) satisfies α iff the polyhedron $\alpha' = [x_y \sim c]$ is included in the zone S.

Non-zenoness and other linear properties are specified by GOAL observers (see Sect. 4). In *safety* mode, the tool detects scenarios leading to states (q, S) in which q designates a **success** or **error** state in at least one GOAL observer. In *liveness* mode, it detects infinite loops which do not pass through states (q, S) in which q designates a **success** state in at least one GOAL observer.

5.3 Interactive Commands and Presentation of Results

The verification tool can be also used in interactive mode, as a simulation tool. The user triggers transitions one by one and examines the reached states, as in normal *Object*GEODE interactive simulation [19]. However, the simulator states that the user gets are symbolic states (q, S) of the same form as above.

The discrete part of the state can be examined using the usual mechanisms of the *Object*GEODE Simulator (watches, the `print` command), while the part S referring to clocks can be examined with an additional command, `clocks`, which prints all or a part of the constraints on clocks which define the polyhedron S. For example, the output of this command during the simulation of the SpaceWire system can look like:

```
> clocks
0 <= iface1!state_machine!c <= 10440
0 <= iface2!state_machine!c <= 7780
0 <= iface1!state_machine!c -iface2!state_machine!c <= 2660
```

Since the time span between two events in interactive simulation is not obvious (and sometimes not even possible) to derive from the clock conditions of S, there is a mechanism to measure time spans: the chronometer. A chronometer behaves like a *clock*, except that it is introduced from the simulation console. The command `addclock` creates a new chronometer, which can be consulted after several simulation steps to check the lower and upper limits of the elapsed time interval.

6   Conclusions and Future Work

We have presented a method for verifying quantitative temporal properties on SDL systems, by mapping the semantics of SDL to that of *timed automata* [2], for which powerful timing analysis methods exist. In order for the mapping and analysis to work, the SDL designer must use only a limited set of constructs for modeling timing conditions in the SDL specification: *timers* and *linear conditions* on **now** having a particular form.

However, the restrictions imposed by the mapping are looser than the restrictions imposed by most SDL verification tools, so a larger set of SDL models can be correctly analyzed with the methods presented here. We also argue that the analysis is more precise in the sense that it gives a better coverage of the possible behaviors of the system.

The use of *linear conditions* on **now**, together with a notion of transition *urgency* that we take from timed automata, make it possible to model more precisely systems presenting non-determinism from the temporal point of view. Incompletely specified systems and systems interacting with a non-specified environment are in this category. *Urgencies* have been proposed first for timed automata in [3,4], and later for SDL in [6].

The verification method presented in the first part of the paper is implemented in a tool discussed in the final part. We examine the practical aspects of modeling temporal properties so that they can be verified by the tool, as well as the internal working of the tool.

From the experience we have with our tool, including the SpaceWire protocol used as an example throughout this paper, we can conclude that the tool is able to give a better coverage of the behavior of the system, compared to classical SDL analysis tools. Further experimentation is taking place within the INTERVAL project[5].

Further work is needed in order to improve the tool: in particular cases, the use of transition urgencies may lead to non-convex clock polyhedra. Handling non-convex polyhedra poses theoretical difficulties, as signaled in [18]. Our tool can handle such polyhedra, but the performances are not always satisfactory.

Another direction for our efforts is to include the MSC language in our timed verification method. MSC'2000 offers the possibility to specify timing information, which could be exploited by our tool either in order to check temporal

---

[5] *Formal Design, Validation and Testing of Real-Time Telecommunications Systems*, European project IST-1999-11557.

properties specified in MSC, or in order to generate timed MSC traces for simulation scenarios.

Acknowledgements

The authors thank the European Space Agency for providing the SpaceWire specification, on which the example from Sect. 3 is based.

References

1. B. Algayres, Y. Lejeune, and F. Hugonnet. GOAL: Observing SDL behaviors with GEODE. In R. Braek and A. Sarma, editors, SDL '95 with MSC in CASE. Elsevier Science B.V., 1995.
2. R. Alur, C. Courcoubetis, and D.L. Dill. Model checking in dense real time. *Information and Computation*, 104(1):2–34, 1993.
3. S. Bornot and J. Sifakis. Relating time progress and deadlines in hybrid systems. In *International Workshop HART'97*, volume 1201 of *LNCS*. Springer-Verlag, 1997.
4. S. Bornot, J. Sifakis, and S. Tripakis. Modeling urgency in timed systems. Technical report, Verimag, Grenoble, 1998.
5. M. Bozga, J.C. Fernandez, L. Ghirvu, S. Graf, J.P. Krimm, L. Mounier, and J. Sifakis. IF : An intermediate representation for sc SDL and its applications. In R. Dssouli, G.v. Bochmann, and Y. Lahav, editors, SDL '99. The Next Milenium. Proceedings of the 9th SDL Forum, Montreal, Canada, 1999. Elsevier.
6. Marius Bozga, Susanne Graf, Alain Kerbrat, Laurent Mounier, Iulian Ober, and Daniel Vincent. SDL for real-time: What is missing? The 2nd Workshop on SDL and MSC, 2000.
7. M. Diefenbruch, E. Heck, J. Hintelmann, and B. Müller-Clostermann. Performance evaluation of SDL systems adjunct by queueing models. In R. Braek and A. Sarma, editors, *Proceedings of SDL Forum '95*. Elsevier Science B.V., 1995.
8. A. En-Nouaary, R. Dssouli, and F. Khendek. From timed scenarios to SDL: specification, implementation and testing of real-time systems. In R. Dssouli, G.v. Bochmann, and Y. Lahav, editors, SDL '99. The Next Milenium. Proceedings of the 9th SDL Forum, Montreal, Canada, 1999. Elsevier.
9. Y. Gurevich. Evolving algebra: The lipari guide. In E. Börger, editor, *Specification and Validation Methods*. Oxford University Press, 1995.
10. Thomas A. Henzinger, Peter W. Kopke, Anuj Puri, and Pravin Varaiya. What's decidable about hybrid automata? *Journal of Computer and System Sciences*, 57(1):94–124, 1998.
11. John E. Hopcroft and Jeffrey D. Ullman. *Introduction to Automata Theory, Languages, and Computation*. Adisson-Wesley, 1979.
12. ITU-T. Annex F (11/00) to recommendation Z.100 – Specification and Description Language (SDL) – Formal definition of SDL – to be published.
13. ITU-T. Recommendation Z.100 (11/99) - Specification and Description Language (SDL).
14. I. Ober. Extending SDL with timed automata concepts. Technical report, VERILOG, 1999.

15. J.-L. Roux. SDL performance analysis with *Object*GEODE In A. Mitschele-Thiel, B. Müller-Clostermann, and R. Reed, editors, *Workshop on Performance and Time in SDL and MSC*, Erlangen, Germany, February 1998. Friedrich-Alexander Universität, Erlangen-Nürnberg.
16. SpaceWire Working Group.   SpaceWire — Serial point-to-point links.   European Space Agency document UoD-DICE-TN-9201, Issue D, May 2000. http://www.estec.esa.nl/tech/spacewire.
17. TELELOGIC A.B., Malmö, Sweden. T*elelogic* TAU SDL *Suite Reference Manuals*, 1999.
18. Stavros Tripakis. *The Formal Analysis of Timed Systems in Practice*. PhD thesis, Joseph Fourier University, Grenoble, 1998.
19. VERILOG, Toulouse, France. ObjectGEODE *4.1 Reference Manuals*, 1999.

A General Approach for the Specification of Real-Time Systems with SDL*

Ralf Münzenberger, Frank Slomka, Matthias Dörfel, and Richard Hofmann

University of Erlangen-Nuremberg
Department of Computer Networks and Communication Systems
Martensstr. 3, 91058 Erlangen, Germany
{rfmuenze, slomka, doerfel, rhofmann}@informatik.uni-erlangen.de

Abstract. In contrast to protocols of the network or transport layer the protocols for medium access have to consider the timing behavior of the communication medium. Although SDL is a widely used language for the specification of communication systems, in most cases time critical parts are not considered. In this paper, a design pattern is discussed that allows the specification of time critical functionality sucg as multiplexers or Quality-of-Service (QoS) schedulers. In many applications such services are running in a synchronous manner with the communication medium. A notation for timing aspects is needed for the specification of this behavior which itself is only possible in a sensible way with a formal model of time. Clocks are used to define the term real-time in a formal way, leading to the specification of timing constraints, for example sending data packets in deterministic time intervals within a communication system. In a case study from the mobile communication area, the design pattern was used to specify the MAC-Layer including time critical parts.

1 Introduction

The specification of large, distributed communication systems is a well known application area of the specification and description language (SDL). SDL is mainly designed for the development of communication protocols. Many language constructs support the requirements needed by protocol development, such as non-deterministic choices, sending and receiving signals asynchronously and the specification of protocol timers.

Nearly all protocols specified with SDL are at layer 3 of the OSI reference model or above. Layer 2 functionality such as medium access is often implemented in a mixed hardware/ software architecture to achieve the performance

---

* This work has been funded by the Deutsche Forschungsgemeinschaft (DFG) under grant HE 1408/4-3 as a part of the program Rapid Prototyping of Embedded Control Systems with Real-Time Constraints. We thank Lennard Kerber for his critical comments and ideas; Günther Peitz and Georg Sandhaus of Tenovis for their support during the development of the DECT MAC-Layer and providing the Tenovis DECT transceiver modules; additionally Kai Lampka for his work on the specification of the DECT MAC-Layer.

R. Reed and J. Reed (Eds.): SDL 2001, LNCS 2078, pp. 203–222, 2001.
© Springer-Verlag Berlin Heidelberg 2001

goals. The separate consideration of hardware and software has led to the use of different design methodologies for the two parts. Especially, the separated consideration often leads to errors during the integration phase. To improve the design process, an integrated specification language is needed where all aspects can be formulated in an adequate way. Because of the wide usage of SDL at the system level, it is desirable to use SDL also for the hardware/software parts of the system. To reach this goal, SDL needs a new time model which has to enable the definition of clocks, the specification of timing constraints and the specification of synchronous behavior.

The current time model of SDL allows the use of timers as a way of defining a timing behavior. Unfortunately the value given to a timer has no unit. This leads to additional informal specifications given in comments or in separate documents to determine which unit and resolution to use for the respective timers. For formal verification and integrated hardware/ software code generation it is important that this information is an integral part of the system specification. Even the reception of signals sent by expired timers is asynchronous because the consumption of the timer signal has no relation to its expiration time. This is in contradiction to the former required deterministic and synchronous behavior of real-time systems.

The usage of the construct *now* as an alternative to timers is not sufficient to implement deterministic real-time behavior because *now* only accesses the system clock variable which has an implementation dependent behavior. The impossibility to specify the behavior of the system clock in a formal and code generator independent way clearly rules it out. The only valid assumption about *now* is the relation $now_n <= now_{n+1}$ in two consecutive statements in the same transition. Neither the size of a time step nor the linearity of the clock is defined by this.

Without a formal model of time, it is not possible to specify all aspects of real-time systems non-ambiguously. Currently, only the functional behavior can be specified, or, in other words, only a system. Due to the aforementioned reasons a formal model of clocks and timing constraints was developed. The timing constraints are subdivided into two different aspects: First, a timing requirement describes a requirement to the system that has to be validated only during the design and the test phase of the system. Second, a timing condition describes a requirement which has to be augmented repeatedly during normal execution of the system and which triggers specified actions depending on validity.

Based on this insight, a design pattern was developed to specify real-time systems. It enables the designer to specify synchronous reaction to medium access components, deterministic real-time behavior and layer 2 functionality as in packet multiplexers or schedulers. It can also be used to build a complex system, such as an IP based switch, by connecting different instances of the same pattern. This paper focusses on the specification of real-time systems with SDL. In [15] strategies for hardware/software partitioning and implementation of such systems are shown.

2   Related Work

Typically, in SDL [8] the timer mechanism and the construct **now** are used to specify the timing behavior of the system. As mentioned, this is not sufficient to specify timing constraints in general. Also the use of the timer active expression or priority input for a timer signal is not a solution. For the specification of timing constraints, even SDL-2000 [7] does not include any enhancements. For this purpose, even SDL-2000 [7] does not include any enhancements. The description of timing requirements is considered in SDL* [13], an extension of SDL to specify non-functional aspects such as resource constraints or cost constraints in annotations. The SDL extension described in [10] enables the specification of a predefined set of timing requirements based on signal chains, which are sequences of signals inside the SDL system. Complex timing requirements are not considered in both SDL extensions.

Message Sequence Charts (MSC) [9] is a language to describe the system behavior by scenarios. MSCs are used during the analysis phase for analyzing the behavior of the system in early design phases and to validate the behavior of the implemented system against the specification. Functional behavior of the system cannot be specified sufficiently. During the last four years a few extensions to MSC were defined to specify temporal aspects. Timed MSC [14], which has been integrated in the upcoming standard MSC-2000, and Performance MSCs [3] consider the specification of timing constraints restricted to special aspects of the system. Properties are not mentioned for the clock used to measureme the specified values; hence again, there is the problem of having deterministic clocks with a reliable behavior usable for real-time systems. The same arguments (except the functional deficit) hold for QSDL [11]: an extension to SDL based on queuing models.

Despite the temporal specification gap, there are two case studies which deal with the specification and implementation of real-time telecommunication systems without considering timing constraints. While in [16] the design of a base station for mobile communications (DECT) is discussed, the case study in [1] deals with the specification of an ATM network interface card. However, both approaches avoid the specification of complex timing behavior of layer 2 protocols. In [16] only the layers without time-critical parts have been specified in SDL. All layers with needed real-time capability have been implemented using fixed hardware and software parts of an existing PBX framework. The case study presented in [1] considers the specification of the whole system in SDL and discusses different implementation alternatives with respect to hardware/software partitioning. The missing consideration of timing constraints leads to the mentioned problems of having a large overhead for the runtime support system (RTSS) in the implementation.

The case studies shown above clearly prove the demand for specification support for timing constraints especially for medium access layer applications. Generally it is efficient to use design patterns for repeatedly arising problems. An SDL design pattern is an SDL module which can be used in different contexts for different applications with only minor modifications. The concept especially of SDL patterns and examples for well known protocols are described in [4] and [5]. None of these papers contains a pattern that considers any timing aspects.

3 Specification of Systems with SDL

3.1 Communication Systems

The specification of communication systems is a typical application domain for SDL. These systems can be subdivided into systems based on connection oriented services and based on packet oriented services. In connection oriented systems (such as conventional telephone systems, wireless telephone systems — GSM, DECT, UMTS, or ATM networks) a negotiation phase is needed to setup a connection before data can be transmitted.

The conventional telephone system based on line switching and wireless telephone systems, implicitly guarantees sufficient bandwidth for each connection even if it is not used. By selecting appropriately installed lines, Quality-of-Service (QoS) requirements (such as maximum tolerable delay) can be guaranteed under every circumstance.

To achieve the same guaranteed quality in packet oriented networks like ATM, a virtual connection has to be reserved explicitly during the negotiation phase. To do this, the ATM protocol stack contains elements for connection setup and resource management. The implementation of layer 2 elements in the network infrastructure has to consider the general QoS requirements (such as the fixed packet delay in each network device, an efficiently used time-slotted medium and a guaranteed data packet rate). The increasing demand for continuous media applications with real-time traffic such as teleconferencing, interactive multimedia games or voice over IP requires more and more powerful networks which can only be implemented in reasonable time if the timing requirements are already considered in the specification.

3.2 Specification of Stream-Oriented Services

The scheduler is a central component of layer 2 to realize QoS requirements in stream-oriented networks. It has to be fast enough to serve the data rate of the output interface and to guarantee a maximum delay. In addition it has to be on time if the medium access is time slotted.

Different queueing algorithms are currently used in internet devices for scheduling, e.g. virtual clock [19], weighted fair queueing (WFQ) [18], or class based queueing (CBQ) [17]. To schedule real-time traffic, each packet belongs to a reserved connection with a guaranteed QoS or gets best-effort service. These algorithms uses at least one virtual time function to compute the virtual time of a data packet in consideration of the QoS traffic parameters, generates a relative ordering of the data packets via their timestamp(s), and select the next packet for transmission.

The SDL diagram in Fig. 1 shows an obvious approach to specify such a scheduler. The procedures *SortInto* and *SortOut* together contain the scheduling algorithm. After the process has received K signals $Input1(Packet)$ an output signal is sent. From the K packets the procedure *SortOut* selects the one for output.

This specification has several disadvantages: Signals can only be sent if the process input queue is not empty. It cannot be specified to send a signal at a particular point of time. If the input data rate is bursty, the scheduler cannot serve the communication medium efficiently because the maximum duration between two output signals is not bounded.

To fulfill these points a data packet has to be sent at least after the duration T_{out}, independently of the number of signals in the process input queue. This has to be done also if there are no signals in the input queue, but buffered data packets inside the process.

Summarizing, the input packet stream and the output packet stream of the scheduler have to be decoupled.

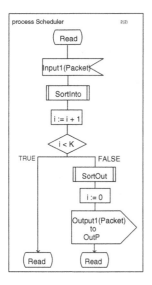

Fig. 1. SDL process Scheduler

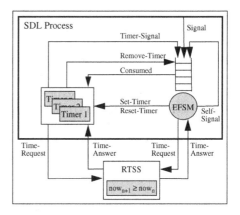

Fig. 2. SDL process with timers

3.3   Specification of Real-Time and Timing Constraints

The specification of temporal aspects is fundamental to describe the complete behavior of real-time systems. SDL contains the predefined expression *now* to get the actual time. As shown in Fig. 2 *now* represents a request (*Time-Request*) from the extended finite state machine (ESFM) to the system clock of the run-time support system (RTSS) to get the point of time (*Time-Answer*). If in the same transition *now* is called twice, the return value of *now* is system dependent and implementation dependent, but has to follow the condition: $now_{n+1} \geq now_n$.

This temporal semantic of *now'* is not sufficient to specify real-time systems, for example to send a data packet on time.

In SDL processes, timers are a powerful method to specify protocol timers. They can be set (*Set-Timer*) and reset (*Reset-Timer*). The timer becomes active after it is set and becomes inactive when the timer signal is consumed or the timer is reset. A timer gets the actual point of time from the RTSS and follows the same rules as discussed above. If the set time has arrived, the timer sends a signal (*Time-Signal*) to the process input queue. The input queue sends a signal (*Consumed*) to the timer, after the signal was consumed from the process. If a timer is reset, an already existing timer signal is removed (*Remove-Timer*) from the input queue. The specification of timing constraints with timers is insufficient in general, because after a timer has fired, the SDL process does not get this information immediately. As a result of the asynchronous semantics of sending and receiving a timer signal, the specified duration of the timer does not correspond with the specified timing constraint, because it consists of two parts: the specified timer value and the delay to receive the timer signal. The delay is indeterminable, it is dependent on the number of signals in the input queue and the time needed to perform them.

If the timer signal is the only priority input for a state, the timer transition is triggered immediately after the timer has fired, but only if no other transition is performed by this point of time. Otherwise the timing constraint is not fulfilled. In addition, the restriction to use only one timer in each state is not sufficient in general as the case study in Sect. 5 shows.

A second possibility to prevent the asynchronous semantics of sending and receiving a timer signal is the timer active expression. This predefined boolean expression evaluates to *true* if the corresponding timer is active otherwise to *false*. The timer active expression can be used in an enabling condition or in a continuous signal to trigger a transition immediately after the timer becomes inactive. But it cannot be used to specify timing constraints in general, such as to perform the same send action every 10 ms: first, the timer still has no unit, so to set the timer value to 10 has no physical meaning; second, (as discussed above) the temporal semantics of *now* is not sufficient.

4   Extensions for the Specification of Real-Time

4.1   Functional Approach

In Sect. 3.2 the problems with specifying stream-oriented services with SDL are discussed. A solution of the problem is shown in Fig. 3, the functional SDL specification of a QoS scheduler. It is divided into three main parts, a *Control*, an *Input* and an *OutputTransition*. The continuous signal concept (*BuffCnt* > 0) is used to decouple the input and the output data packet stream. This is needed to handle continuous, varying, and bursty traffic. Continuous signals have lower priority than input signals. A transition with a continuous signal is only triggered if the boolean condition is true and the input queue contains no signals that can be received in the state leading to the transition. The two procedures *SortInto*

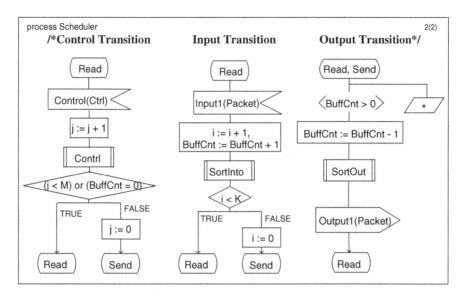

Fig. 3. Design pattern Decouple Module to specifiy a QoS scheduler

and *SortOut* contain the whole functionality of the scheduling algorithm including buffering. The constants K and M specify how many signals (*Control(Ctrl)* and *Input1 (Packet)*) can be received before the signal *Output1 (Packet)* is sent. The variable *BuffCnt* specifies the number of data packets currently buffered for scheduling.

The *ControlTransition* is responsible for all control information of the scheduler, e.g. the QoS traffic parameter for each reserved connection. The procedure *Contrl* contains this functionality. The constant M specifies how many control signals (*Control(Ctrl)*) the Scheduler is able to receive ($j < M$) and stay in the state *Read* on the condition there is not a data packet that is buffered (*BuffCnt* = 0) for scheduling. If no packet is buffered, the process can receive any number of control signals.

The *InputTransition* receives all data packets to be scheduled. After buffering K data packets the state *Send* is reached. Now, it is only possible to trigger the *OutputTransition*. In the state *Send* the continuous signal condition (*BuffCnt* > 0) is always true because at least the *InputTransition* was triggered once. After an output signal is sent the state *Read* is reached. In this state, the *Control* or *InputTransition* is triggered depending on the type of signal in the input queue. The *OutputTransition* can also be triggered in the state *Read* to prevent blocking up the sending of signals if the input queue is empty.

The SDL process in Fig. 3 is a solution for the above discussed decoupling problem. It is a design pattern [5] because it can be used to specify applications that collect and distribute data. This so called Decouple Module can be used for many systems, e.g. routers, control systems, or mobile communication systems. The two procedures *SortInto* and *SortOut* hide the whole functionality needed for a dedicated application (except control).

If a complex protocol for the control part is needed, the described design pattern is not sufficient. In order to have its own state machine the *ControlTransition* has to be specified in its own module. To prevent implementation overhead the SDL service concept is used as shown in Fig. 4. The input and output transition is specified as a service (*InputOutput*) with its own

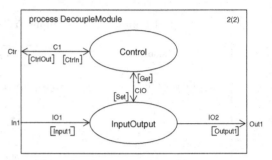

Fig. 4. Design pattern Decouple Module in general

state machine and the same is done with the control transition (*Control*). The two services can communicate via process variables (synchronous communication) and/or via the signal mechanism (asynchronous communication).

In SDL-2000 the service concept is no longer available. Instead of services, process agents can contain further agents that execute alternating. The design pattern Decouple Module needs an agent for the *Control* service and an agent for the *InputOutput* service. In order to share the constants M and K and the variable *BuffCnt* between the contained agents they need to be specified in the containing process agent.

If signals for scheduling are buffered in the process *DecoupleModule*, the maximum duration between two output signals depends on the values of the constants M and K and no longer on the number of signals in the input queue. But it is not feasible to send an output signal at a specified point of time. Therefore the specification of timing conditions is needed.

4.2   Temporal Approach

Clocks. In order to describe timing constraints a time measure is needed. An instrument called clock measures time and displays the measured value called the reading of the clock $C(t)$. The aim of this section is to discuss different mathematical models of clocks. The basic idea is to express the reading of the clock $C(t)$ by mathematical functions. In Fig. 5 examples of different clock models are shown. A detailed description of these clocks including the mathematical background can be found in [12].

The ideal clock $C_i(t)$ is a linear function of time. In order to approximate the physical properties of clocks more precisely, a continuous real clock $C_c(t)$ requires more parameters: The start-offset is the reading of the clock at time $t_0 = 0s$ and the accuracy of the clock is the gradient, often denoted in data sheets with a unit of parts per million (ppm). The efficient change of accuracy describes changes of working conditions that are influencing the clock, e.g. temperature variations. In Fig. 5 the efficient change of accuracy of the continuous real clock is zero and the start-offset is $1s$.

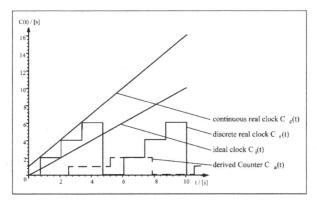

Fig. 5. Functional behavior of clocks

In contrast to the clocks discussed above, the temporal resolution implemented in real systems is finite. The discrete real clock $C_r(t)$ has a granularity to describe the discrete changes of the reading of the clock. Due to the finite memory space to store the reading of the clock, the range of value has an overflow. In the example in Fig. 5 the discrete real clock has a range of value of $8s$ and a granularity of $2s$. Typically a watch has an overflow after 12 or 24 hours and a granularity of $1s$.

To handle interlocking slot schemes in mobile communication systems, clocks are needed to be derived from a reference clock. This concept is called derived counter $C_a(t)$. The reading of the derived counter changes once after the discrete real clock has changed the number of times *ticksofreferenceclock*. The delay time defines the delay between the changing of the reference clock and the incrementing of the devived counter. The *ticksofreferenceclock* of the derived counter shown in Fig. 5 is 2, the granularity is 1, the delay time is $0.2s$, and the range of value is 3.

In order to specify the physical properties of a clock and to be able to use different clocks in an SDL system, each SDL system gets at least one clock of its own. The physical properties of a clock are specified as an annotation in SDL* as shown in Fig. 6. SDL* is an extension of SDL to describe non-functional system aspects, such as resource constraints or cost constraints, as annotations [13]. A timing annotation starts with the key word $define followed by the kind of the annotation (*Clock*) and the name of the clock (*ProcessClock*). The physical properties of the clock are specified in braces.

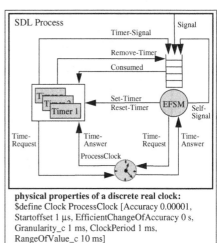

physical properties of a discrete real clock:
$define Clock ProcessClock [Accuracy 0.00001,
Startoffset 1 μs, EfficientChangeOfAccuracy 0 s,
Granularity_c 1 ms, ClockPeriod 1 ms,
RangeOfValue_c 10 ms]

Fig. 6. Clock specification with SDL*

The EFSM as well as the timers can get the time from this clock in the same way as it was discussed in Sect. 3.3.

In conformance to the hierarchical concept of SDL, clocks can be assigned to any entity of the hierarchy. All processes and services of a block/system can use a block/system clock. An example of an SDL system that has one system clock (*ReferenceClock*1) and five process clocks is shown in Fig. 7. There are three discrete real clocks: *Clock*1 and *Clock*2 in *ProcessA* and *Clock*1 in *ProcessB*. *ReferenceClock*1 is the reference clock for the *DerivedCounterA* in the derived counters, all clocks are running independently of each other.

A new concept to access more than one clock is needed because SDL only allows the access to one system clock with the construct *now*. In order to con-

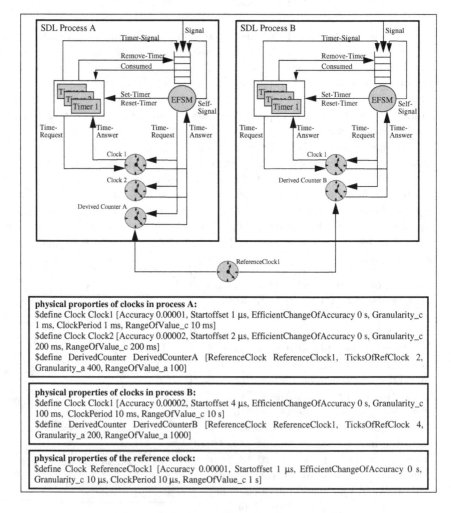

Fig. 7. Specification of clocks with SDL* on different hierarchy

form with the SDL standard, the access to the clocks is described by the abstract data type *Clock* shown in Fig. 8. The abstract data type *Clock* defines the operators that are needed for this purpose: The first parameter of the operator *mkClock(...)* is a reference to the SDL* clock annotation, e.g. *Clock*1 of *ProcessA* defined in Fig. 7 in *ProcessA*. The name of this clock is specified as an SDL syntype. The operator *getTime(...)* accesses a discrete real clock and the operator *setClock(...)* sets it to a certain value. In a similar way *getTicks(...)* and *setCounter(...)* can be used for a derived counter. The operators *setTimer(...)* and *resetTimer(...)* have the same semantics as the SDL timer constructs, discussed in Sect. 3.3. In addition to each timer, a correlated clock is referenced and used to fire the timer. Most operators have a parameter *TimeUnit* to specify the unit of the given value, e.g. ms, μs, or ns, but the default unit of Clock is second.

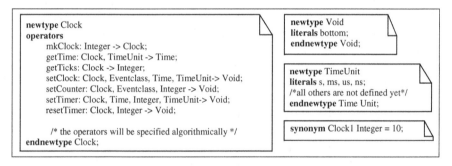

Fig. 8. Abstract data type *Clock*

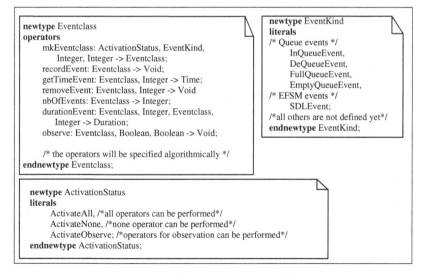

Fig. 9. Abstract data type to specifiy timing requirements

This approach does not distinguish between SDL sorts *Time* and *Duration*. Every reading of a clock is the *Duration* since the starting point of *Time*, such as 00:00 Uni-versal Coordinated Time (UTC), January 1, 1958, or since the last overflow.

Timing Constraints. A timing constraint consists of a timing bound and a logical condition to describe the temporal behavior of complex systems: The timing bound describes the required execution time of system parts. It is a boolean expression that specifies the relation $(<, \leq, >, \geq, =, \neq)$ between the required duration defined by a start event and an end event and the real duration. An event is a particular change of system properties at a point of time, e.g. states, values of variables, or level of the input queue. The timing bound is *false* if the required duration is violated. In communication systems a timing bound has not to be fulfilled in every case, e.g. if there are no data packets buffered in the scheduler nothing can be sent. To specify such conditions, a boolean expression describing the history of the system is needed. At least this logical condition is defined by the relation between an event and the end event of the timing bound. The logical condition is *false* if the preconditions for the timing constraint are not met. In this case the timing constraint is always *true* because the timing bound does not apply. If the system does not fulfill the timing bound and the logical condition is *true* the timing constraint is *false*.

If the scheduler discussed before has to perform a data rate of 1 Kbyte/s, it has to send a data packet of one byte every 1 ms. The timing bound of this timing constraint is dependent on the number of data packets stored in the process buffer for scheduling. If the buffer is empty, the timing bound does not have to be fulfilled because no data packet can be sent.

As practical experience has shown, timing constraints have to be subdivided into timing requirements and timing conditions. A detailed description is given in the following sections.

Timing Requirements. A timing requirement is the specification of a timing constraint that has to be observed for validation during design and test phases of the design cycle. They are not considered further after the installation of the final system. Different performance evaluation methods for validation of such requirements are state of the art. These methods are subdivided into monitoring, simulation, and analytical approaches. In all these cases, the mathematical clock model for the validation has to be chosen carefully to approximate the system behavior sufficiently. In addition to this, the use of different clock models can cause errors, e.g. if for simulation continuous real clocks and for analytical validation ideal clocks have been used. This has to be considered during the definition and the utilization of engineering models for system development.

The abstract data type concept is used to specify timing requirements in SDL. The abstract data type *EventClass* shown in Fig. 9 contains different operators to describe complex timing requirements and to observe the fulfilling of a timing requirement online. An *EventClass* handles all events of the same kind, e.g. all events *writeintotheinputqueue* or all events *aparticularprogrampositioniscalled*.

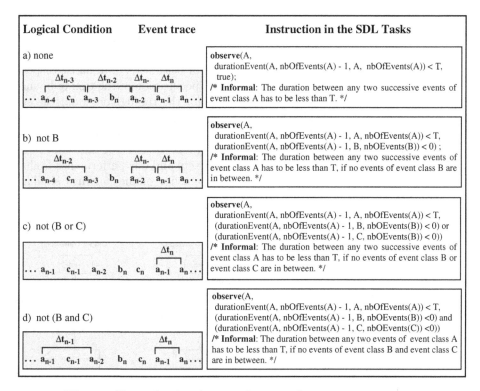

Fig. 10. Examples for the specification of timing requirements

The operator $mkEventclass(\dots)$ has four parameters for the initialization: *ActivationStatus* specifies which operators will be implemented in the target (executable) system or can be used by simulation, for example *ActivateAll* specifies that all operators can be performed and *ActivateObserve* specifies that only the operators which are needed for observation are available. This makes it very comfortable to leave out unneeded operators in the final version. Timing requirements can depend on events of the RTSS, e.g. a signal is written into the input queue (*InQueueEvent*), and events of the EFSM. The SDL type *EventKind* denotes this determination. The last two parameters of the operator $mkEventclass(\dots)$ are a reference to a clock specified in an SDL* annotation and the maximal number of buffered events. The operator $recordEvent(\dots)$ calls the operator $getTime(\dots)$ of the abstract data type *Clock* to get the timestamp of an event of the EFSM and buffers it in the internal data structure of the event class. Events of the RTSS are recorded in conjunction with the RTSS, e.g. if such an event occurs, their timestamp is automatically saved in the data structure of the event class. $getTimeEvent(\dots)$ can be used to get the timestamp of a recorded event. After $removeEvent(\dots)$ is called, a recorded event is deleted. The operator $nbOfEvents(\dots)$ returns the index of the last recorded event of the event class. The duration between two recorded events from the same or from different event classes is calculated by $durationEvent(\ldots)$.

The operators discussed above are necessary for specifying timing requirements that can be described by the operator $observe(\ldots)$ in an SDL specification. The two boolean parameters are the specification of the timing bound and the logical condition. As discussed above, for the timing constraint the system requirement is always fulfilled, if the logical condition is $false$ or if both the timing bound and the logical condition are $true$. The operator $observe(\ldots)$ can be linked to a monitor for performance evaluation. We use our hardware monitoring system ZM4 to record timing requirements and our performance evaluation system SIMPLE to analyze the system behavior offline [6]. In a similar way $observe(\ldots)$ can be used for simulation.

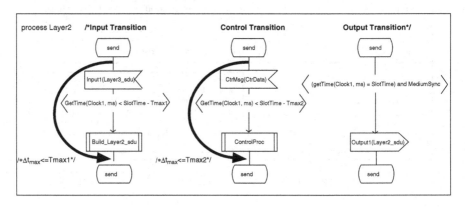

Fig. 11. Specification of a timing condition

In Fig. 10, some characteristic examples of timing requirements and their specification in SDL are shown. In each case, the event classes A, B, C, and D handle a set of events of the same kind. As a special case, all start and end events of the timing bound are members of the event class A with $a_n \in A$. The timing bound is $true$ if two successive events of A are less than the duration T. The operator $durationEvent(\ldots)$ compared with the required duration T specifies the timing bound in all cases. The events $b_n \in B$, $c_n \in C$, and $d_n \in D$ are used to specify the logical conditions. The logical condition in example a) is always true because any other events can occur between two successive events of $EventClassA$. In examples b) to d) $durationEvent(\ldots)$ and the relations or and and describe the logical conditions. Only a few language operators and the new operators $durationEvent(\ldots)$ and $nbOfEvents(\ldots)$ are needed to specify such important logical conditions.

Timing Conditions. A timing condition is a kind of timing constraint that actively controls the behavior of the system. In opposition to timing requirements, a timing condition needs to be part of the implementation. A timing condition consists of a timing bound, a logical condition, and a reference to the code that has to be performed if the timing bound and the logical condition are *true*. A typical example is a communication system which performs the medium access in a timed division multiplex scheme (TDMA). A TDMA frame is subdivided into slots. Such a system has to send a data packet exactly in a defined slot.

In SDL-2000, it is possible to access the operators of abstract data types in enabling conditions and continuous signals. So the pattern *DecoupleModule* can be modified to specify the above mentioned timing condition. As shown in Fig. 11, the output transition of the process *Layer*2 is triggered every time the reading of *Clock*1 has the value of the beginning of the assigned slot (*SlotTime*). This happens periodically because the overflow of the discrete real clock is the duration of the TDMA frame. The variable *MediumSync* denotes that all communication partners are already synchronized. The enabling condition construct prevents triggering the *InputTransition* or the *ControlTransition* in the case that the *SlotTime* is reached. The enabling condition ($getTime(Clock1, ms) <$ $SlotTime - Tmax1$) of the *InputTransition* takes into account the longest duration $Tmax1$ of the *InputTransition*. In the same way ($getTime(Clock1, ms) <$ $SlotTime - Tmax2$)) the longest duration $Tmax2$ of the *ControlTransition* is considered.

The timer active expression can be used in continuous signals or enabling conditions to trigger transitions after the timer has fired and has become inactive. As discussed in Sect. 3.3 this is not sufficient to specify timing conditions in general. To allow this, each timer needs access to a clock whose physical properties are described as shown in Sect. 4.2.

The example shown in Fig. 11 needs three timers to specify the timing condition. The continuous signal contains a time expression to trigger the *Output Transition* in the case the *SlotTime* is reached. The timer has to expire if the reading of the clock is *SlotTime* ms. The time expressions of both enabling conditions refer to different timers. The timer of the *InputTransition* has to expire if the reading of the clock is *SlotTime* ms - *Tmax1* ms and the timer of the *ControlTransition* has to expire at *SlotTime* ms - *Tmax2* ms.

The specification of the timing condition based on timers that have an extended semantic and the operator $getTime(\dots)$ are identical in respect to triggering the *OutputTransition* at the defined time. But it is more elegant and intuitive to make a decision in dependency on the reading of the clock than to use timers. In the case the timing condition is periodically the timers have to be set every time again after their expiration. This needs not to be done if the operator $getTime(\dots)$ is used. In addition, the automated synthesis of the system specification shown in Fig. 11 to an implementation is easier.

5 Case Study: Specification of the DECT MAC-Layer

The specification of the MAC-Layer [2] of the wireless communication system DECT will be discussed in this section to demonstrate the usability of our ap-

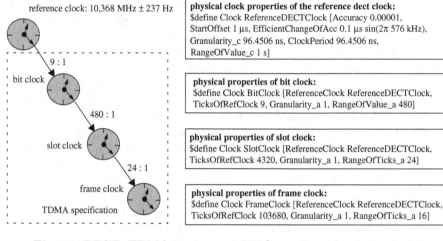

Fig. 12. DECT: TDMA timing and SDL* specification of the clocks

proach.[1] The case study focusses on the main functions of a DECT fixed part
to specify the multiplexing scheme of the logical channels with SDL. Functions
such as connection management, measurement of radio signal strength, and handover algorithms are not yet considered. The DECT example was used because
the system behavior is well known and we already had experience developing
DECT systems. In addition, this case study covers most real-time problems of
future mobile communication systems because DECT is very similar to GSM or
UMTS. A detailed description of hardware/software partitioning alternatives of
this SDL specification and their implementations on a rapid prototyping board
can be found in [15].

    The DECT protocol specification describes layer one, two and three of the
OSI reference model for wireless speech and data services. It is designed to provide large cordless private branch exchange (PBX) installations or wireless LANs
and is also available for domestic consumers. The radio fixed parts (base stations)
of the system are connected to a PBX via a standard telephone line interface
(analog or digital). Portable parts, mobile phones, or laptops with a DECT
interface communicate with the telephone system via the air interface. DECT
supports seamless handover to change the radio fixed part being connected to a
portable part, or to switch the radio channel if the quality of a connection gets
worse. This may occur when the user leaves the area of the current radio fixed
part or other radio signals interfere with the radio channel.

[1] The case study is supported by an industrial cooperation with the DECT development group of the Tenovis GmbH.

5.1   Specification of the Clocks for the TDMA Timing

The air interface is realized with TDMA to support high traffic load. A 10 ms frame is subdivided into 24 time slots: 12 time slots for the uplink and 12 time slots for the down-link result in 120 channels on 10 radio frequencies. In each time slot, a packet of 480 bits can be sent. A bit, slot, and frame clock is needed to send the data packets on time. The frequency of these clocks has to be derived from the reference clock (10.368 MHz) as shown in Fig. 12 where the factor 9 leads to the bit clock. Additionally the SDL* clock annotations are shown. The reference clock is described as a discrete real clock while the bit, slot, and frame counters are derived from this reference clock.

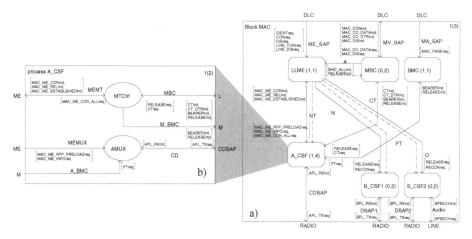

Fig. 13. DECT MAC: a) SDL specification overview, b) process A_CSF

5.2   SDL Specification

The specification of the DECT MAC-Layer shown in Fig. 13 a) is split into six different SDL processes: The complete MAC-Layer is managed by the process Lower Layer Management Entity (LLME). Multi Bearer Controller (MBC) and Broadcast Message Controller (BMC) perform the handover and broadcast management. The cell site function (CSF) performs the control of the radio links. The process A_CSF performs the setup of the so called bearer - the radio link - while the other two CSF processes perform speech data transmission. Every bearer supports different logical channels identified by message types, e.g. Mt-messages for bearer setup or Pt-messages for paging.

The specification of the multiplexing scheme of the logical channels performed by the process A_CSF is discussed in the following. As shown in Fig. 13 b) A_CSF is divided into two services with synchronous communication between them: Service AMUX manages the multiplexing and the access to the radio module while service MTCtrl contains the MAC protocol. The pattern Decouple Module has been used to specify the strict DECT timing conditions that must be fulfilled only by the AMUX service. In the same way the CSF functions B_CSF1 and B_CSF2 for transmitting speech can be specified.

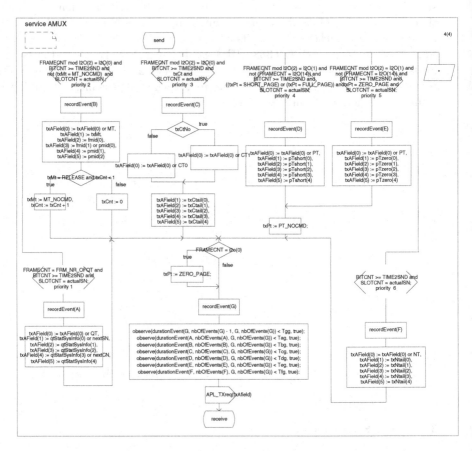

Fig. 14. DECT sending: MUX algorithm

The AMUX service shown in Fig. 14 consists of six transitions to multiplex the logical channels. With the help of the continuous signal construct, the timing condition of each logical channel is specified. The correct timing behavior of the MAC-Layer can easily be described and validated using the above discussed derived counters for bit-, slot-, and frame-timing and the operators of the abstract data type *Clock*. For the sake of clarity a macro is used for the operator *getTicks*(...), e.g. *FRAMECNT* to get the reading of the derived frame counter. Depending on the desired values of these clocks the desired transition fires synchronously. The used radio module has a fixed transmission latency. This has to be taken into account for the calculation of the transitions firing point. In the discussed specification, the constant *TIME2SND* describes this duration including the runtime of the transitions. To validate that each send transition has finished within $TIME2SND$ the operators $observe(\ldots)$, $recordEvent(\ldots)$, etc. of the abstract data type $EventClass$ are used.

6 Conclusion

In this paper a new approach is presented that extends SDL with the ability to specify timing aspects. The specification of timing aspects is important to support the development of communication systems with QoS requirements, such as mobile telephone systems. The formal time model consists of clocks and timing constraints. Based on this model, a design pattern to specify Layer 2 protocols was developed. The design pattern decouples the data input stream from the data output stream. Different data rates of the input and output data stream can be considered. For the specification of synchronous medium access it is possible to send signals at a particular point of time.

A typical example that shows that timing enhancements are needed to specify real-time systems is the DECT MAC-Layer discussed in the case study. Currently, we are starting a cooperation with an industrial partner to specify and implement time critical parts of UMTS-mobiles. A key point of this cooperation is to prove the usefulness of our SDL extensions and the acceptance by industrial engineers.

References

1. J.M. Daveau, G. Marchioro, A.J. Jerraya. Hardware/Software Co-Design of an ATM Network Interface Card: a Case Study. 6th International Workshop on Hardware/Software Codesign, IEEE Computer Society Press, Seattle, 1998.
2. ETSI. ETS 300-175-3, Digital Enhanced Cordless Telecommunications (DECT); Common Interface (CI); Part 3: Medium Access Control (MAC) layer. ETSI, 1996.
3. N. Faltin, L. Lambert, A. Mitschele-Thiel, F. Slomka. An Annotational Extension of Message Sequence Charts to Support Performance Engineering. In A.Cavalli, A. Sarma (eds.), SDL'97 Time for Testing, SDL, MSC and Trends, 8th SDL Forum, Elsevier Science, 1997.
4. B. Geppert, R. Gotzhein, F. Rößler. Configuring Communication Protocols Using SDL Patterns. In A.Cavalli, A. Sarma (eds.), SDL'97 Time for Testing, SDL, MSC and Trends, 8th SDL Forum, Elsevier Science, 1997.
5. B. Geppert, F. Rößler, R. L. Feldmann, S. Vorwieger. Combining SDL Patterns with Continuous Quality Improvement: An Experience Factor Tailored to SDL Patterns. In Proceedings of the 1st Workshop of the SDL Forum Society on SDL and MSC, Berlin, June 1998.
6. R. Hofmann, R. Klar, B. Mohr, A. Quick, M. Siegle. Distributed Performance Monitoring: Methods, Tools, and Applications. IEEE Trans. Parallel and Distributed Systems. Vol. 5, No. 6, June 1994.
7. ITU-T. Z.100 (11/99), ITU, Specification and Description Language. ITU, Specification and description language (SDL), November 2000.
8. ITU-T. Z.100, Appendix I. ITU, Specification and Description Language. ITU, 1993.

9. ITU-T. Z.120, Message Sequence Chart. ITU, 1996.
10. P. Langendörfer, H. König. Specifying and Validating Quality of Service Requirements Using Signal Chains. Workshop on Performance and Time in SDL and MSC, Internal report IMMD-VII-1/98, University of Erlangen-Nuremberg, February 1998.
11. B. Müller-Clostermann, M. Diefenbruch. Queueing SDL: A Language for the Functional and Quantitative Specification of Distributed Systems. Workshop on Performance and Time in SDL and MSC, Internal report IMMD-VII-1/98, University of Erlangen-Nuremberg, February 1998.
12. R. Münzenberger, F. Slomka, M. Dörfel, R. Hofmann. A New Time Model for the Specification, Design, Validation and Synthesis of Embedded Real-Time Systems. Submitted to IEEE Transactions on VLSI Systems.
13. A. Mitschele-Thiel, F. Slomka. Codesign with SDL/MSC, In K. Buchenrieder, Al Sedlmeier (eds.), International Workshop on Conjoint Systems Engineering (CONSYSE'97), IT-press, 1999.
14. I. Schieferdecker, A. Rennoch, O. Mertens. Timed MSCs - an Extension to MSC'96. In A. Wolisz, I. Schieferdecker, A. Rennoch (eds.), Formale Beschreibungstechniken für verteilte Systeme, GI/ITG-Fachgespräch, Berlin, June 1997.
15. F. Slomka, M. Dörfel, R. Münzenberger. Generating Mixed Hardware/Software Systems from SDL Specifications. Accepted for the International Symposium on Hardware/Software Codesign, Codes 2001.
16. H.-J. Vögel, W. Kellerer, S. Sarg, M. Kober, A. Beckert, G. Einfalt. SDL based prototyping of ISDN-DECT-PBX switching software. In Proceedings of the 1st Workshop of the SDL Forum Society on SDL and MSC, Berlin, June 1998.
17. I. Wakeman, A. Ghosh, J. Crowcroft, V. Jacobson, S. Floyd: Implementing Real-Time Packet Forwarding Policies using Streams Proceedings of the USENIX Technical Conference, Louisiana, New Orleans, Jan. 1995.
18. H. Zhang. Service Disciplines for Guaranteed Performance Service in Packet-Switching Networks. Proceedings of IEEE, Vol. 83, No. 10, October 1995.
19. L. Zhang. Virtual Clock: A New Traffic Control Algorithm for Packet-Switched Networks. ACM Transactions on Computer Systems, Vol. 9, No. 2, May 1991.

Timed Extensions for SDL*

Marius Bozga[1], Susanne Graf[1], Laurent Mounier[1], Iulian Ober[2],
Jean-Luc Roux[2], and Daniel Vincent[3]

[1] VERIMAG, Centre Equation, 2 avenue de Vignate, F-38610 Gières
Marius.Bozga@imag.fr
[2] Telelogic Technologies, 150 rue Nicolas Vauquelin, F-31106 Toulouse Cedex
Iulian.Ober@telelogic.com
[3] France-Telecom R&D, DTL, 2 avenue Pierre Marzin, F-22307 Lannion Cedex
Daniel.Vincent@rd.francetelecom.fr

Abstract. In this paper we propose some extensions necessary to enable the specification and description language SDL to become a more appropriate formalism for the design of real-time and embedded systems. The extensions we envisage concern both roles of SDL: first, in order to make SDL a better real-time *specification* language, allowing to correctly simulate and verify real-time specifications, we propose a set of *annotations* to express in a flexible way assumptions and assertions on timing issues such as execution durations, communication delays, or periodicity of external inputs; second, in order to make SDL a better real-time *design* language, several useful real-time programming concepts are added. In particular we propose to extend the basic SDL timer mechanism by introducing new primitives such as cyclic timers, interruptive timers, and access to timer value. All these extensions rely on a clear and powerful time semantics for SDL, which extends the current one, and which is based on *timed automata with urgencies*.

Keywords: SDL, time semantics, timed automata, urgencies.

1 Introduction

The ITU–T Specification and Description Language (SDL, [10]) is increasingly used in the development of real-time and embedded systems. For example many recent telecommunication protocols (such as RMTP-II [18] or PGM [17]) integrate such real-time features in their architecture, and these non-functional aspects are essential in the expected behaviour of the application. This kind of system imposes particular constraints on the development language, and SDL is a suitable choice in many respects: it is formal, it is supported by powerful development environments integrating advanced facilities (such as simulation, model checking, test generation, code generation, etc.), and thus it can cover several phases of the software development, ranging from analysis to implementation and on-target deployment.

---

* This work is supported by the INTERVAL IST-11557 European project on timed extensions for SDL, MSC and TTCN.

R. Reed and J. Reed (Eds.): SDL 2001, LNCS 2078, pp. 223–241, 2001.

It appears however that several important needs for a real-time systems developer are not covered by SDL. These problems range from pure *programming issues* (such as the lack of useful primitives commonly available in real-time operating systems) to *specification issues*, (such as the difficulty to describe in a appropriate way the assumptions under which the system is supposed to be executed). Clearly, the needs are not the same for both uses of the language, and, in many cases, the *programming side* has been given priority in the supporting tools to the detriment of the *specification side*.

Several proposals already exist to extend SDL with real-time features. We can mention for example the work carried out on performance evaluation [8,13,15,12], on schedulability analysis [4], or on real-time requirements [11]. In this paper we are more concerned with the use of SDL as a specification language for real-time systems and its application for formal validation. In particular, one of the important questions we address is what kind of real-time features should be modeled in SDL and at which level of abstraction.

The simplest use of time which is frequent in communication protocols, is the use of timeouts (whose value is often meaningless) in order to avoid infinite waiting. The time semantics of SDL, together with the fact that timeouts are notified via a signal in the message queue of the process, corresponds exactly to this use: no guarantee can be given when the signal arrives and is dealt with, but it is after some finite time. Nevertheless, when SDL is used as a programming language, it is often done with much more restricted assumptions on the possible time behaviour in mind, and, if they are correct, the implemented system will behave as expected. As such assumptions are not (and cannot be) expressed explicitly, the specification cannot be validated: the verification using the standard SDL progress of time may invalidate even apparently time independent safety properties.

A typical workaround used for obtaining a convenient result at simulation time consists of using on one the hand timers to force minimal waiting, and, on the other hand, a very restricted interpretation of time progress, allowing it only when the system is not active. This "synchrony hypothesis" is in general as unrealistic as the standard semantics. The correct assumption would be that certain tasks will be executed timely, whereas for others this cannot be guaranteed and the correctness of the system must be verified even if they take longer than expected.

We propose a solution to reconcile these two extreme choices that relies on a more flexible time semantics for SDL, based on timed automata with urgencies [5]. In particular, urgencies give a very abstract means to express *assumptions* on the environment and on the underlying execution system, such as action durations, communication delays, or time constraints on external inputs. From the user point of view, all these "non functional" extensions are offered in a uniform way by means of *annotations* on the SDL specification.

The propositions presented in this paper are the results of the INTERVAL IST project and preliminary work of its partners. The aim of INTERVAL is to take into account real-time requirements during the whole development process of real-time systems and to define consistent extensions to the languages SDL, MSC and TTCN.

The remainder of the paper is organized as follows: in section 2, we give an overview on the problems occurring when using SDL for real-time systems, concerning both programming and specification aspects, in section 3 we propose extensions to improve SDL as a real-time specification language and in section 4 we propose necessary programming concepts. All new concepts are illustrated by examples illustrating their use and proposed syntax. Finally, in section 5 we draw some conclusions and give some perspectives.

2 Real-Time SDL: What Is Missing?

SDL has the double aim of being on the one hand a high-level *specification* formalism, meaning that it must abstract from certain implementation details, and on the other hand a *programming* formalism from which direct code generation is possible. These two roles of the language seem sometimes conflicting, as the needs at the different levels are not the same in general.

It is important that SDL can fully play this double role of being an implementation and a specification language, and all information needed for both uses of SDL must be expressible, but also in such a way that these two concerns are clearly separated. This feature is particularly crucial when dealing with real-time systems, in which non-functional elements need to be taken into account even in the early stages of the design.

We summarize here the main difficulties currently arising when trying to use SDL for the design and validation of real-time systems.

2.1 Real-Time Semantics

First of all, the semantics of SDL, as presented in Z.100, is very abstract in the sense that it allows no assumptions to be made about progress with respect to time: actions take an indeterminate amount of time to be executed, and a process may stay an indeterminate amount of time in the current state before taking one of the next fireable transitions. This notion of time that is external, unrelated to the SDL system, is realistic for code generation, in the sense that any actual implementation of the system conforms to this abstract semantics. However, for simulation and verification, this total absence of controlability of time is not satisfactory: timer extents do not have any significance besides defining minimal bounds, any timer that gets in a queue may stay there for any amount of time, with the consequence that hardly any real-time property holds on models based on this abstract semantics.

A simulator that would use the semantics of time as described in Z.100, would not be able to make any assumption on the way time progresses, and therefore many unrealistic executions will be present in the resulting graph. As a result, the simulator would not guarantee elementary properties such as: *when a timer expires, it will be consumed by the concerned process in a* reasonable *amount of time* (whatever the notion of reasonable is). Even worse, according to Z.100, *when two timers are set in the same transition (for example in two consecutive tasks), the timer with the lower delay is not always consumed first*[1].

In practice, existing simulation and verification tools have foreseen means for limited control over time progress. However, the control over time they propose is in general quite limited and moreover these annotations are tool dependent whereas they are really part of the complete specification in the sense that they describe assumptions on the system environment. The fact that designers and design languages neglect a clear description of relevant properties of the environment in which the system should be executed, is a frequent source of errors.

2.2   How to Note Non-functional Aspects?

The development of a complex real-time protocol usually needs to consider several preliminary stages, during which some abstract or incomplete descriptions are produced. In order to properly validate (and document) these early designs, general assumptions on both the "environment" of the system and on its "non-functional" aspects have to be taken into account. For example, such assumptions concern:

- the expected duration of some internal task (which might be either informal or fully specified),
- the periodicity of some inputs triggered by the environment,
- or even the expected behaviour of the communication channels used within the system (these channels may be reliable or not, assumptions may be made on communication times, etc.).

Of course, some of these assumptions can already be partly included in the specification, either directly in SDL (for example using timers for explicit waiting) or using some separate formalisms offered by the verification tools (like the GOAL language [2] proposed in *Object*Geode to specify external observers). However, none of these two solutions is satisfactory: the first one leads to a specification in which external and non-functional assumptions do not appear as such, and this is obviously not desirable for code generation (these timers need not to be implemented), whereas the second one is restricted to a particular tool. Our objective is to provide a more suitable framework, based on standardized *annotations* on SDL specifications and compatible with the real-time semantics we propose.

2.3   High Level Synchronisations and Other Real-Time Primitives

Clearly, SDL has several characteristics that are attractive for real-time system designers: asynchronous communication is a first class language feature, a specification is organized in a logical hierarchy that can be mapped in many ways to different physical configurations of software modules (and SDL code generators

---

[1] Which timer is consumed first will depend on the inputs attached to states.

usually provide this feature), external code may be called from SDL, making it possible to use system libraries directly in SDL.

Several synchronisation mechanisms that are commonly employed in real-time systems should be usable as a concept at the SDL level. In particular, SDL timers are rather limited: the only available primitives are **set** and **reset** operations, the **active** function (which allows to determine if a given timer is running or not), and timeouts are always transmitted in the form of signals in the input buffer of the process.

2.4 Deployment Information

Z.100 asserts that the agents composing a system are executed truly in parallel. In the context of the very weak time semantics of SDL (only minimal waiting time can be enforced, and any action or message transmission takes either zero or an arbitrary amount of time), this simplifying assumption is possible, because any mapping onto a set of processors and any notion of atomicity will lead to the same functional behaviour, and any time behaviour is included in the semantics.

However, the introduction of a more precise notion of time introduces also global constraints, so that different degrees of atomicity or different mappings on processors will lead to different time behaviours, and — if the functional behaviour depends on real-time constraints — even to different functional behaviours. An obvious example is the fact that, if a set of processes are executed interleaved, their execution times must be added, whereas if they are truly parallel the global execution time will be the maximum of the individual execution times.

One must obviously be very careful, and try to make assertions on global execution times that use no or very abstract assumptions about the architecture on which the system is executed. However, it is not always necessary to introduce much knowledge about the architecture:

- First of all, there exist time dependent properties which are not architecture dependent: for example, often the safety of a protocol may depend on the relative values of a set of timers, the expiration of which are used as implicit signals between processes. This is a common use of timers which cannot be expressed in the present SDL time semantics, and still does not need any architecture indication.
- In a system where time is consumed either in communications or in requests to external systems (such as a distant data base or anywhere else in the environment), the execution times will not depend on the mapping of processes onto processors if the execution time of all activities consume a negligible amount of time and can be safely simplified to zero. Also, in the case where time is consumed within the system, but within a single process per processor, analysis of execution time is still possible in the same way.
- In the case where time is consumed in several parts of the system, it would be sufficient to indicate which parts of the system are executed in parallel and which ones are not. The distinction between block agents, process agents

and sub-agent gives some limited possibility to indicate such an architectural information.

Going one step further, scheduling policies also can influence the properties of the system in critical hard real-time systems. Moreover, there exist important advances in the synthesis of schedulers [3], where scheduling policies are expressed mainly in terms of dynamic priorities.

3   Extending Specification Aspects

As we mentioned above, what is missing in SDL to improve real-time systems *specification* is both a flexible time semantics together with more facilities to express some "non functional" parts of the system or its environment. Our proposal is to solve these two problems uniformly – from the user point of view – by offering the possibility to *annotate* the specification in a standardized way. In particular, we propose to distinguish between two types of annotations:

- *assumptions*, which express *a priori* knowledge or hypotheses about the environment or the underlying execution system (system architecture, scheduling policy, etc). The use of assumptions is twofold: first, they might be necessary for the verification of properties which do not hold otherwise; second, they might be used for code generation, both to guide some implementation choices or to add specific code in order to check their correctness at run-time or during the test phase.
- *assertions*, which express expected (local) properties on the system components. Such properties have to be proved on the specification, during the verification phase, possibly taking into account some of the assumptions.

The annotations we propose concern respectively, control over time progress by means of *urgencies*, *durations* and *periodicity* of actions, and flexible *channel* specifications.

3.1   Urgencies

A very abstract – and still very powerful – manner for making realistic assumptions on the time environment of a system is by means of transition urgencies [5]: a transition is "urgent" if it is enabled and will be taken or disabled before time progresses. Three types of urgency assumptions (*eager*, *lazy* and *delayable*) are enough to control the progress of time with respect to the progress of the system:

- **eager** transitions are urgent as soon as they are enabled: they are assumed to be executed "as soon as possible"; in a simulation state *time does not progress* as long as there are enabled eager transitions.
- **lazy** transitions are never urgent: enabled **lazy** transitions do not inhibit time progress in any simulation state.

- **delayable** transitions are a combination of eager and lazy transitions: they become urgent when time progress disables them. They are supposed to be executed within some interval of time in which they are enabled. A delayable transition usually has an enabling condition depending on time, such as **now** $\leq x$ or **now** $- x \leq y$ (where x and y are numerical values) and *time may progress* in all simulation states in which this transition is enabled *as long as* **now** $\leq x$ (or **now** $- x \leq y$). When the extreme point of the interval is reached the transition becomes urgent.

  Notice that the attribute "delayable" is not primitive: a delayable transition with a guard **now** $< x$ (for instance) can always been replaced by two transitions, a lazy one with the same guard and an eager one with guard **now** $= x$. In particular, in SDL specifications in which explicit time guards (other than timeouts) are not used, explicit delayable transitions are not useful. However, whenever a task or a communication is assumed to take some time specified by an interval, this is expressed by a delayable transition in the semantics model.

Expressed in terms of urgencies, the semantics of time in Z.100 considers all transitions as lazy: time progress is not a constraint at all, whatever transitions are enabled. Nevertheless, most SDL tools implement an eager semantics: transitions are fired as soon as they are enabled without letting time progress. It appears in practice that none of these two extreme interpretations of time progress in isolation allows to obtain satisfactory models of real-time systems. It is often appropriate to mix these two extreme views of time progress:

- one would like to consider some of the inputs as lazy (which is the standard point of view). When such a transition is enabled, the system can choose to react immediately or to wait. In the case of an *external* input laziness denotes the absence of knowledge about the possible arrival time of the input; for *internal* inputs laziness can be used to interpret internal action durations or internal propagation delays as unconstrained or unknown.
- On the other hand, one would like to consider some of the inputs as eager to express that the system cannot ignore them when they are enabled and must react immediately without waiting. This can be used to guarantee an immediate response to *critical* events such as timeout expirations or other priority inputs.

  Some care must be taken here, as one can easily write system specifications in which time is blocked "forever" because at least one eager transition is always enabled (this is called a *Zeno behaviour*). This is particularly problematic when all transitions are assumed eager.

Default choices, depending on the type of the system and on the type of transitions, can be envisaged in order to provide a user-friendly way of specifying how urgencies are associated with system transitions.

Example 1. A generic situation is presented in Fig. 1. The informal action *open* must be executed in the interval defined by timers *t_early* and *t_late*, as soon as

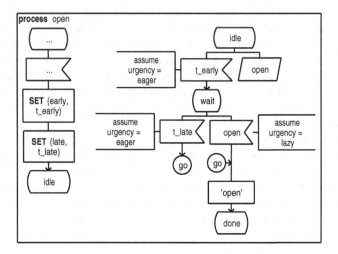

Fig. 1. Urgency assumptions - an example

the external signal *open* is received. Thus, if the *open* signal arrives too early, it must be saved; if it does not arrive in time, the action is executed anyway at the expiration date of *t_late*. In order to obtain a realistic time behaviour of such a system, timeout consumptions have to be considered as eager: the action *open* is assumed to be executed at time *t_late* at latest. In order to specify all possible behaviours, the (external) *open* signal input must be considered as lazy. Considering it as eager prevents the timer *t_late* expiring, because this transition is always enabled in state *wait* (in the absence of an environment process that sends only a limited amount of *open* signals). In the case where the *open* signal is sent from within the system, the consuming transition might be considered as eager (meaning that one can assume that it will be executed as soon as possible within its allowance interval). In this case one may want to *verify* if the signal always arrives in the given interval, and how this depends on the time constraints of the other parts of the system.

Timer semantics. Timers are the most used primitive to observe the progress of time. The behaviour of a timer can be sketched by the automaton given in Fig. 2. Basically, it switches between *inactive* and *active* states depending on the set and reset actions performed on it. When active, once the expiration time is reached, it will *expire* and the timeout signal becomes available to the corresponding process instance. Previous interpretations of Z.100 have considered this expire transition as *lazy*, which means that one cannot make any assumption on the maximum time elapsed since its last setting.[2] In other words, nothing can prevent the automaton from remaining in the *running* state after the expiration time.

---

[2] Editor's note: The text in Z.100 itself has remained unchanged since SDL-88 apart from the change of the word *process* to *agent* for SDL-2000.

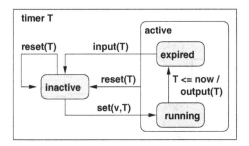

Fig. 2. Timer behaviour

The current Z.100 semantics has a more reliable timer concept ensuring the availability of the timeout signals *exactly* at expiration time. This can be done simply by considering the implicit expiration transition as *eager*. Note that this cannot be captured by the urgency annotation described before, because it relies on an implicit semantic choice which cannot be explicit at the specification level. Also, this is not too restrictive: it still allows the situation in which the timeout is really consumed at a later point of time as the consumption time depends on the urgency of the consuming transition.

Example 2. A second generic process is illustrated in Fig. 3. The informal actions *action_1*, *action_2*, *action_3* must be executed *precisely* at moments, respectively *T0-d1*, *T0-d2*, *T0-d3*, where *T0* is given as parameter to the process. Let us assume first that the informal actions correspond to external commands which need just to be initiated by the process (for example actioning some external devices). That means it is reasonable to assume that the process is essentially idle, waiting for the timer expiration, and reacts immediately. Unfortunately, even if the SDL description of the process seems intuitive and concise, based on the standard semantics of SDL, no assumption can be made about the time at which the actions will be executed. Considering the timeout transitions as lazy, the actions are executed *not earlier than* the required moment. The eagerness of the timeout consumption transitions expresses the assumption that actions are executed at the moment at which the timer expirations are available.

3.2   Durations

SDL does not foresee the possibility to impose or to assume any quantitative restriction on the duration that an action may take to be executed or how long a process may stay idle in a state before executing one of the enabled actions. Also, concerning the amount of time that a signal may need to travel through a channel, only two cases can be distinguished: either zero time or an arbitrary amount of time. Nevertheless, in real-time designs, such execution times may not only influence the performance, but also the functional behaviour of the system.
Currently, it is possible in SDL to describe minimum and maximum durations by means of explicit waiting using timeouts and a notion of "invalid state" to

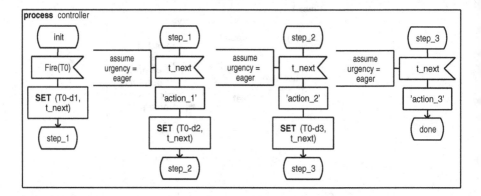

Fig. 3. Timeout urgencies - an example

mark the executions taking more than the maximum execution time as "uninteresting". This is used frequently in SDL specifications, but it is a bad solution as it is cumbersome and it uses a programming construct in order to indicate assumptions about the environment. Such a specification cannot be used directly for (automatic) code generation, since it is difficult to detect the nature of these timeouts.

We suggest to explicitly indicate such durations using predefined annotations. For instance, we propose to use either interval constraints (e.g. delay=[9.0,11.0]) or mean-values plus jitter (e.g., delay=[10.0±5%]) attached to the corresponding action or channel.

Moreover, such annotations could be enriched with probability laws. This extension is mandatory for performance evaluation, but still not sufficient, as we also need a model of available computation resources. Note that from a functional verification point of view probabilities are not necessary since all the behaviours have to be analyzed independently of their probability: verification wants to ensure absence of errors and not just "low probability" of them.

3.3   Periodicity

Many telecommunication applications (such as multimedia services) are expected to cope with large streams of data arriving with high and continuous rates. In practice, components of such applications are designed to fit into particular environments, able to deliver multiple inputs at given frequencies.

Therefore, periodicity of inputs tends to be an important feature that has to be expressed at the specification level. Similar to duration, we propose to annotate external inputs with interval constraints (or mean-values plus jitter) describing the expected period (where applicable). Such annotations are not only mandatory for verification but they also clearly improve the readability of the specification because they allow systematic description of the relevant characteristics of the assumed environment.

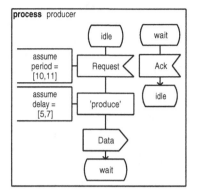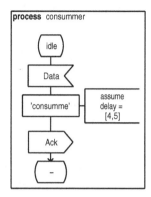

Fig. 4. Delay and periodicity assumptions

Example 3. A typical producer/consumer system is illustrated in Fig. 4. The producer reacts to external *request* signals and sends *data* signals to the *consumer*. When the *consumer* receives *data*, it sends an acknowledgment back to the *producer*. Several annotations are used here. First, we assume that *request* signals come from the environment at the given rate of one signal every 10 or 11 time units. Also, both data production and consumption times are supposed to be not negligible, the former being between 5 and 7 and the second between 4 and 5 time units.

Notice that, contrary to any description using timers or the global time **now**, here it is obvious that a production cycle can be greater than the period of *request* signals. This may be considered as problematic, as there exist scenarios in which the *request* signals accumulate indefinitely in the input queue of the *producer*. At this point, the use of probabilities may become important to decide if the design is acceptable or not.

3.4 Flexible Channel Specifications

In SDL, only reliable FIFO channels exist as a primitive concept. In its use as a specification language, however, a channel is often considered to be part of the environment of the implemented system and may represent an abstraction of a whole network. Many protocols are supposed to implement reliable communication through unreliable channels or networks. For verification it is therefore necessary to consider the situations where channels lose or re-order messages.

Thus, by flexible channel specification we typically mean that messages can be lost, re-ordered or delayed to some extent, leading to a well-defined set of channel types. It is possible to describe any of these channel types by means of additional SDL processes, but it is not necessarily desirable to do so, for several reasons. First of all, these processes serve only for simulation and not for implementation. Also, in some (rare) cases it may be problematic as the transported signals do not carry the pid of the original sender anymore. Finally, having a predefined set of channel attributes is an advantage for simulation and verification tools as this knowledge can be directly exploited by appropriate techniques and lead to more efficient algorithms (notice for example that an interesting set of properties is decidable for finite state machines communicating through lossy channels whereas they are undecidable in the case of communication through reliable channels [1]).

We propose annotations on channels allowing specification of a propagation delay in a similar way as execution times and distinguish between fully reliable (which never lose messages) and unreliable channels (where arbitrary losses are possible), and between ordered and unordered ones. Here again, for performance evaluation purposes, it is possible to extend these annotations to include a probability law which specifies a distribution of message losses or a degree of re-ordering.

Example 4. Figure 5 gives an example of the use of annotations to denote propagation delays and reliability of channels. Notice that the default option for a channel is to be ordered and reliable in conformance to the standard SDL semantics.

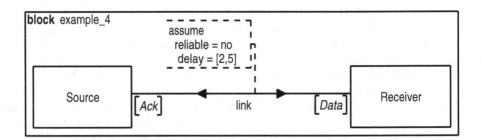

Fig. 5. Channel assumptions

4   Extending Programming Aspects

We make here two concrete proposals to improve SDL as a *programming* language for real-time systems. These proposals concern respectively the extension of the timer concept and the introduction of standardized packages to provide at the SDL level useful synchronization and atomicity primitives.

4.1   Extension of Timer Concepts

Timers play a central role in SDL to control and observe time progress. They are however limited with respect to what is commonly offered in real-time systems, and we propose to extend them in several directions.

Interruptive timers and signals. SDL timeouts are always received in the form of asynchronous signals. For general-purpose time dependent code this is usually fine, but it is difficult to write real emergency procedures using asynchronous timeouts. To ensure that a piece of code is executed immediately, or in a specified interval, after the expiration of a timeout, the SDL designer must first make sure that the corresponding agent (process) is idle when the timeout message is received, otherwise, the agent may consume the timeout message from the input queue only when it finishes its current job, which may be too late. This obliges the designer to artificially restructure the system, whereas in the implementation this task can be left for implementation where it may even be useless because of the existence of interrupts. Therefore, SDL needs a notion of emergency timer, whose expiration is taken into account *immediately* by the receiving agent. Emergency actions which interrupt the normal execution of an agent were already introduced in SDL-2000 with the advent of exceptions. What we need is an extension of this exception mechanism to be triggerable by system time.

We propose to define interruptive timers using an optional attribute (called *interruptive*). The behaviour of interruptive timers will rely on an extension of the exception mechanism already existing in SDL. More precisely, when an interruptive timer expires, instead of sending a timeout signal via the input port of the process, it will raise a timeout-exception in the concerned process. The handling of this exception is left to the user. However, special care is required to clarify what happens if an interruptive timer wants to interrupt a transition which is required as atomic. In particular, when an interruptive timer expires in a service, while another service (within the same process) is running and executes a time-consuming job.

Example 5. In Fig. 6 we illustrate the use of an interruptive timer. When set, it receives as a parameter the maximum allowed execution time T. Then, the process enters a loop executing some time-consuming job *"decode"*. Two cases are possible: either the entire job is finished in time, or the maximum allowed time is reached while the process is processing a *"decode"*. In the second case, an interruptive timer *alarm* is needed in order to break the normal execution flow to abort the execution of *decode* and to stop the process immediately.

Note that for the sake of readability we use the same notation for normal timeouts and interruptive timeouts. Nevertheless, the meaning is quite different: whereas the former denotes a normal SDL transition from a state, the latter is a shorthand notation for an exception handling at this state (and implicitly *in all implicit states and actions within the state scope*).

Timer consultation. The second extension we propose concerns the ability to consult the expiration time (which is of the predefined sort **Time**) assigned to a timer. In order to do so, our proposal is simply to add a predefined **value** operator on timers (similar to the existing **active** construct).

Such a primitive reduces the current distinction made in SDL between timers and other variables. In particular, the value of the timer can be passed via communication signals from one process to another, used to compute the remaining duration until its expiration, or used to set another timer depending on it, etc.

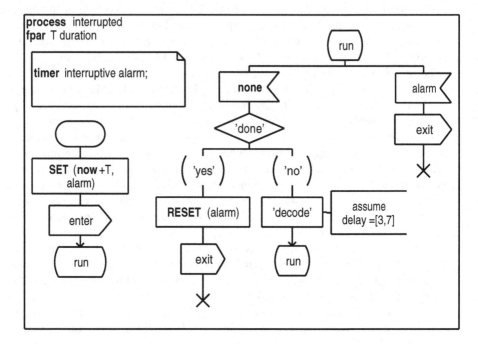

Fig. 6. Interruptive timer example

Obviously, the same result can be obtained using variables of sort **Time** instead of – or in addition to – timers. But the use of an explicit timer concept (with a set of well defined primitives) makes the specification more readable and the mapping into an implementation easier.

Cyclic timers. Finally, the third extension concerns cyclic timers. The idea is to eliminate the current limitation in SDL, where timers are one-shot and have to be explicitly reset in order to model a periodic behaviour. This could be done by simply considering an optional timer attribute (called **cyclic**) which fixes the nature of the timer at its declaration. When a cyclic timer is explicitly set in the specification, its period is computed (by subtracting the value of **now** from the expiration time). After that, at the expiration time (when the corresponding timeout signal is sent) this timer will be implicitly set again using the period and the value of **now** at expiration. This will continue until either an explicit reset or set occurs (the later restarts the whole behaviour, possibly with a new period). Finally, note that interruptive and cyclic attributes can be safely combined (the same timer can be both cyclic and interruptive).

4.2   Atomicity

There are no atomicity and synchronization primitives foreseen in SDL, the idea being that these are implementation details which should not be mentioned at the SDL level. The fact is that in implementation oriented SDL descriptions, they are necessary and several SDL users have expressed the need for native SDL constructs for synchronization [9], especially to achieve atomicity and mutual exclusion. The reason is that such constructs are often used in the specification and the implementation of real-time systems such as those developed with SDL.

The current practice in SDL is to use calls to external code (OS primitives) in order to achieve these functionalities. This approach has at least the following obvious inconvenience: the SDL specification, which is supposed to be high-level, becomes unnecessary configuration and platform dependent, and the external code cannot be handled properly by simulation and verification tools.

The need stated above can be addressed without making first-order extensions to SDL. Synchronisation behaviour can be expressed in terms of existing primitives of SDL, such as asynchronous signal exchange and remote procedures. We propose the use of (standard) libraries for this purpose, as is the case for other languages.

Example 6. For example, a semaphore may be specified in SDL as a process type exporting two (empty) procedures P and V, implementing the usual operations on semaphores. The specification of such a semaphore type is shown in Fig. 7. The only prerequisite for this implementation to work is that the atomicity of P and V are preserved for a same instance of semaphore. This prerequisite is ensured by the execution semantics of SDL. Moreover, concurrent wait operations P which arrive after the semaphore is already blocked on a wait are implicitly sequentialized by the save operation from the *busy* state.

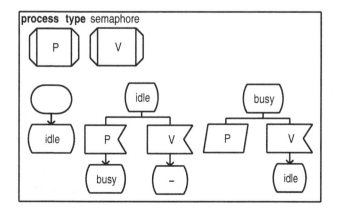

Fig. 7. Semaphore example

5   Conclusion and Perspectives

We have defined a number of real-time extensions of SDL in order to obtain a specification and modeling language really adequate for real-time systems based on general ideas and reflections we have proposed in [7]. The proposed extensions include:

- a more flexible time-progress semantics for SDL, including as particular cases both the time progress semantics of Z.100 and the time-progress semantics used in verification tools;
- powerful and flexible annotations to express non-functional aspects: both very abstract ones necessary at an early design stage, and concrete ones useful for code generation, code annotation and performance analysis;
- some primitives which to make specifications simpler, easier to understand, and to better separate the implementation oriented and specification oriented features;
- implementation oriented features like emergency timeouts which are mandatory for the use of SDL as a real-time modeling language.

These extensions have been submitted as a first draft to the ITU-T (Study Group 10) within the framework of Question 7 *Time expressiveness and performance annotations on ITU-T modeling languages.* All these extensions are based on a unified and sound semantic framework which can be integrated smoothly into the existing semantics of SDL in which time progress is mainly "unconstrained". In particular, almost all of them require only local restrictions of the current non-deterministic time semantics.

We have already started to extend existing SDL tools in order to deal with these extensions and they have already proved their usefulness in real applications. We have implemented a translation of SDL with our proposed time extensions into communicating timed automata with urgencies which are the input language of the IF toolset [6]. IF has been developed at VERIMAG for the purpose of prototyping timed extensions of SDL-like languages. In particular, it includes a model-checker based on the Kronos-tool [19] allowing verification of quantitative time requirements. In addition, the proposed timed extensions have been implemented also in the *Object*Geode simulator [14]. For both tools we obtain good results, both in what we can express and in what analysis we can perform on annotated models. As an example, we are currently modeling real-time multicast protocols [17,18] using these annotations.

The main approaches which need to be compared with ours are QSDL[8,13] and UML related real time extensions (such as UML-RT[16]). QSDL (Queuing SDL) is a performance analysis oriented extension of SDL using a resource mapping and annotations of tasks in terms of "computation power" from which execution and waiting times are computed depending on some scheduling policy which can be user defined. The underlying semantics considers all non annotated transitions as taking zero time and all transitions are considered urgent, respectively delayable if they may take a variable amount of time. To this extent this framework is compatible with ours, but is exclusively performance analysis oriented.

The foreseen real-time extensions of UML are all of a very syntactic nature. It is not clear if there will be a semantic time model at all. However, it is interesting to note that non-functional annotations concerning Quality of Service (such as throughput of channels, execution times,...) are planned and can take likewise the form of *requirements* and *assumptions*.

The proposed extensions are not completely satisfactory from the user point of view. More extensions may well appear to be useful. In particular, there is a strong need for an expressive notion of *deployment diagram*, allowing definition of the mapping of processes to resources, scheduling policies and QoS annotations. The proposed annotations on the SDL level will then have 3 different sources: some of them may be user defined, especially at an early design stage; some of them will be generated from information extracted from such a deployment diagram (which goes far beyond the resource mapping of the QSDL proposal); finally, some will be obtained from analysis results of other parts of the system, especially in a compositional verification approach, which is the only one able to deal with large specifications.

References

1. P. Abdulla, A. Bouajjani, and B. Jonsson. On-the-fly Analysis of Systems with Unbounded, Lossy Fifo Channels. In A .Hu and M. Vardi, editors, *Proceedings of CAV'98 (Vancouver, Canada)*, volume 1427 of *LNCS*, pages 305–318. Springer, June 1998.
2. B. Algayres, Y. Lejeune, and F. Hugonnet. GOAL: Observing SDL Behaviors with GEODE. In *Proceedings of SDL FORUM'95*. Elsevier, 1995.
3. K. Altisen, G. Gößler, and J. Sifakis. A Methodology for the Construction of Scheduled Systems. In Mathai Joseph, editor, *Proceedings of FTRTFT 2000*, number 1926 in LNCS, pages 106–120. Springer-Verlag, September 2000.
4. J.M. Alvarez, M. Diaz, L.M. Llopis, E. Pimentel, and J.M. Troya. Integrating Schedulability Analysis and SDL in an Object-Oriented Methodology for Embedded Real-Time Systems. In R. Dsoulli, G.v. Bochmann, and Y. Lahav, editors, *Proceedings of SDL-FORUM'99 (Montreal, Canada)*, pages 241–256. Elsevier, June 1999.
5. S. Bornot, J. Sifakis, and S. Tripakis. Modeling Urgency in Timed Systems. In *International Symposium: Compositionality - The Significant Difference (Holstein, Germany)*, volume 1536 of *LNCS*. Springer, September 1997.
6. M. Bozga, J.Cl. Fernandez, L. Ghirvu, S. Graf, J.P. Krimm, and L. Mounier. IF: An Intermediate Representation and Validation Environment for Timed Asynchronous Systems. In J.M. Wing, J. Woodcock, and J. Davies, editors, *Proceedings of FM'99 (Toulouse, France)*, volume 1708 of *LNCS*, pages 307–327. Springer, September 1999.
7. M. Bozga, S. Graf, A. Kerbrat, L. Mounier, I. Ober, and D. Vincent. SDL for Real-Time: What is Missing ? In *Proceedings of SAM'00: 2nd Workshop on SDL and MSC (Grenoble, France)*, pages 108–122. IMAG, June 2000.

8. M. Diefenbruch, E. Heck, J. Hintelmann, and B. Müller-Clostermann. Performance evaluation of SDL systems adjunct by queueing models. In R. Braek and A. Sarma, editors, *Proceedings of SDL Formu'95*. Elsevier Science B.V., 1995.
9. Interval Consortium. Requirement Analysis Report. Technical Report D11, Interval Deliverable, October 2000.
10. ITU-T. Recommendation Z.100. Specification and Description Language (SDL). Technical Report Z-100, International Telecommunication Union – Standardization Sector, Genève, November 1999.
11. S. Leue. Specifying Real-Time Requirements for SDL Specifications – A Temporal Logic-Based Approach. In *Proceedings of the Fifteenth International Symposium on Protocol Specification, Testing, and Verification PSTV'95*. Chapmann & Hall, 1995.
12. M. Malek. PerfSDL: Interface to Protocol Performance Analysis by means of Simulation. In R. Dsoulli, G.v. Bochmann, , and Y. Lahav, editors, *Proceedings of SDL-FORUM'99 (Montreal, Canada)*, pages 441–455, 1999.
13. A. Mitschele-Thiel and B. Müller-Clostermann. Performance Engineering of SDL/MSC Systems. In A. Mitschele-Thiel, B. Müller-Clostermann, and R. Reed, editors, *Workshop on Performance and Time in SDL and MSC (Erlangen, Germany)*, February 1998.
14. I. Ober, B. Coulette, and A. Kerbrat. Timed SDL Simulation and Verification - Extending SDL with Timed Automata Concepts. Technical report, Telelogic Technologies Toulouse, 2000.
15. J.-L. Roux. SDL Performance Analysis with ObjectGeode. In A. Mitschele-Thiel, B. Müller-Clostermann, and R. Reed, editors, *Workshop on Performance and Time in SDL and MSC (Erlangen, Germany)*, February 1998.
16. B. Selic and J. Rumbaugh. Using Uml for Modeling Complex Real-Time Systems. Whitepaper, Rational Software Corp., March 1998.
17. T. Speakman, D. Farinacci, S. Lin, and A. Tweely. PGM Reliable Transport Protocol Specification. Internet draft, IETF, 1999.
18. B. Whetten, M. Basavaiah, S. Paul, T. Montgomery, N. Rastogi, J. Conlan, and T. Yeh. The Reliable Multicast Transport Protocol, version 2 (RMTP-II). Internet draft, IETF, 1998.
19. S. Yovine. KRONOS: A Verification Tool for Real-Time Systems. *Software Tools for Technology Transfer*, 1(1+2):123–133, December 1997.

ASN.1 Is Reaching Out!

John Larmouth

Salford University Professor
1 Blueberry Road, Bowdon, Altrincham, Cheshire WA14 3LS, England.
j.larmouth@salford.ac.uk

Abstract. This paper takes a light-hearted look at the life of ASN.1 and it relationships with other languages from the time of its birth to the year one of the third millennium – the information age.·

1 Birth and Marriages

ASN.1 [1,2,3,4,5], SDL [6], and TTCN [7,8] were all born in the 1980s, in the hey-day of Open Systems Interconnection (OSI). SDL would claim a birth-date of 1976, ASN.1 of 1984, and TTCN a year or so later. ASN.1 was conceived by Xerox putting seminal ideas into their Courier protocol, but the birth pangs accompanied X.400 [9] into the world, a twin brother, and the dominant partner in child-hood. X.400 was soon to be joined from the same parentage by X.500 [10], which quickly established itself on the worldwide stage, and rampaged through the worlds of security and e-commerce. It should have been certified at birth!

ASN.1 was a brilliant child, widely adopted and embraced by many long-dead OSI Standards, and by some that still retain major importance today.

In its youth it was wooed and wedded by SDL and TTCN, forming a comfortable liaisons with those Recommendations that have stood the test of time. Together they have formed important partnerships that have not only brought civilization to the world of protocol specification, but have (perhaps more importantly!) brought profits to many tool-vendors and "I sleep-well-a'-nights" to many implementors.

To quote from the SDL Forum Web-site: "The use of the object model notation of SDL-2000 in combination with MSC, traditional SDL state models and ASN.1 is a powerful combination that covers most aspects of systems engineering."

The combination of SDL, TTCN, and ASN.1 is powerful indeed. SDL will be well-known to everyone in this audience as providing what is probably still today the most powerful and complete means of specifying (and with SDL tools, implementing) the required behaviour of communicating systems. TTCN provides the means for the specification and implementation of test suites.

As maturity develops, both TTCN and SDL are developing new acquaintances and alliances, but ASN.1 remains a major support and platform for their activities.

This presentation will identify some of the exciting new developments in the second youth of ASN.1. The ASN.1 platform for SDL and TTCN is quite

R. Reed and J. Reed (Eds.): SDL 2001, LNCS 2078, pp. 241–249, 2001.
© Springer-Verlag Berlin Heidelberg 2001

dramatically extending its reach into legacy protocols and is developing new relationships with the fledgling XML [11] world, and perhaps soon with UML.

This in turn gives new opportunities for SDL and TTCN, and for their supporting tools. Together they can enable what were hitherto non-ASN.1 protocols (such as Bluetooth - surely to be the buzz-word of the next decade) to be easily and precisely specified, and their implementations rapidly produced and tested.

There has been a long feud between the Montagues and Capulets (character-based protocols with ad hoc BNF versus binary-based protocols with formal notations and tools), but the "middle-way" of XML-based encodings now seems an attractive proposition to many.

Will the children of the 1980s live to long old-age in 2080? Time will tell, but in human terms, they are now at their prime - mature, with plenty of real-world experience, but still young enough to adapt an develop. Viva SDL, ASN.1, and TTCN!

2   Protocol Specification through the Ages

Computer communications technology has been developing for approximately the last 1.5 billion seconds, with major advances every 150 million seconds.

We can identify a very clear stone-age era with crude syntaxes and very rudimentary tools, a bronze-age era when tools were sharpened and refined, through to the present day of advanced technology and clear insights.

There are, of course, three main thrusts to protocol development: syntax specification, procedure specification, and testing methodologies and suites. As this is a presentation on ASN.1, please excuse concentration on the former.

The earliest work (Montague's stone-age) was quite crude and simple: fields of encodings were fixed length, always present, and never repeating. Pretty pictures drawn on the walls of caves sufficed to specify all that was needed.

Capulet's stone-age was not much better: simple command lines with some sort of terminator, ASCII (a dinosaur if there ever was one) encoded, but Capulet matured to the bronze-age very rapidly through the insights of Bachus and Naur, but has not progressed much since then.

Montague progressed from simple pictures to "tabular notation", based on a "type, length, value" (TLV) approach, showing the first real signs of an awareness of the problems of "extensibility" - the need for easy interworking between deployed version 1 systems and enhanced version 2 systems developed many years later.

But it was still another 150 million seconds or so before realization dawned that a clear separation of abstract syntax specification from encoding specification provided real advantages of re-usability, and for good profits for tool vendors.

And it took another three hundred million seconds before there was realization that extensibility did **not** require a TLV style of encoding. The world could have very efficient, compact encodings, and **still** have interworking between deployed systems and new systems with minimal effort, and in a bug-free manner.

We will ignore the rather short Iron Age and the Agricultural and Industrial Revolutions and proceed to the modern era. There is now much better understanding of the problems to be solved in protocol design, and the options that are possible, with significant recent advances in understanding ways of achieving extensibility without prejudicing compact encodings.

Tools are widely available with increasing flexibility in what they can support.

3 So What Does ASN.1 Offer?

There are many well-known features of ASN.1, some of which have been present from its earliest beginnings and are well known:

- A simple notation for abstract syntax definition, independent of the complexities of encoding (transfer syntax) design.
- Full support in all encoding rules for interworking between version 1 and version 2 systems, and for vendor-specific extensions of a base protocol through the open type, information object class, and relational constraint mechanisms.
- Easy mappings with good tool support from abstract syntax definition to C, C++, Java, on a very wide range of platforms, enabling rapid development of implementations incorporating much reusable and bug-free code.
- Highly robust and mature TLV-style encodings giving plenty of support for extensibility.
- Highly compact encodings for low-band-width applications, but still with extensibility support, and available at the touch of a button from an increasing number of ASN.1 tools.
- Mature notation and tools deployed in a very wide range of industries from telecommunications through to intelligent transportation systems, parcel delivery, and news networks.
- Fully canonical encoding rules for specialised applications, such as those supporting digital signatures and certificates, used by many security-related protocols, such as SET.
- Strong links to system design and testing methodologies such as SDL and TTCN.

4 And What Are the Main ASN.1 Developments Today?

There are two pieces of important new work which are near completion as extensions to ASN.1, and which are expected to increase considerably the range of applications in which it is employed. This will not only increase the market sector for ASN.1 tools, but also for associated tools such as SDL and TTCN.

The two developments are very different, and provide very different opportunities and challenges, but interestingly the greatest benefits may come from a synergy between the two developments.

The first to be described is the Encoding Control Notation (ECN) [12]. The second is the XML Encoding Rules and other support for XML. There is also strong interest in linking ASN.1 with UML, but that work is still in its infancy, and will not be discussed further in this presentation.

With these together we can add the bullets:

- The ability to use ASN.1+ECN tools to support any protocol, whether initially defined using ASN.1 or not.
- The ability for ASN.1 tools to interact with XML tools, including standard Web browsers, to input and display ASN.1 values in a very human-readable form.

5   Encoding Control Notation (ECN)

The origins of ECN were in a dinner conversation during an ISO/ITU-T meeting in Lannion between Frank Schramm (Siemens) and Colin Wilcock (Nokia). The idea (like most good ideas) was very simple.

It had been known for some time that the ASN.1 notation could easily be used to describe the abstract syntax pretty well any protocol. Put another way, with the mapping of ASN.1 type definitions to C, C++ and Java data-structures, ASN.1 could be used to define an API for any protocol, hiding all details of "auxiliary fields" (length and choice determinants, mechanisms for extensibility, etc) and details of actual encoding.

Unfortunately, the application of the standardized encoding rules to such a specification rarely produce the bits-on-the-line that these "legacy protocols" were using.

With the development of the ASN.1 Packed Encoding Rules (PER) [13], it was often possible to get the correct encodings, but only by including these "auxiliary fields" within the ASN.1 definition. This lost most of the advantages of information hiding that ASN.1 offered, and put the onus of generating correct length and choice determinants (and other aspects of encoding) back onto the application. This was very unattractive!

The ECN work aimed to provide a notation for encoding control that could not only specify the way in which primitive types in ASN.1 would be encoded, but would also enable auxiliary fields to be added in a separate specification. These goals have been achieved.

Put simply, it is now possible to:

- Specify any existing "legacy" protocol using ASN.1, including in the ASN.1 only fields that carry application semantics, hiding all aspects of encoding detail.
- Specify separately (with considerable flexibility) the details of the encoding rules to be applied to produce any desired form of encoding.

5.1   Application to Bluetooth

ECN has been applied in the complete definition of the Bluetooth Service Discovery Protocol [14], and in the partial definition of a number of other protocols such as Hiperlan and Tetra. Here we concentrate on the Bluetooth Service Discovery protocol.

Service Discovery is at the heart of Bluetooth, enabling a mobile device moving into a new geographic environment to discover the existence of other Bluetooth devices in that environment. It is key to the fully automatic configuration that is needed for communication by domestic devices.

The protocol has many similarities with features of ASN.1 BER and PER, but of course uses totally different encodings at the detailed level. Many messages are encoded in a fixed-length PER-style, but contain an "attribute id" and "attribute value" field that corresponds very closely to an ASN.1 "open-type".

Bluetooth uses a tabular notation to define attribute ids and types, but also provides for vendor-specific additions of new types using constructors in much the same way as ASN.1 types are defined. For these types, Bluetooth uses a BER-like TLV structure, but of course with different encodings.

It has similar concepts of universal class tags for the "T" part (but without user over-ride of tag values), and of a number of built-in primitive types and constructors.

Encoding rules have been written using ECN that can be applied to **any** ASN.1 type definition that can be written using the Bluetooth primitive types and constructors. This provides facilities that enable tool-support for encoding Bluetooth application, and means that the Bluetooth Service Discovery Protocol can now be regarded as an ASN.1-based protocol, with important implications for tools that support ASN.1.

For ECN to be useful, it is not necessary for standardised protocols to be re-defined using ASN.1+ECN. The necessary ASN.1+ECN specifications can be provided by ASN.1 tool vendors to (for example, Bluetooth) implementors, with the correctness of the ASN.1 and ECN warranted and supported by the tool-vendor in the same way as correct encodings are normally warranted and supported by the tool. This enables implementation teams to use a single familiar tool for all protocols.

5.2   Examples of ECN

Examples of two pieces of ECN used in the specification of the Bluetooth data element sequence are:

```
bluetooth-tag-encoding #TAG ::=
        {ENCODING SPACE SIZE 8
        EXHIBITS HANDLE "Bluetooth tag" AT {0..7} }

length-delimited-repetition
        {< REFERENCE >} #REPETITION ::=
                {ENCODING
                        {REPETITION-SPACE
                        SIZE variable-with-determinant
                        MULTIPLE OF octet
                        USING length } }
```

5.3 Status of ECN

The base standard is ITU-T Rec. X.692 | ISO/IEC 8825-3. It has completed its ISO FCD ballot, and final text is now ready for ITU-T Last Call. It is expected to undergo only editorial changes, and to be finally approved before the end of this year.

There is also an amendment giving extensive support for a range of extensibility mechanisms. The FPDAM ballot for this is underway in ISO, and final approval is also expected before the end of the year.

Syntax checkers are available, and a more complete tool is expected to become available during the summer.

6 XML Support

There are several pieces of XML work planned, but the most mature, and arguably the most important, is the provision of ASN.1 XML Value Notation, and XML Encoding Rules (XER).

The addition of XML Value Notation means that within any ASN.1 module, in test specification, or as input and output to ASN.1 tools a new value notation can be used which is XML mark-up.

The tags used are based on the identifiers in ASN.1 SEQUENCES and SETS, and on type reference names where necessary. An example is:

6.1 An Example of ASN.1 XML Value Notation

With the ASN.1 type:

```
Invoice ::= SEQUENCE
        {number            INTEGER,
        name               UTF8String,
        details            SEQUENCE OF
                SEQUENCE {
                        part-no     INTEGER,
                        quantity    INTEGER},
        charge             REAL,
        authenticator      BIT-STRING}
```

we can have the value:

this-invoice ::=
        <Invoice>
                <number>32950</number>
                <name>Funny name with <</name>
                <details>
                        <part-no>296</part-no>
                        <quantity>2</quantity>
                        <part-no>4793</part-no>
                </details>
                <charge>397.65</charge>
                <authenticator form=hex>
                        EFF8 E976 5403 629F
                </authenticator>
        </Invoice>

For the XML Encoding Rules, it is a UTF8 encoding of a value of the outer-level type that is transmitted down-the-line, with a short XML preamble and postlude.

6.2   Advantages of the XML Value Notation

This new ASN.1 facility provides:

- The ability to send value of ASN.1 PDUs to browsers for easy monitoring or display of received messages.
- The ability to generate ASN.1 PDUs (probably for conversion by an ASN.1 tool to PER encodings before transmission) from an XML tool.
- A more readily understandable (although more verbose) value notation for input and output of values to and from ASN.1 tools, for monitoring and testing, and for the specification of test suites.
- A human-readable text-based (but verbose) encoding format suited to browsers, which can be easily converted to or from a PER encoding by ASN.1 tools.

This is quite powerful support in itself, but combined with ECN provision, it enables, for example, the input and display of Bluetooth Service Discovery Protocol messages (for testing and monitoring purposes) using standard XML tools. The whole can be much greater than the two parts.

There is further XML work planned in the area of mapping XML schema languages and ASN.1 schemas (type definitions) but the precise form of this (nor the precise user demand) is not yet clear. This work will probably begin to mature during 2002, and is not discussed further.

6.3   Status of the XML Work

The work involves two new amendments (one to ITU-T Rec. X.680 | ISO 8824-1 and one to ITU-T Rec. X.681) to support the new value notation, and one new (quite small) standard (ITU-T Rec. X.693 | ISO 8825-4) to define use of the XML Value Notation to provide the XML Encoding Rules (XER).

These are all complete and under FPDAM and FCD ballot, and are expected to be finally approved in late 2001. ASN.1 tools supporting the new value notation and XER are expected to become available during this summer.

7   Conclusion

ASN.1 has come a long way from the ten page X.409 produced in 1984, and is showing an ability to adapt to changing user requirements which should hopefully give it a long life yet, and reinforce its suitability as a base platform for communications protocol specification.

The recent merger of the main standards bodies involved in ASN.1 work (ITU-T SG7) and in SDL and TTCN work (ITU-T SG10) to form the new (ITU-T SG17) should enable integrated support for these new features across the complete spectrum of protocol specification, system definition, and testing, both at the standards level and within tools, to the benefit of both standardisers and implementors.

We have finally reached the Information Age! (April 2001).

References

1. Recommendation X.409 - part of the X.400 series (1984) - withdrawn when X.208/X.209 published, ITU Geneva 1985.
2. [Recommendation X.208 (11/88) | ISO/IEC 8824], *Specification of Abstract Syntax Notation One (ASN.1)*, ITU Geneva 1989 - obsolete.
3. [Recommendation X.209 (11/88) | ISO/IEC 8825], *Specification of Basic Encoding Rules for Abstract Syntax Notation One (ASN.1)*, ITU Geneva 1989 - obsolete.
4. [Recommendation X.680 (12/97) series | ISO/IEC 8824], *Information technology - Abstract Syntax Notation One (ASN.1): Specification of basic notation (X.680 | 8824-1); Information object specification (X.681 | 8824-2); Constraint specification (X.682 | 8824 - 3); Parameterization of ASN.1 specifications | 8824-4)*, ITU Geneva 1998.
5. [Recommendation X.690 (12/97) series | ISO/IEC 8825], *Information technology - ASN.1 encoding rules:*
   Specification of Basic Encoding Rules (BER),
   Canonical Encoding Rules (CER) and
   Distinguished Encoding Rules (DER) (X.690 | 8825-1);
   Specification of Packed Encoding Rules (PER) (X.691 | 8825-2);
   Encoding Control Notation (ECN) (X.692 | 8825 - 3);
   Parameterization of ASN.1 specifications (X.693 | 8824 - 4),
   ITU Geneva 1998 (except ECN - to be published).
6. Recommendation Z.100 (11/99), Specification and Description Language, ITU Geneva 2000.
7. [Recommendation X.292 (09/98) | ISO/IEC 9646-3], *Data Communication Networks: OSI conformance testing methodology and framework: Tree and Tabular Combined Notation (TTCN)*, ITU Geneva 1999.

8. [Recommendation Z.140 series | ETSI ES (see [15])
9. [ITU-T Recommendation X.400 | ISO/IEC 10021-1], Message handling Systems: Message handling services - Message handling system and service overview, Geneva 1999.
10. [ITU-T Recommendation X.500 | ISO/IEC 10021-1], Information technology - Open Systems Interconnection - The Directory: Overview of concepts, models and services, Geneva 1997.
11. http://xml.coverpages.org/xml.html
12. see [Recommendation X.692 | ISO/IEC 8825-3] in [5].
13. see [Recommendation X.681 | ISO/IEC 8825-2] in [5].
14. http://www.bluetooth.com/developer/specification/ Bluetooth_11_Specifications_Book.pdf
15. A. Wiles, *ETSI Testing Activities and the Use of TTCN-3*, this voluume.

Distributed Systems:
From Models to Components

Fabrice Dubois[1], Marc Born[2], Harald Böhme[3], Joachim Fischer[3],
Eckardt Holz[3], Olaf Kath[3], Bertram Neubauer[3], and Frank Stoinski[3]

[1] France Telecom R&D, 2, Avenue Pierre Marzin
22300 Lannion, France
`fabrice.dubois@rd.francetelecom.fr`
[2] GMD FOKUS, Kaiserin-Augusta-Allee 31,
D-10589 Berlin, Germany
`born@fokus.gmd.de`
[3] Humboldt-Universität zu Berlin, Dept. of Computer Science,
Rudower Chaussee 25, D-12489 Berlin, Germany
`{boehme|fischer|holz|kath|neubauer|stoinski}@informatik.hu-berlin.de`

Abstract. Advanced design methods are needed to fulfill the increasing requirements of telecommunication service development. For a design method the relevant concepts for the application domain have to be defined, a supporting notation has to be declared and finally rules have to be developed to map design models to supporting runtime environments. The ITU-T has followed this route by defining concepts for the design of distributed telecommunication applications and supporting notations for these concepts. In the past, the ITU-T has defined several languages and notations to support structural and behavioral descriptions of distributed telecommunication systems, namely ODL, SDL-2000 and MSC-2000. With the rise of the component age, an additional technique (DCL) is under development that enables component based manufacturing of distributed systems. Beside these languages, the ITU-T recognized the common need for open, component aware object middleware platform standards as the runtime environment for these systems. This contribution is about integration.

1 Motivation and Introduction

A dedicated and efficient design methodology contributes significantly to a reduction of the time to market distributed applications and telecommunication services. An appropriate treatment of all kinds of communication aspects lies in the very nature of the targeted application domain. These aspects span from functional requirements on object interactions over quality-of-service issues to security properties. Taking into account the broad acceptance of object middleware technology, middleware platforms provide an ideal implementation environment for such designs.

Conceptually, an appropriate design method may be split into three separate parts:

R. Reed and J. Reed (Eds.): SDL 2001, LNCS 2078, pp. 250–267, 2001.
© Springer-Verlag Berlin Heidelberg 2001

- A concept space, that contains all relevant entities that conceptually reflect the elements of the problem domain and the information for the description of these elements and their relations;
- a concrete notation to visually specify models using the elements of the concept space;
- and a set of rules to specify the mapping of models onto middleware technologies.

Following this approach, a concept space is independent of a specific design notation. Design models can be developed in different notations but are based on the same concepts. Design information can then be exchanged on the basis of the common concept space. Second, both the notation and the concept space are independent of a specific runtime-environment. The same design can be mapped onto different environments. This enables a high flexibility and is also important for the aspect of re-use of component design models.

In the past, the ITU-languages ODL [1] and SDL [3] as well as the proposed language DCL [19] represented different aspects within the design of distributed systems. No common concept space for them has been defined, an integration is only possible by pair-wise direct mappings based on notational concepts ([12,13]). This is also valid for the relation of ITU languages to languages of other communities, for instance between ODL and OMG-Component IDL or SDL/MSC and UML [5].

This paper discusses the concepts of and the relations between selected ITU-languages and the current trends for their future development and refinement taking into account corresponding OMG activities. In order to apply different techniques with their specific advantages in a combined approach for the design of distributed systems, the underlying concepts have to be integrated within a common concept space. This can be achieved in 3 main steps:

- the identification of those concepts of each technique, that are relevant to a combined approach;
- the definition of the semantics for a combined concept space;
- the realization of the integration within a common meta-model framework by applying a suitable technology.

Consequently, major parts of the components that form a distributed system will be generated from design models that are based on the common meta-model and target specific object middleware platforms. Therefore, an example of such a platform will be presented here as well.

2 Modelling Distributed Systems

2.1 ODL for Modelling of Software Components

Starting from the basic reference model of Open Distributed Processing (ODP), the TINA community has taken up this general approach to define dedicated computational modelling concepts and a supporting notation Object Definition

Language (ODL) that was standardized later by ITU. As a result ITU-ODL offers necessary key concepts necessary to design distributed telecommunication applications as Computational Objects (COs) communicating via multiple well defined interfaces. Interfaces are described in terms of signatures defining the elements for potential interactions a CO instance may participate in. ODL distinguishes between operational and stream interactions. Because ODL was designed to be technology independent, different language mappings to technology dependent modelling or implementation languages have to be defined for a complete model based generation of software components. The design process should also include the distribution, installation, and configuration of produced binary components, taking into account concrete distributed processing environments. Some of the mappings (Fig. 1) will be discussed in the following.

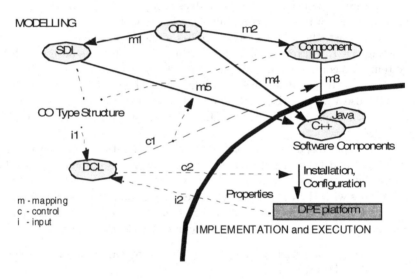

Fig. 1. Mapping and code generation scheme for ODL

Since ODL describes only structures and signatures of the application, the traditional way is to proceed directly to the implementation. This requires language mappings from ODL into the implementation language, reflecting the specified structure and signature information. Such a language mapping to C++ has been developed and is part of Z.130 (m4).

ODL does not support the behaviour specification and the binding functionality. It does also not provide mechanisms for type instantiations and the configuration of systems based on components. Furthermore, signal communications cannot be directly described. Moreover, practical experiences have shown that the support of stream flow types in ODL is not adequate. The concept of flow types should be substituted by the usage of standardized formats as MP3, MPEGII or others.

2.2   SDL for Object Behaviour and Object Configuration Description

In order to allow validation of an application's behaviour before its implementation and to perform automated code generation and testing, a suitable description of the computational behaviour for the involved COs is needed. This description should be an abstract one since in some cases only the externally visible behaviour should be specified, without prescribing any implementation details. If the behaviour of the COs can be expressed by state machines, SDL is a suitable language to do that, otherwise different languages should be selected. With SDL-2000, the ODL-to-SDL mapping (**m1**) can be defined in a much more structural equivalent way than with SDL-92/96 in the current Z.130.

We concentrate here on those aspects that differ from the mapping defined in Z.130.

- The package and package nesting concepts of SDL-2000 now provide a direct way to reflect ODL Name scopes (modules). The same concepts are applied to map definitions local to an interface definition in ODL: a special package definition in SDL contains all entities defined locally to the ODL interface definition and is defined in the same scope as the interface and with the same name as the interface.
- Data types of ODL are be mapped to SDL data types as follows:
  1. Simple types are mapped onto specialization of simple SDL data types;
  2. Constructed data types are defined as SDL structures or choices;
  3. Sequences are mapped to SDL vector types or powersets;
  4. Exceptions are directly mapped to SDL exception definitions. The names of the exception members are not relevant in SDL and are omitted;
- Operational interface definitions in ODL are mapped to SDL interface definitions. Operations are mapped to remote procedures, exceptions can be declared as raised by the remote procedure. One-way operations are defined as signals at the SDL interface. Attributes are mapped to a pair of get- and set-remote procedures, where the set-remote procedure is only present, if the attribute is not declared as being read only. Additionally, in SDL-2000 interfaces imply the definition of a related Pid type that is used for interface references.
- As capsules for state and behaviour, CO definitions in ODL are mapped to SDL block types. The block type definition gets three gate definitions: one is named *initial* and is defined with the initial interface, the second is named *supported* and is defined with all supported interfaces of the CO and the third is named *required* and is defined with all required interfaces of the CO. The block type definition contains a state machine which is bound to the gate providing the initial interface of the CO.

This skeleton of an SDL-specification can now be enriched with the specification of behaviour (SDL state machines) and specifications of further elements of the application (artifacts). The completed specification serves as input for the verification/validation as well as for the code generation. An important factor for the application of the ODL-SDL combination is the availability of tools. In the past tools for the mappings **m1** and **m5** have been developed and applied in different projects.

2.3   Component IDL for Describing Potential Object Collaborations

ODL was defined before the OMG standardization activities for an architecture and description technique for components with multiple interfaces (Component IDL, [7]) reached their final stage. As a result Component IDL allows in addition to ODL:

- the enabling of behaviour contracts for a collaboration of CO type[1] instances based on supported and required interface declarations;
- a mechanism to manage the CO collaboration life cycle (finding/creation/configuration/binding);
- a navigation mechanism to detect and access required functionalities of a CO instance (interface access).

Following the traditional CORBA approach [10], a mapping (**m3**) from Component IDL to programming languages is provided. However, this mapping is not given directly, instead Component IDL is mapped onto CORBA IDL where multiple language bindings are available. In general a mapping from ODL to Component IDL is straight-forward because the Component IDL concepts are refinements of ODL concepts. The major missing features in Component IDL concern the specification of streams and stream interfaces. It seems to be interesting to investigate whether a Component IDL extended by these concepts could be a successor version of ODL.

2.4   DCL for Distribution and Configuration
of Software Components

In the last few years, information technology has been moving from centralized homogeneous computer systems to more flexible and efficient distributed networks. Binary software components, which are distributed across the network, interact to fulfil the system's need. The languages introduced until now support the development and implementation of CO types. CO types reside in software components and are incarnated as COs during their execution. However, up to the time of writing, there is no tool-supported notation providing comprehensive support for the description of the deployment and configuration of software components. There is still a lot of manual and cost-intensive work needed to deploy, install and maintain distributed applications at the customer's target environment.

The development of DCL (Deployment and Configuration Language) has been started in order to bridge the gap between the development phase of a

---

[1] Note, the OMG term component does not directly correspond to the term (software) component used here, but is reflected by the term CO type.

distributed system and its deployment to a customer's target environment. It will facilitate the automation of the deployment process of distributed applications and help service developers and providers to reduce costs.

DCL specifications are independent of a specific notation used during the system development phase. However, the CO type structure as a result of this phase serves as input for DCL specifications (**i1**). DCL fulfils the following purposes:

- Specification of the partitioning of the developed CO types into binary software components (Fig. 2). This specification is required to control the code-generation from computational specifications (**c1**). Code generation produces binary software components executable on suitable computing nodes.
- Specification of properties of CO types and binary software components and the requirements they have on the distribution and the target environment. Examples of such requirements are the need for a certain amount of memory on a target node or the co-location requirement for certain CO types.
- Specification of the initial COs of a distributed software system and their initial connections. This specification is needed to establish the initial configuration of a distributed system automatically.
- Specification of a concrete target environment (DPE) in terms of the available capabilities (computing nodes, network links) and their properties (**i2**). This specification is needed to perform the assignment of software components to target nodes, their installation and configuration in an automated way (**c2**).

It is expected that the standardization process in the ITU-T on DCL will lead to a suitable notation as well as to an exchange format for DCL-specifications. This is needed since DCL serves as a bridge between distributed systems development and the deployment and usage of such systems, which normally takes place in different domains with a heterogeneous infrastructure.

2.5   Language Mapping Problems

The current approach to combine different techniques for the development or design reveals a series of problems. First, not all concepts of ODL have a directly corresponding concept in the other techniques. The most prominent examples here are streams and stream interfaces. The only possible way of mapping is to map these interfaces to implicit operational interfaces for the management of the stream flows, which are declared by the stream interfaces. A generation of concrete operations (and their signatures) can only be realized for a concrete platform which itself has to be standardized before.

A much harder problem concerns the inconsistency of some concepts of the different domains and their mappings. Especially, it cannot be expected that the mapping **m4** is equivalent to the concatenation of mapping **m2|m3** or **m1|m5**. Many of the mappings are described informally, independent of each other and often in a tool-dependent way. For developments that explicitly require the application of different design and code generation techniques therefore an additional manual adaptation of the artifacts of the development process is necessary.

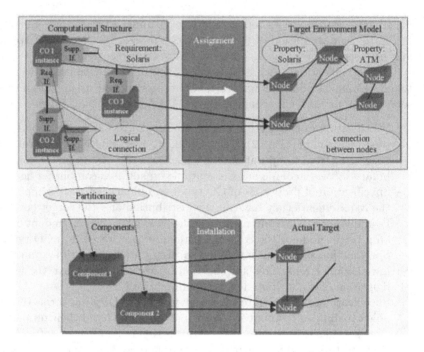

Fig. 2. Explicit distribution configuration

Two principle ways are open to overcome these problems:

- Complete redefinition of all these languages based on a common semantic formalism; or
- the definition of a common conceptual framework in combination with rules mapping between this framework and concrete techniques.

The first one is obviously irrelevant. The latter alternative will not imply modifications to existing languages and promises to improve the consistency of the transformations between concrete models.

3   Common Concept Space

The starting point for the definition of the common concept space is the basic reference model of ODP and its telecommunications domain specific application within the Computational Modelling Concepts as defined in [11].

3.1   Structural Concepts

The central terms here are *interface*, *object* and *component*. An *object* is defined in [2] as a model of an entity, that encapsulates state and behaviour and is distinct from any other *object*. To make the term *object* clearer, a distinction should be made between the type of an object (in the sense of an *object template* in [2]) and an *object* itself as instance of such a type. Because we focus here on the functional decomposition of a distributed system, we specialize the term *object type* by referring to it as *computational object type* (*CO type*). We use the term CO to denote instances of those types. COs define units of *distribution*, which interact with their environment (that is other computational objects) via well defined interfaces. Interactions between COs are transparently supported by a distributed infrastructure.

In contrast to CO types the term *component* is used to refer to units of *deployment* in a distributed system. Components contain implementation artifacts (such as the classes of an object-oriented programming language), that realize the behaviour of one or more CO types. Consequently, there is an association between one or multiple CO types as the unit of distribution and a component as the unit of deployment. Besides containing CO type implementations, a component provides additional interfaces and accompanying implementations for the component life cycle management. A component can be deployed onto nodes, but there needs a runtime environment. Nodes are the processing entities within the target distributed environment. The definition of a component here corresponds directly to the definition in [17]. Since ODP does not deal with the concept of deployment in detail, there is no directly corresponding entity to a component as part of the ODP viewpoint languages. However, an analogy can be seen to the definition of a basic engineering object (BEO), which needs support from a distributed infrastructure in the same way a component does, and to the definition of a cluster (grouping of BEOs). However, the requirement of ODP, to have an 1:n relation between COs and BEOs, is a contradiction to the component definition. In our concept space, there is a n:m relation, meaning that different components may contain the implementations for a CO type and that one component may contain implementations of many CO types. Practical experiences have shown, that such a n:m relation as introduced here is more realistic.

In the remaining part of this contribution, we mainly focus on the specification of CO types and CO type behaviour. According to [2], a CO communicates with its environment (that is, other COs) at its interaction points. Those interaction points may be uniquely referenced. From the computational perspective, interactions are classified into the following three kinds:

- operational interactions relate to remote method invocations (RPC style);
- signal interactions refer to asynchronous sending and reception of atomic information (information publishing style);
- stream interactions are a continuous sending/reception of information (continuous media delivery style).

An *interface* (type) defines the signature of an interaction point. In contrast to the definition of a computational interface given in [2], a single interface type here allows to combine all three interaction kinds. By doing so, the design of the computational entities of a distributed system turned out to be more simple and intuitive. As will be shown later, the separation of the interaction kinds is a technology issue and therefore an item for the mapping onto specific middleware platforms. There are two different types of relations between CO types and interface types:

- *supports* relation: a CO may provide instances of an interface supported by it;
- *requires* relation: a CO makes use of an instance of an interface required by it.

With these relations, we do not refer to configuration aspects, which deal with COs and interfaces. Instead we constrain configuration definitions to be defined only between CO types and interface types for which a corresponding *supports* and *requires* relation is specified. The signature of an interface type is defined as the set of its interaction elements. Interaction elements are distinguished with respect to the interaction kind they belong to:

- Operations and attributes for operational interactions;
- Consumed and produced signals for signal interaction;
- Sourced or received media sets for stream interaction.

Interaction elements in turn are also defined by signatures. An operation signature consists of the operation name and a set of parameters, each of them having a type and a direction specification (*in, out, inout*). Furthermore, operations may specify terminations in the form of return types or exceptions. The difference to other existing interface definition languages is the data type system to be applied. Although a design method must be open to different type systems, it has to be concrete with respect to data types to allow for an unambigous mapping onto specific middleware environments.

Concerning the signature of signals, we distinguish between the concept signal and the information the signal carries. The type of the information is described in terms of values. A signal declares one or more values as being carried by it.

A signature of a media set is given by an aggregation of media, where each medium is interpreted according to one or more appropriate media types. A medium here refers to the continuous provision of information, were the information is formatted in conformance to one of its realizing media types. Given that, a medium characterizes the information delivered, while a media type characterizes the format of the information delivery. Commonly known media types are MIME types [20].

The concepts introduced so far are pure type information, namely signatures and potential structures and therefore are referred to as structural concepts. Consequently a specification of an application given in these terms forms the structural view.

3.2   Configuration Concepts

Besides these structural definitions, we introduce also concepts to describe configuration aspects of CO types. The term configuration here refers to mechanisms that allow access to instances of supported interfaces also allow the storing of references to required interfaces at a CO instance. The concept *port* is used to denote both, the access points and the points to store interface references. Ports are uniquely identifiable in the context of a CO type. Since there can be potentially an infinite number of interface instances supported by a concrete CO at runtime, ports can be declared as being single or multiple ports. The property **single** implies that only a one interface reference at a time can be registered or obtained at that port, whereas a port with the property **multiple** allows dynamic registration and multiple interface references to be obtained. The specification of the configuration of ports belonging to a CO type forms the configuration view. Currently, only a single configuration is foreseen per CO type. This definition completes the view on a CO type in addition to the structural view. It requires further study, whether it is feasible and practical to allow the definition of more of these configurations. In particular, we currently do not see an application case where multiple configurations are necessary.

3.3   Interaction Concepts

Special concepts are required to allow the specification of properties, rules for and constraints on interactions in certain contexts. The main concept to support this is the *binding*. Bindings are associations between instances of interfaces supported by COs. A binding is a prerequisite for an interaction: interactions between COs may occur via the bound interface instances only. A binding always defines a (common) subset of the signatures of the interfaces involved in the binding and by doing so it specifies the interaction elements that may be used for interactions in the context of the binding (binding context).

Bindings can be established implicitly or explicitly. In order to establish a binding, it is required that the subsets of the signatures of the interfaces involved in the binding are complementary to each other. Rules for a definition of the conditions under which signatures are complementary can be found in [2]. An example is the requirement that the interface types defining the signatures are in a subtype relation.

For both kinds of bindings, rules can be specified to determine the characteristics of the interactions in the binding context. The set of binding rules for the server and client side of a specific binding is referred to as the binding contract for that binding. Binding rules themselves are defined as constraints (logical expressions) over special types. We call these types Quality of Service (QoS) types. Their attributes may be assigned values according to desired QoS characteristics or policies. Examples for these attributes include security levels, bandwidth, response time or transaction policies. In order to restrict the QoS types available for the specification of a binding rule, the types must be associated with the interface types involved in the binding.

The concept of QoS types declared for specific interface types is known from other QoS specification approaches as [21]. However, the approach to define rules for bindings that are possibly different in different contexts is more dynamic. Predicates are used to identify a binding case at runtime. Instances implementing

CO types are checked whether or not they fulfil predicates attached to binding cases. Taking the results of this check, a CO instance wanting to participate in a binding selects an appropriate binding case and therefore the binding contract to be used at runtime.

A specification given in terms of policies and rules on interactions and binding of interfaces is called an interaction view. Together with the other two views (structural and configuration) it provides a sufficient set of concepts to form a black box model of the system by concentrating on communication aspects only.

3.4   Implementation Concepts

As described in Sect. 1, the design process should lead to software components that implement CO types. Therefore internal aspects regarding the implementation of CO types also have to be covered. Similar to the RM-ODP [2], our design approach defines an object as encapsulation of state and behaviour, but until now it does not address how this behaviour is provided. While the term CO type refers to an abstract entity, in concrete distributed systems the expected behaviour of objects is realized by programming language elements, such as classes in a programming language. For that reason, a design method must consider the relationships between abstract, referable objects andthe concrete implementation language code, which implements the behaviour of such objects. Questions of interests here are:

- What are the structural elements implementing a COs behaviour?
- What information can be considered as being a COs state?
- What is the relation between implementation elements and the interaction elements at the interfaces, and which element is responsible for implementing a particular operation?
- What part of the state information is needed to provide the behaviour of a certain interaction element?

To answer these questions, additional concepts are introduced. The programming language elements realizing parts of the behaviour of CO types are called artifacts. A CO type is implemented by a set of those artifacts. Interaction elements of the supported and required interfaces are associated with artifacts implying that the artifact implements those elements: the aritfact consumes a signal or provides behaviour of an operation. State information is described by storage types. Storage types have attributes whose values form the current state. Such storage types can be assigned to the association between an interaction element and an artifact. In this way it is expressed, that the part of a COs state which is covered by the storage type is required for the implementation of the selected interaction element. In our common concept space the implementation view can be considered as the artifacts, their relations to CO types and interaction elements as well as storage types, and their relation to realizations of interaction elements.

The concepts we presented here are to be understood as either refinements of ODP computational language concepts (CO type and CO, interface or the different interaction elements) or extensions to the ODP computational concept space, like the implementation concepts.

3.5 Deployment Concepts

Since ODP itself does not provide any explicit concepts relating a CO type to the concrete realization of its behaviour, ODP also does not provide sufficient means for the description of a collection of CO type implementations together with their life cycle management: there is no means to describe the term component as unit of deployment. However, components obviously have static and dynamic requirements upon the execution environment that hosts them. The definition of such requirements is the main concern of the additional view, the deployment view. There is an ongoing international project performed in the EURESCOM programme which deals with the definition of deployment aspects for components ([19,18]).

Our common concept space does not cover the other ODP viewpoints. The ODP enterprise viewpoint focuses on the requirement analysis for the system to be developed. This is within the software engineering process prior to the design, even if the development process is an iterative approach. The information viewpoint deals with the structure of information to be manipulated by the system to be developed. Further investigation is needed whether or not – and how – to cover these aspects within the common concept space. Currently only the reflection of information entities in signatures of the structural view and the state descriptions in the implementation view are included.

4   The MOF Approach

The extensive application of models and modelling techniques within the scope of the development and use of CORBA-systems lead to the requirement for a unique and standardized framework for the management, manipulation and exchange of models and meta-models. This need has been addressed by the OMG with the Meta-Object-Facility (MOF, [4]). The architecture of MOF is based on the traditional four-layer approach to meta-modelling:

- *Layer M0 – the instances*: information (data) that describes a concrete system at a fixed point in time. This layer consists of instances of elements of the M1-layer.
- *Layer M1- the model*: definition of the structure and behaviour of a system using a well defined set of general concepts. An M1-model consists of M2-layer instances.
- *Layer M2 – the meta-model*: The definition of the elements and the structure of a concept space (that is, the modelling language). An M2-layer model consists of instances of the M3-layer.
- *Layer M3 – the meta-meta-model*: The definition of the elements and the structure for the description of a meta-model.

Although further levels seem to be necessary (for example to define of the elements to specify the M3-layer), MOF as well as other meta-modelling frameworks stops at M3. This is due to the application of a so-called hard-wired meta-meta-model: Elements and structure of the M3-layer are directly derived from a well-known formalism, which in the case of MOF is object-orientation. In fact, the MOF concepts are defined using MOF itself. The MOF-model consists of the following concepts for the definition of meta-models:

- *Classes*: Classes are first-class modelling constructs. Instances of classes (at M1-layer) have identity, state and behaviour. The structural features of classes are attributes, operations and references. Classes can be organized in a specialization/generalization hierarchy.
- *Associations*: Associations reflect binary relationships between classes. Instances of associations at the M1-layer are links between class instances and do neither have state nor identity. Properties of association ends may be used to specify the name, the multiplicity or the type of the association end. MOF distinguishes between aggregate (composite) and non-aggregate associations.
- *Data types*: Data types are used to specify types whose values have no identity. Currently MOF comprises the CORBA data types and IDL interface types.
- *Packages*: The purpose of packages is to organize (modularize, partition and package) meta-models. Applicable mechanisms here are generalization, nesting, import and clustering.

The MOF specification defines these concepts as well as supporting concepts in detail. Because there is no explicit notation for MOF, the UML [8] notation has been deliberately used to visualize selected concepts. The interface to the conceptual framework of MOF is formally defined in the CORBA-IDL modules **Model** (meta-model specific interfaces) and **Reflective** (generic interfaces). All interfaces in **Model** directly or indirectly inherit from **Reflective** interfaces. The MOF interfaces allow:

- stepwise creation of a new meta-model in the MOF by creating new objects;
- modification of an existing meta-model in the MOF;
- extraction of information from a meta-model using query- and traversal functions;
- a request to be made to validate the meta-model.

In addition, functions are defined to produce an external representation of a meta-model (externalize) or to create a meta-model in MOF from such an external representation (internalize). Currently two specific mappings from MOF to external formats are defined:

- *MOF-IDL-mapping*: This mapping generates the IDL-specification for a meta-data service from a MOF-meta-model specification. This service (a repository) can be used to store or manipulate models (M1-layer), that conform to the meta-model. An example for such a service is the UML CORBA facility generated from the UML-meta-model, but also the MOF itself (generated from the MOF-meta-model).

- *XMI (XML based model interchange [6])*: This mapping defines rules to derive an XML Document Type Definition (DTD) [16] from a meta-model in MOF and to externalize meta-model data (i.e. a model) from MOF into an XML document structured according to that DTD and vice versa.

The advantage of the MOF-approach is to make the definition of (meta)-models independent of the concrete application domain of the models and to provide a concise and unique set of concepts for the definition of meta-models. Moreover, multiple meta-models can be managed by the MOF. Further investigations have to show how relations between meta-models can be utilized as a basis for a transformation of models based on these meta-models.

The integration of MOF into CORBA enables applications to access meta-information about application objects during runtime (using the repository interface). The XMI as well as the proprietary externalize/internalize functions enable an application of MOF and MOF-based repositories in a scope wider than CORBA.

5   Towards a Common Concept Space Based Integration of ITU and OMG Techniques

Starting from the informal definition of the common concept space, we have applied the concepts defined by MOF to specify a meta-model that captures the concepts of our concept space. The meta-model consists of a set of packages, one for each of the views defined in Sect. 3. According to the current application domain of the concept space [18], we chose to base all definitions on the meta-model for CORBA as defined in [9]. The complete meta-model is stored in the Meta Object Facility.

This approach allows us to apply the MOF functionality. A consequent continuation of these principles is to construct such models from concrete specifications present in one of the techniques discussed in Sect. 2. Currently, we have realized prototype front ends for ITU-ODL, Component IDL, CORBA IDL and a subset of UML. These prototypes have shown that the concepts of the concept space are powerful enough to serve as a bridge between these techniques. Further investigations now concentrate on the definition of a rule set to cover behavioural aspects (in SDL-2000) and the deployment aspects (in DCL). On the other hand, back end tools can be developed that transform models into skeletons of concrete software components, possibly based on a specific middleware technology. A set of mapping rules has already been defined targeting the environment presented in Sect. 6. The main advantage of this approach is that – instead of mappings between different pairs of techniques as presented in Sect. 2 – for each technique only a mapping from and to the common concept space is needed. Relations between concrete elements of each of the techniques are implied by mappings between the techniques and the common concept space. As a result, different techniques can be applied in combination to manipulate the same model.

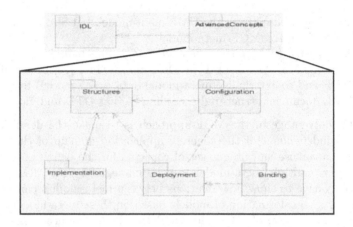

Fig. 3. Common Concept Space as Meta-Model

6   The HU-NTT Platform

The goal of the HU-NTT platform is to provide a flexible solution for content providers to:

- describe, store and catalogue binary media information;
- assemble the stored them to high quality multimedia information content;
- deliver this content to consumers with respect to static and dynamic network conditions and with respect to the facilities of the used terminal at the consumer side.

This CORBA-based platform was developed in a joint project between HU Berlin and NTT R&D Tokyo ([14,15]). For the verification of the applicability of the model driven design technique for software components the concepts (especially operational and stream oriented interactions) of our concept space have been mapped to this middleware platform.

6.1   Mappings of Concepts

The concept space allows all three interaction kinds (operational, signal, stream) to be part of an interface type. For the mapping onto a certain middleware platform, a separation of the interaction kinds is needed since they are mapped onto different and independent technology concepts.

1. **Operational interactions** are directly be mapped to the CORBA IDL concept of an operation. Operational ports are mapped to the CORBA server resp. client concept. The **binding** of operational ports since this is automatically established by the ORB.
2. To map the **signal concept**, oneway operations as defined by the CORBA standard are not sufficient. When using this approach it would be very difficult to apply multiplicity constraints on the signal ports of an interface type.

As signals are a concept that is based upon an asynchronous communication style, a similar concept in the CORBA domain is the Event Service resp. Notification Service. The Event Service defines a programatic model enabling suppliers and consumers of events to register at a special entity, called the event channel. The event channel handles multiplicity in a way that all suppliers of events of a certain event channel can send events to the channel and, depending on the implementation of the channel, a single consumer, some consumers or all consumers receive this event from the channel.

To find a mapping of signal interaction onto the CORBA concept of an event channel, first an appropriate structural concept in the CORBA domain has to be found to map signal types. An intuitive mapping uses CORBA IDL value types. Value types share the characteristics of a CORBA interface by supporting an interface for operational interaction and the characteristics of CORBA data structures by being able to be marshalled into a byte stream. Each signal type in a model defined using the concept space is mapped to a CORBA value type. **Signal producing ports** are mapped to event supplier interface types and **signal consuming ports** are mapped to event consumer interface types that are standardized by the event channel specification.

The **binding** of two signal interaction points at runtime is realized by the creation of a new event channel and the registration of the signal port implementations at that event channel. Instances of a CORBA value type that represent the signal type from the concept space are sent to the event channel by the CORBA event supplier and received through the channel by the CORBA event consumer.

3. The mapping of the **stream concept** onto a CORBA platform requires additional services built on top of the CORBA platform. The **interaction endpoints** for stream data at source and sink are realized by the mechanisms of associated network flow and stream flow endpoints. **Network flow endpoints** are specific to a certain network and network protocol type. They are created at runtime by factories (Network Devices), which in turn are also network and network protocol type specific. Network flow endpoints are able to connect to a compatible network flow endpoint using network and network type specific configuration parameters. These configuration parameters are negotiated at runtime between the network flow endpoints.

Stream flow endpoints are specific to the type of source resp. sink in the model. They are created by factories at runtime (Stream Devices). A source stream flow endpoint delivers data continuously according to the specified media set of the model to a connected sink stream flow endpoint. This logical connection between the two stream flow endpoints is physically realized by the connection of the two network flow endpoints, which are co-located with associated sink or source stream flow endpoints.

The HU-NTT platform utilizes an independent singleton, the Stream Coordinator to setup stream flows. Besides the **binding** between stream interaction elements, the Stream Coordinator is responsible for:

   − initiating the creation of appropriate stream flow resp. network flow endpoints;

- negotiating the network flow resp. stream flow endpoint connection configuration;
- setup of the nodal binding within the two network nodes;
- activation and deactivation of the stream flow and network flow connections.

Network Device, Stream Device, Network Flow Endpoint and Stream Flow Endpoint are CORBA objects providing operations to bind two stream interaction elements.

7   Summary

With this contribution we have pointed out some of the major shortcomings arising currently when multiple design techniques are used in combination for the development of distributed applications. We have presented how a common concept space can help to better integrate these techniques. Furthermore, the application of the Meta Object Facility enabled us to implement these ideas. A tool supported bi-directional mapping between concrete models and their MOF representation has turned out to be the important integration technology. The applicability of this approach has been demonstrated by the development of specific code generators, that produce – out of a MOF representation of a model that instantiates the concepts of the common concept space – software components executable on top of a CORBA middleware environment.

This meta-modelling method turns out to be an interesting technology in the context of modelling and language standardization within the ITU-T. It allows definition of a common concept space for the different techniques relevant in this community. This approach contributes significantly to the challenge of the integration of object-oriented design and component based manufacturing in the context of distributed and telecommunication systems.

References

1. ITU-T Recommendation Z.130: ODL – Object Definition Language, Geneva, 1998
2. ITU-T Recommendation X.901-X904: 1995, Open Distributed Processing – Reference Model Part 1-4
3. ITU-T Recommendation Z.100: Specification and Description Language, Geneva, 2000
4. OMG: Meta Object Facility Specification, OMG-Doc. formal/2000-04-03
5. OMG: Unified Modelling Language Specification v1.3, OMG-Doc. formal/2000-03-01
6. OMG: XML Metadata Interchange (XMI) Specification, OMG-Doc. formal/2000-11-02
7. BEA Systems et. al.: CORBA Components – Volume I; OMG Doc. orbos/99-07-01BEA Systems et.al.: CORBA Components – Volume III; OMG Doc. orbos/99-07-03
8. OMG: OMG Unified Modeling Language Specification, V. 1.3; OMG Doc. ad/99-06-08

9. BEA Systems et.al.: CORBA Components – Volume III; OMG Doc. orbos/99-07-03

10. OMG: The Common Object Request Broker: Architecture and Specification, Revision 2.3.1; OMG formal/99-10-07

11. TINA-C: Computational Modeling Concepts, 1998.

12. Fischer J., Fischbeck N., Born M., Hoffmann A., Winkler M.: Towards a behavioral Description of ODL, TINA'97 conference.

13. Born, M.; Hoffmann, A.; Li, M.; Schieferdecker, I.: Combining Design Methods for Service Development, FMOODS'98, Florence, Italy.

14. Kath, O.; Takita, W.: OMG A/V Streams and TINA NRA: An integrative Approach; Proc. of TINA '99

15. Tsuchiya, Y.; Takita, W.; Kath, O.; Stoinski, F.: Multimedia Contents Mill – A Platform for Authoring and Delivery of Interactive Multimedia Contents. Proc. of SoftCom 2000, Split .

16. W3C: Extensible Markup Language (XML), version 1.0, Febr. 1998

17. Szyperski, C.: Component Software – Beyond Object-Oriented Programming; Addison-Wesley '99

18. Bonnet, D.; Efremidis, L.; Malavazos, V.: ‚Cooling the Hell of Distributed Applications' Deployment", Proc. of IS&N 2000

19. EURESCOM Project P924: "Deliverable 2: Notation and Semantics for Deployment and Initial Configuration", EURESCOM '01

20. Internet RFC 2046: Multipurpose Internet Mail Extensions (MIME) Part Two: Media Types

21. Frolund, K.: QML: A Language for Quality of Service Specification; Hewlett-Packard Laboratories

Deriving Message Sequence Charts
from Use Case Maps Scenario Specifications[*]

Andrew Miga[1], Daniel Amyot[2], Francis Bordeleau[1],
Donald Cameron[3], and Murray Woodside[1]

[1] Carleton University, Ottawa, Canada
{miga, cmw}@sce.carleton.ca, francis@scs.carleton.ca
[2] Mitel Networks, Kanata, Canada
Daniel_Amyot@Mitel.com
[3] Nortel Networks, Ottawa, Canada
dcameron@NortelNetworks.com
http://www.UseCaseMaps.org

Abstract. A set of scenarios is a useful way to capture many aspects of the requirements of a system. Use Case Maps are a method for scenario capture which is good for describing multiple scenarios, including scenario interactions, for developing an architecture, and for analysing architectural alternatives. However, once a component architecture is determined, Message Sequence Charts are better for developing and presenting the details of interactions, and provide access to well-developed methodologies and tools for analysis and synthesis. This paper considers what must be specified in UCM scenarios and the architecture to make it possible to derive MSCs automatically, and it describes our experience in executing these transformations within a prototype tool, the UCM Navigator.

1 Introduction

A scenario specification technique called Use Case Maps [11,12] is part of a new proposal to ITU-T for a User Requirements Notation (URN) [13]. The role of the UCM notation is to capture functional requirements and it has been baptized URN-FR, while another and complementary component for non-functional requirements [14] is called URN-NFR. UCMs capture functional requirements in terms of *causal scenarios* that link sequences of responsibilities to (external) events. These scenarios may also be bound to underlying abstract components.

UCMs have been useful in describing a wide range of systems, including Wireless Intelligent Networks [2,16,26], agent systems [15], Wireless ATM [7], GPRS [3], and others discussed in [5,25]. As suggested by the I.130 and Q.65

[*] This research was supported by Nortel Networks, Mitel Networks, and by the Industrial Partnerships program of the Natural Sciences and Engineering Research Council of Canada. Discussions with John Visser, Jim Hodges, Jacques Sincennes, Luigi Logrippo and Gunter Mussbacher were helpful in developing the ideas.

R. Reed and J. Reed (Eds.): SDL 2001, LNCS 2078, pp. 268–287, 2001.

methodologies [17,20] and several UML-based approaches [4], the process of creating specifications and standards is generally composed of three major stages. At Stage 1, services are described from the user's point of view in prose form and with use cases. The focus of the second stage is on control flows between the different entities involved, represented using Message Sequence Charts (MSCs) [19]. Finally, Stage 3 aims to provide (informal) specifications of protocols and procedures. Formal specifications are sometimes provided (in SDL [18] for example), but overall they still suffer from a low penetration, especially in North-America.

In such methodologies, scenarios are often used as a means to model system functionality and interactions between the entities such that different stakeholders may understand their general intent as well as technical details [27]. Use Case Maps are used in Stage 1, and to bridge the conceptual gap into Stage 2 descriptions [5]. UCMs are used to capture user (functional) requirements when very little design detail is available, *without reference to messages or component states*. In Stage 1 documents, UCM scenarios may or may not be bound to any particular components for execution. The organization and architecture of components can be introduced into the map when moving towards Stage 2 documents. One of the strengths of UCMs at this level is their ability to show a number of scenarios together, and to reason about architecture and behaviour over a set of scenarios. Once appropriate architectural decisions are taken, UCMs can be used to guide the generation of MSCs to complete Stage 2 descriptions. In turn, MSCs can be used for the synthesis and the validation of component-based behavioral models in SDL or similar languages [1,22]. Many such synthesis techniques are studied and compared in [6].

This paper builds on previous work [2,9] and describes research on a well-defined transformation from a subset of the UCM notation to MSC-96, along with preliminary results in implementing the transformation. This transformation enables the rapid and consistent generation of MSCs from UCMs, and the extraction of simple end-to-end scenarios from complex multilevel UCMs. These MSCs can further be refined for Stage-2 like documents (where the specifics of messages becomes more relevant), used for system understanding, and used for functional testing of more detailed models and of implementations. The UCM notation is briefly reviewed and illustrated in Sect. 2. The UCM/MSC relationship is further studied in Sect. 3, with a particular emphasis on scenario variables in Sect. 4. Section 5 explains the proposed transformation, which is then illustrated with an example in Sect. 6. Finally, Sect. 7 presents our conclusions.

2   Use Case Maps

Use Case Maps visually describe causal relationships between *responsibilities* superimposed on organizational structures of abstract *components*. Responsibilities represent generic processing (actions, activities, operations, tasks, etc.). Components are also generic and can represent software entities (objects, processes, databases, servers, etc.) as well as non-software entities (actors or hardware). The relationships are said to be causal because they link causes (preconditions and triggering events) to effects (postconditions and resulting events) by arranging responsibilities in sequence, as alternatives, or in parallel. Essentially, UCMs show related and interacting use cases in a map-like diagram, whereas UCM *paths* show the progression of a scenario along a use case. Scenario interactions are shown by multiple paths through the same component and by one path triggering or disabling another. In UML terms, UCMs fill the gap between requirements described as (natural language) use cases and detailed behavioral based on components and messages (e.g. sequence, collaboration, and statechart diagrams). They exhibit several advantages over UML activity diagrams, as discussed in [4].

Fig. 1. Simple Use Case Map

The scenario in Fig. 1 represents a simplified call connection initiated at a start point labelled req. The system first checks whether the call should be allowed (responsibility chk) and then verifies whether the called party is busy or idle (vrfy). In both cases here, we assume that the call request goes through as no alternative is provided. The system status then is updated (upd) and a resulting ringing event occurs (ring). Additional UCM notation elements for alternatives, concurrent paths, submaps, path interactions, dynamic components, dynamic responsibilities, etc. are described in Appendix A.

2.1   UCMs, Messages, and Architectural Reasoning

UCMs are useful for describing features at an early stage, even when no components are defined, and then developing a scaffolding of components to "execute" the scenarios. Alternative architectures can be developed for the same UCM, for early architectural reasoning. For instance the UCM path from Fig. 1 is bound to two users connected through an agent-based architecture in Fig. 2a, whereas Fig. 2b uses a more conventional architecture based on Intelligent Networks (IN).

UCMs are more robust over architectural changes than the corresponding MSC. For instance, Fig. 2c is an MSC capturing the scenario in Fig. 2a in terms of message exchanges. Figure 2d is a potential MSC for the same scenario in an IN-based architecture. In this last MSC, complex protocols or negotiation mechanisms may be involved between the Switch and service nodes (SN), resulting in additional messages, and the Switch may be involved as a relay involving refinement of the relationship between req and chk. These figures make use of

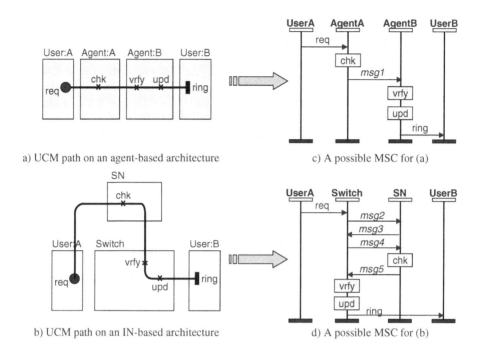

a) UCM path on an agent-based architecture

c) A possible MSC for (a)

b) UCM path on an IN-based architecture

d) A possible MSC for (b)

Fig. 2. UCM path bound to two different component structures, and potential MSCs

abstract messages (msg_x, in italic characters) that indicate which entities need to communicate in order to ensure the flow of causality found in the UCM.

The UCM view is a useful reference point which remains stable over changes related to messages, protocols, communications constraints and structure. UCMs avoid early commitments to detailed structures and messages and they remain focused on intended functionality and on reusable causal scenarios, within an evolving context.

2.2   UCM Navigator

The UCM notation is supported by a freely available editing tool: the *UCM Navigator* [21,25]. Among other features, this tool supports the path and component notations found in Appendix A, and it maintains various kinds of bindings (plug-ins to stubs, responsibilities to components, sub-components to components, etc.). Also, it allows users to navigate much like a Web browser, and to visit and edit the plug-ins related to stubs of all levels (see screen captures in Fig. 5). Editing operations maintain syntactic correctness, by either inserting correct elements, or transforming existing paths.

The UCM Navigator saves, loads, exports and imports UCM as XML files, which are valid according to a UCM Document Type Definition (DTD) [13].

This DTD describes the current formal definition of UCMs, which is based on hypergraphs, and the UCM Navigator ensures that syntactic and static semantic rules are satisfied.

The tool can also export UCM figures in three formats: Encapsulated Post-Script (EPS), Maker Interchange Format (MIF), and Computer Graphics Meta-file (CGM). Flexible reports can be generated as PostScript files ready to be transformed into hyperlinked and indexed PDF files. Multiple platforms are currently supported: Solaris, Linux (Intel and Sparc), HP/UX, and Windows (95, 98, 2000 and NT).

3   Relationships between UCM and MSC

To develop Message Sequence Charts from UCMs, we need to analyze how their respective concepts relate to each other. Here is a comparison of their main concepts:

- **Abstraction**. UCMs describe scenarios in terms of causal sequences of re-sponsibilities, which are identified at a high level of abstraction by a label and a brief textual description. The description may abstract away some inter-component communication. On the other hand, MSCs describe scenarios in terms of sequences of inter-component messages, actions, and methods.
- **Components**. A UCM defines components in terms of the role they play in a scenario (by means of a short textual description), and the responsibilities they provide. Components are represented by rectangles which can express layered and peer-to-peer relationships by position. An MSC represents com-ponents (called instances) using timelines, which express only the existence of separate locations. UCM components are optional whereas MSC instances are mandatory.
- **Alternative and concurrent sub-scenarios**. Both provide for a main scenario with a set of alternatives. Alternative paths are combined in a UCM diagram using the OR-fork segment connector. Dynamic stubs may also be used to describe alternatives. MSCs use the inline alternative box in basic MSC (or the OR notation in HMSC). Similar concepts exist for concurrent scenarios.
- **Scenario interactions**. UCMs have explicit notation for scenario interac-tions, as shown in Appendix A; this feature does not exist in MSC.

For the purpose of the UCM-to-MSC transformation, we establish a one-to-one relationship between valid UCM scenarios and basic MSCs, and between the following elements of the two models:

- a UCM component and an MSC instance (UCM sub-components are not being converted) an unbound UCM triggering (or resulting) event and an MSC message from (or to) the environment
- a UCM path crossing from one component to another and an abstract MSC message

- a UCM start (or end) point that is not bound to a stub input (or output) segment and an abstract MSC message
- a UCM precondition or postcondition and an MSC condition (expressing system state) (partially implemented in the Navigator at present)
- a UCM responsibility and an MSC action
- a UCM OR-fork or a UCM dynamic stub with multiple plug-ins, and multiple basic MSCs. Although the alternative inline box could be used here, having multiple MSCs simplifies the understanding of end-to-end scenarios
- a UCM AND-fork and an MSC parallel inline box (to avoid an explosion in the number of MSCs);
- a UCM loop and an MSC loop box (not implemented in the Navigator at present)
- a UCM timer and an MSC timer, with UCM triggering events as MSC resets and UCM timeout paths as MSC timeouts (partially implemented in the Navigator at present).

It is important to note that in the transition from UCM to MSC, the explicit interactions between scenarios expressed in UCM maps is lost because basic MSCs do not allow to explicitly express interactions between scenarios. However, causal flows in end-to-end scenarios will be preserved.

4   Scenario Variables in UCMs

A Use Case Map describes multiple scenarios, some with separate starting points and others that share starting points but under different types of input data, or different system states. A single scenario is the path traced out by placing a "token" on a particular map start point and by tracing one path through the various choices that are offered. These tokens are assumed to duplicate and merge along AND-forks and AND-joins. A single scenario gives rise to one basic MSC.

A particular scenario can be distinguished by the designer, by defining the conditions that govern it, as the *context* of the scenario. They include the state of the system at the time the scenario is executed, and the data that triggers the start point. Typically the designer can express the context in words, drawn from some previous use case. However, a more precise definition of each scenario context is needed to allow a meaningful set of MSCs to be created. *Scenario variables* have been introduced for this purpose, to govern the choice of alternatives. There are two kinds of *choice points* where a path has alternatives:

- an *OR-fork* allows a path to branch into multiple segments, which may possibly be joined subsequently by OR-joins (or before in the case of a loop)
- a *dynamic stub* is a placeholder for a set of plug-in maps, with the choice of submap depending on the particular scenario.

Without scenario variables, the UCM indicates that all alternatives are possible at every choice point. A symptom of inadequately specified scenario contexts is a combinatorial explosion of scenarios, many of which are meaningless. To illustrate this point, consider the example in Fig. 3. While there are 64 potential combinations of OR-fork choices in the path above and thus 64 potential scenarios, when the branch choices are based on the state of the variable x there are only four valid scenarios.

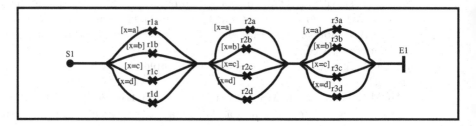

Fig. 3. Example of correlated OR-fork choices

As another example, dynamic stubs are often used to represent specific behaviour of features of a system with different plug-ins for each feature and many different dynamic stubs along a scenario path. Depending on which feature is active (part of the system state), only the corresponding plug-ins for that feature should be selected at each dynamic stub. For example if a path contains 10 dynamic stubs, each of which contains 10 plug-in maps to implement 10 different features, then there would be 1000 possible scenarios of which only a handful could be valid.

A *scenario data model* is added to UCM notation to allow designers to specify the context of an end-to-end scenario, and the logical selection conditions at choice points. This provides:

- precise specification of each scenario context for named end-to-end scenarios through a multi-level UCM design
- MSC generation of a specific scenario
- selection and highlighting of one scenario path in the UCM Navigator, for understanding of a design
- capability to discover invalid scenarios.

4.1 Scenario Variables

Selection conditions at choice points are based on global boolean variables whose values are defined for the scenarios that use the choice point. Future refinements of this most simple data model will allow for enumerated and integer values and local variables scoped to a particular token, map or stub.

There are three steps needed to specify the set of valid scenarios:

1. define the set of global variables for the map
2. specify logical selection conditions at OR-forks and dynamic stubs
3. specify a scenario definition for each valid scenario.

The first two steps are done once for the entire design. The third step, creating the scenario definitions simply sets the variable state so that the proper decisions are made at each branching point.

Usually the global variables emerge from consideration of the set of intended scenarios. They should be identified while the map is being created from use cases, with documentation either by formal notation or in words, of the role and conditions for each scenario. The full set of these scenario contexts is then described by a set of boolean variables, such that for each start point, each scenario from that start point has a unique representation in the values of the variables. The dialog box for defining these variables is shown in Fig. 4. The names may not include logical operators (i.e. +, &, =, and !).

4.2   Scenario Definitions

Scenario definitions are the means by which named end-to-end scenarios can be specified and referenced. Apart from a name and description of the scenario, they contain the starting point of the scenario as well as a set of variable initializations. A variable initialization is a reference to a global variable coupled to the value to which it should be set for a scenario. A list of these initializes the system state to the proper values for a given scenario.

Scenario variables can be set to the values *true* and *false*, for the given scenario. If not set, they have the value *undefined*, which causes logical expressions referencing the variable to evaluate to *false* and a warning to be emitted. In addition, it would be useful to be able to define a variable as *don't care*, within a given scenario; however this has not been implemented yet.

Figure 4 shows the dialog box (at right) that is used to create and edit scenario definitions as well as invoke MSC generation and scenario highlighting operations for defined scenarios. Once scenarios are defined, another dialog box (not shown) can be invoked from a path start point to list the scenarios that begin at that point and invoke MSC generation and path highlighting operations.

Scenarios are organized into groups which are listed on the top left section of the dialog. Individual scenarios are listed in the top right section. They must be given a unique name and can be given a description (same for scenario groups). The bottom section allows for the editing of the currently selected scenario's variable initializations and for invoking operations using the scenario definition. The list of variable initializations is created by selecting an unset variable from the left list and specifying its initial value. It is then removed from the left list and the variable-initial value pair is placed in the center list. Controls allow for creating, modifying and destroying such initializations. The set of variables that need to be initialized for a given scenario is generally a subset of the total list of predefined variables.

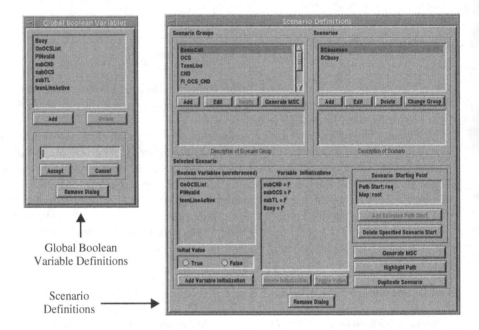

Fig. 4. Dialog boxes for scenario variables and definitions

4.3   Logical Selection Conditions at OR-Forks and Dynamic Stubs

Logical selection conditions are logical expressions designed to produce a boolean output value with a combination of variable references and logical operators (*and* '&', *or* '+', *not* '!', *equal* '=', *not equal* '!=', and brackets ()).

For example, if A, B, C, and D are variable names then ((A+B) &(C!=D)) & !(A=D) is a valid logical expression. So is A=T & B=F as the constants T (*true*) and F (*false*) may be referenced explicitly in conditions.

Syntactic validation is performed at creation time so that invalid conditions are not accepted in the tool. In addition all conditions are stored internally with references to variable objects so that the renaming of a variable causes no problem. Also reference counting on variable objects is performed so that a variable referenced in either a selection condition or a scenario definition cannot be deleted prematurely.

Figure 5 shows a set of screenshots that illustrate the relevant dialog boxes invoked for specifying logical conditions for OR-fork branches (top) and for plug-in maps (bottom). The dialog for editing conditions is the same in both cases and is opened as a subdialog of the OR-fork specification and plug-in choice dialogs. This dialog contains a list of all predefined variables for users to select.

In order to validate if the entire scenario specification operation has been performed correctly, a named scenario can be selected and highlighted with a special path colour throughout a multilevel scenario. This provides immediate feedback to designers in UCM terms. If there are errors either the wrong path or multiple paths will be highlighted. If a branching point is reached where none of the options evaluate to *true*, an error message is given and the highlighting fails.

5 Transformation of UCM Paths into Message Sequence Charts

The UCM-to-MSC transformation generates basic MSCs (MSC-96) from UCMs in two main steps:

- Identification of the valid scenarios in the UCM model (this is where the scenario variables data are required)
- Generate MSC elements according to the correspondences identified in Section 3.

This section expands on this transformation with an overview of the underlying object model and of the scanning algorithm.

5.1 Hypergraph Model

A *hypergraph* model is used to store the use case paths, with hyperedges representing UCM path elements (start, end points, responsibilities, forks, joins, stubs) and nodes being internal objects representing connections between hyperedges. UCM components are separate objects referenced by hyperedges that are inside their boundaries. All of the objects for a given map are contained inside a map object. Connections between parent and child maps are specified by user defined stub bindings, which bind the input and output segments of a stub in a parent map to path start and end points in a submap.

5.2 Recursive Scanning Algorithm

There are two methods of generating MSCs:

- Generating the MSC(s) for one or more predefined scenarios by selecting scenarios from a list
- Generating all possible scenarios that start from a given point by determining all possible combinations of OR-fork choices and plug-in selections and generating a separate MSC for each (hence ignoring scenario variables).

Both methods of generating MSCs use a recursive algorithm to determine all possible combinations of OR-fork branches and plug-in maps. The difference is with use of scenario data and with branch conditions defined, the recursive algorithm only follows those OR-fork branches or plug-in maps whose selection conditions evaluate to *true* at run time. With fully specified selection conditions for a design this should result in only the single proper path taken. Nondeterministic UCM choices (where many conditions evaluate to *true*) will cause multiple MSC scenarios to be generated, one for each alternative.

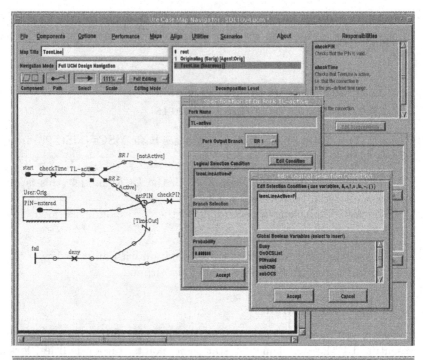

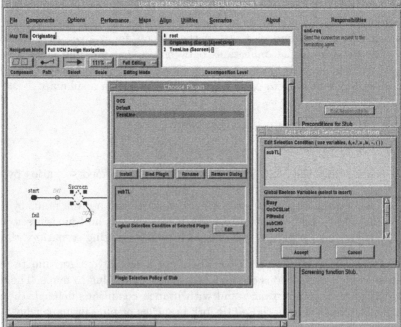

Fig. 5. Specification of logical selection conditions for OR-forks and dynamic stubs

For real designs of any complexity, scenario data must be used as the other method can generate a very large number of MSCs (as already discussed). If errors exist, it is possible multiple scenarios will be generated for a scenario definition as the same recursive algorithm is used. In this manner users are notified that elements are either over- or under-specified.

6 Example

To illustrate the UCM-to-MSC transformation, we extend the simple call request UCM in Fig. 1 to include multiple UCM constructs, embedded maps, and system functionality (*features*).

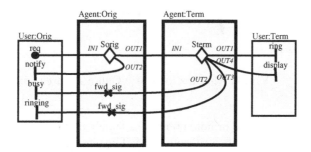

Fig. 6. Basic Call root map

6.1 Basic Call and Feature UCMs

The top-level UCM is the Basic Call of Fig. 6, which contains four components (originating/terminating users and their agents) and two static stubs. The first stub (Sorig) contains the ORIGINATING plug-in whereas the second (Sterm) contains the TERMINATING plug-in. In turn, these two plug-ins have other stubs (Sscreen and Sdisplay) and another level of maps. Each of these latter stubs includes a DEFAULT plug-in (which happens to be the same in both cases) that represents how the basic call reacts in the absence of other features. These plug-ins, generated with the UCM Navigator, are presented in Fig. 7.

The ORIGINATING side has two features (plug-ins) used in Sscreen:

- **Originating Call Screening** (OCS), which checks whether the call should be denied or allowed (chk). When denied, an appropriate event occurs at the originator side (notify). Its binding relationship, which connects the input/output segments of a stub to the start/end points of its plug-in, is $\{< IN1, start >, < OUT1, success >, < OUT2, fail >\}$.

- **TeenLine**, which denies the call provided that the request is made during a specific time interval and that the personal identification number (PIN)

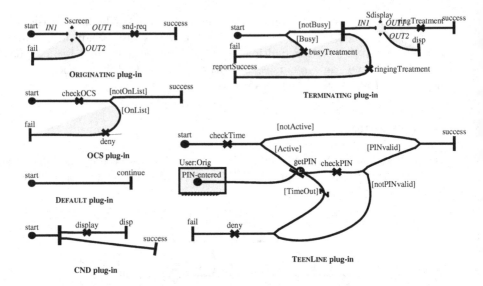

Fig. 7. Plug-ins for Basic Call and three features

provided is invalid or not entered in a timely manner. The zigzag path leaving the timer represents a timeout path. The binding relationship for this feature is also $\{< IN1, start >, < OUT1, success >, < OUT2, fail >\}$.

The TERMINATING side contains only one feature plug-in, used in Sdisplay:

- **Call Number Delivery** (CND), which displays the number of the originating party (display) concurrently with the rest of the scenario (update and ringing). The binding relationship is $\{< IN1, start >, < OUT1, success >, < OUT2, fail >\}$.

Note that the start and end points of these plug-ins are not external events but connectors to the input/output segments of their parent stub.

Adding features to such UCM collections is often achieved by creating new plug-ins for the existing stubs, or by adding new stubs containing either new plug-ins or instances of existing plug-ins. In all cases, the selection policies and pre-conditions need to be updated appropriately.

6.2 Scenario Variables and Conditions

Seven scenario variables were created according to the two categories discussed in Sect. 4:

- branch selection in OR-forks: Busy, OnOCSList, PINvalid, and TeenLineActive
- subscriptions to features, i.e. selection of plug-ins in dynamic stubs: subOCS, subTL, subCND.

Branch selection conditions, found in three plug-ins (OCS, TERMINATING, and TEENLINE), use the first categories of variables. All these conditions reference only one variable, as is and complemented (e.g. PINvalid for [PINvalid] and !PINvalid for [notPINvalid]).

Plug-in preconditions use the second category of variables. They form the *selection policies* found in our dynamic stubs. These policies provide a simple mechanism for feature interaction resolution, which is local to a particular stub. Policies can hence establish precedence of one feature over another. For instance, stub Sscreen contains three plug-ins whose selection is done as follow:

$$subOCS \rightarrow use\ OCS$$
$$!subOCS\ \&\ subTL \rightarrow use\ TEENLINE$$
$$!(subOCS+subTL) \rightarrow use\ DEFAULT$$

Selection policies enable one to derive scenarios that involve multiple features, and hence to visualize desirable and undesirable interactions early in the development process.

6.3   Message Sequence Charts

From the req start point in the Basic Call root map, various scenarios can be generated. The presence of seven variables suggests an upper bound of $2^7 = 128$ such scenarios (assuming all choices are guarded and deterministic), but the path structure actually constrains this number to 15 scenarios. However, this number would increase dramatically as features are added or made more complex, as explained in Sect. 4. Although all these MSCs can be generated, the more valuable and traceable MSCs are the ones produced by explicitly defining contexts using scenario variables. Such scenarios can be well documented, referenced, and studied through the highlighting function of the UCM Navigator.

Scenarios can be defined for the Basic Call (no feature), for one feature at a time, or for multiple features at a time. Figure 8 shows the MSC (MSC-96) corresponding to a successful call connection made by an originating user subscribed to OCS to an available terminating user. Abstract messages (e.g. *m1* and *m2*) are generated where necessary, as discussed in Sect. 2.1. The Z.120 textual representation is output directly from the UCM Navigator, and it can be easily converted to a graphical form by existing MSC/SDL tools or by packages such as [10] (used in this paper). Minor but automatable transformations may have to be done to the textual representation to satisfy identifier conventions (e.g. removal of underscore symbols "_" in messages).

Combinations of features are particularly interesting as the emerging behaviour is often surprising. Figure 9 (left) shows the MSC corresponding to a successful call connection made by an originating user subscribed to OCS to an available terminating user with CND. As a result, three messages are sent in parallel by the terminating agent.

The second MSC in Fig. 9 (right) results from a situation where the originating user has subscribed to both OCS (but the terminating user is not on the

```
mscdocument OCSsuccess;
msc OCSsuccess;
User[Orig] :  instance;
Agent[Orig] : instance;
Agent[Term] : instance;
User[Term] :  instance;
User[Orig] :  out req,1 to Agent[Orig];
Agent[Orig] : in req,1 from User[Orig];
              action 'checkOCS';
              condition [notOnList];
              action 'snd_req';
              out m1,2 to Agent[Term];
Agent[Term] : in m1,2 from Agent[Orig];
              condition [notBusy];
all: par begin;
   Agent[Term] : action 'ringTreatment';
                 out ring,3 to User[Term];
   User[Term] :  in ring,3 from Agent[Term];
all: par;
   Agent[Term] : action 'ringingTreatment';
                 out m2,4 to Agent[Orig];
   Agent[Orig] : in m2,4 from Agent[Term];
                 action 'fwd_sig';
                 out ringing,5 to User[Orig];
   User[Orig] :  in ringing,5 from Agent[Orig];
all: par end;
Agent[Term] : endinstance;
Agent[Orig] : endinstance;
User[Orig] :  endinstance;
User[Term] :  endinstance;
endmsc;
```

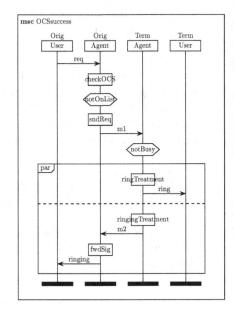

Fig. 8. OCS successful scenario in MSC form

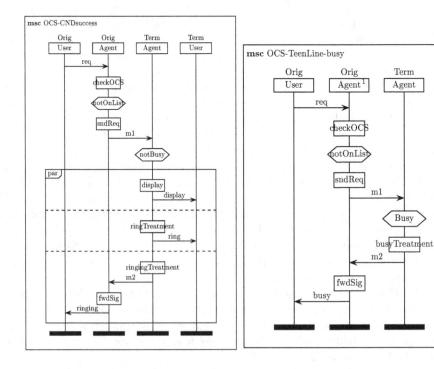

Fig. 9. OCS-CND and OCS-TeenLine MSCs

screening list) and TEENLINE (active, with an invalid PIN). The terminating user is assumed to be busy. In theory, the call should be denied by TEENLINE. However, due to Sscreen's selection policy (described in Sect. 6.2), OCS prevents TEENLINE to be activated. This is a typical feature interaction that needs to be solved at the policy level (and perhaps at the UCM path level as well). MSCs generated from UCMs make it possible to find such unexpected behaviour very early in the design process, without any commitment to complex protocols.

7   Conclusions

This work studied transformation of a UCM into a set of MSCs, intending to jump-start the use of MSCs in design, and link them to requirements analysis. We found that UCMs must be enhanced by introducing a formal *scenario data model*, to identify the intended scenarios from all potential scenarios. The paper has described the concept of scenario variables, how they are used in small but realistic UCM examples, and how MSCs are generated from such UCMs.

Scenario variables capture designer intentions for a given scenario among all those that begin at a given start point; they capture at least some of the preconditions of the scenario as a formal part of the UCM model, and they help to remove or to identify unwanted "emergent possible paths". The intended paths are then transformed into MSCs by establishing simple correspondences between the models.

If the scenarios are the basis of design of tests, the scenario variables also define test cases. One goal in transforming to MSCs is to exploit design and test-generation tools based on MSCs and SDL or other formal languages.

The use of scenario variables in the examples described here is somewhat limited by the size of the examples, and choices are often specified by the value of a single variable. In larger specifications we have found that more complex functions of several variables are often used.

Several enhancements to this work would appear to be useful. The transformation can be expanded to include details which are identified but not modeled in the UCM. Detailed patterns of messages and activities equivalent to sub-MSCs can be defined in the UCM, for responsibilities or for inter-component responsibility transfer, and included in the MSC output either as additional detail within the MSC, or as MSC references. Complex and realistic message exchanges (e.g. as in Fig. 2d, but with meaningful message names and parameters rather than abstract messages) could hence be generated.

The scenario data model could be enriched, without going to a complete data model like ASN.1. The data types described here are restricted to booleans in order to simplify them, to focus the designer on declaring intentions, and to avoid the temptation to write program code at this stage. Modest potential enhancements include:

- better support for pre/postconditions, loops and timers
- enumeration types to distinguish multiple cases, and bounded counters

– operations on scenario variables by responsibilities, to allow counters to be incremented and control flags to be set as a token moves along a path
– token variables could be introduced, instead of just global variables, for scenarios with multiple concurrent tokens.

A clearer and more precise dynamic semantics for UCMs is also desirable in the context of MSC generation. For instance, a stub could create new instances of a plug-in each time it is accessed, or else access the same instance (singleton).

It would not be difficult to transform to HMSCs instead of MSCs. The structure of alternate paths and forks and joins in the UCM could be used to define the HMSC road map, and the linear sub-paths within this structure would define the message sequences within the HMSC.

The transformation machinery described here is also being adapted to create test cases and performance models from UCMs [24], and to exploit UCM scenario information in generating SDL models [8,23].

References

1. Abdalla, M.M., Khendek, F. and Butler, G.: "New Results on Deriving SDL Specifications from MSCs". In: *SDL'99, Proceedings of the Ninth SDL Forum*, Montreal, Canada. Elsevier (1999)
2. Amyot, D. and Andrade, R.: "Description of Wireless Intelligent Network Services with Use Case Maps". In: *SBRC'99, 17° Simpósio Brasileiro de Redes de Computadores*, Salvador, Brazil (May 1999) 418–433
3. Amyot, D. and Logrippo, L.: "Use Case Maps and LOTOS for the Prototyping and Validation of a Mobile Group Call System". In: *Computer Communication*, 23(12) (May 2000) 1135–1157
4. Amyot, D. and Mussbacher, G.: "On the Extension of UML with Use Case Maps Concepts". In: *<<UML>>2000, 3rd International Conference on the Unified Modeling Language*, York, UK (October 2000), LNCS 1939, 16–31
5. Amyot, D.: "Use Case Maps as a Feature Description Language". In: S. Gilmore and M. Ryan (Eds), *Language Constructs for Designing Features*. Springer-Verlag (2000) 27–44
6. Amyot, D. and Eberlein, A.: "An Evaluation of Scenario Notations for Telecommunication Systems Development". In: *9th Int. Conference on Telecommunication Systems (9ICTS)*, Dallas, USA (March 2001)
7. Andrade, R.: "Applying Use Case Maps and Formal Methods to the Development of Wireless Mobile ATM Networks". In: *Lfm2000: The Fifth NASA Langley Formal Methods Workshop*, Williamsburg, Virginia, USA (June 2000)
8. Bordeleau, F.: *A Systematic and Traceable Progression from Scenario Models to Communicating Hierarchical Finite State Machines*. Ph.D. thesis, School of Computer Science, Carleton University, Ottawa, Canada (August 1999)
9. Bordeleau, F. and Cameron, D.: "On the Relationship between Use Case Maps and Message Sequence Charts". In: *2nd Workshop of the SDL Forum Society on SDL and MSC (SAM2000)*, Grenoble, France (June 2000)
10. Bos, M. and Mauw, S.: "A LaTeX macro package for drawing Message Sequence Charts". Version 1.47, http://www.win.tue.nl/~sjouke/mscpackage.html (1999)
11. Buhr, R. J. A. and Casselman, R. S.: *Use Case Maps for Object-Oriented Systems*, Prentice-Hall (1996)

12. Buhr, R. J. A.: "Use Case Maps as Architectural Entities for Complex Systems". In: *IEEE Transactions on Software Engineering*, 24(12) (Dec. 1998) 1131–1155

13. Cameron, D. et al.: *Draft Specification of the User Requirements Notation*. Canadian contribution CAN COM 10-12 to ITU-T, Geneva (November 2000)

14. Chung, L., Nixon, B. A., Yu, E. and Mylopoulos, J.: *Non-Functional Requirements in Software Engineering*. Kluwer Academic Publishers (2000)

15. Elammari, M. and Lalonde, W. (1999) "An Agent-Oriented Methodology: High-Level and Intermediate Models". In: *Proc. of the 1st Int. Workshop. on Agent-Oriented Information Systems (AOIS'99)*, Heidelberg, Germany (June 1999)

16. Hodges, J. and Visser, J.: "Accelerating Wireless Intelligent Network Standards Through Formal Techniques". In: *IEEE 1999 Vehicular Technology Conference (VTC'99)*, Houston (TX), USA (1999)

17. ITU-T: *Recommendation I.130, Method for the characterization of telecommunication services supported by an ISDN and network capabilities of ISDN*. CCITT, Geneva (1988)

18. ITU-T: *Recommendation Z.100, Specification and Description Language (SDL)*. Geneva (1999)

19. ITU-T: *Recommendation Z.120, Message Sequence Chart (MSC)*. Geneva (1999)

20. ITU-T: *Recommendation Q.65, The unified functional methodology for the characterization of services and network capabilities including alternative object-oriented techniques*. Geneva (2000)

21. Miga, A.: *Application of Use Case Maps to System Design with Tool Support*. M.Eng. thesis, Dept. of Systems and Computer Engineering, Carleton University, Ottawa, Canada (October 1998)

22. Mansurov, N. and Zhukov, D.: "Automatic synthesis of SDL models in use case methodology". In: *SDL'99, Proceedings of the Ninth SDL Forum*, Montreal, Canada. Elsevier (1999)

23. Sales, I. and Probert, R. L.: "From High-Level Behaviour to High-Level Design: Use Case Maps to Specification and Description Language". In: *SBRC'2000, 18° Simpósio Brasileiro de Redes de Computadores*, Belo Horizonte, Brazil (May 2000)

24. Scratchley, W. C.: *Evaluation and Diagnosis of Concurrency Architectures*. Ph.D. thesis, Dept. of Systems and Computer Engineering, Carleton University, Ottawa, Canada (November 2000)

25. *Use Case Maps Web Page and UCM User Group* (since March 1999) http://www.UseCaseMaps.org

26. Yi, Z.: *CNAP Specification and Validation: A Design Methodology Using* LOTOS *and UCM*. M.Sc. thesis, SITE, University of Ottawa, Canada (2000)

27. Weidenhaupt, K., Pohl, K., Jarke, M., and Haumer, P.: "Scenarios in System Development: Current Practice". In: *IEEE Software* (March/April 1998) 34–45

N.B. Most UCM papers are available at http://www.UseCaseMaps.org/pub/

Appendix A: UCM Quick Reference Guide

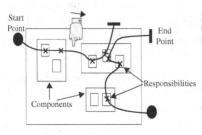

Start Point

End Point

Responsibilities

Components

Imagine tracing a path through a system of objects to explain a causal sequence, leaving behind a visual signature. Use Case Maps capture such sequences. They are composed of:

- **start points** (filled circles representing preconditions and/ or triggering causes)
- causal chains of **responsibilities** (crosses, representing actions, tasks, or functions to be performed)
- and **end points** (bars representing postconditions and/or resulting effects).

The responsibilities can be bound to **components**, which are the entities or objects composing the system.

A1. Basic notation and interpretation

(a) OR-join   (b) OR-fork

[yes]

[no]

(c) Permissible routes assumed identified

Indicate routes that share common causal segments. Alternatives may be identified by labels or by conditions ([guards])

A2. Shared routes and **OR-forks/joins**.

1:N   N:1   N:M

(a) AND-fork   (b) AND-join   (c) Generic version

A4. Concurrent routes with **AND-forks/joins**.

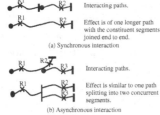

R1   R2
Interacting paths.

R1   R2
Effect is of one longer path with the constituent segments joined end to end.

(a) Synchronous interaction

R1   R2   R3
Interacting paths.

R1   R2
            R3
Effect is similar to one path splitting into two concurrent segments.

(b) Asynchronous interaction

A3. Path interactions.

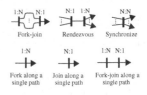

1:N   N:1   N:1 1:N   N:N
Fork-join   Rendezvous   Synchronize

1:N   N:1   1:N N:1
Fork along a   Join along a   Fork-join along a
single path   single path   single path

A5. Variations on AND-forks/joins.

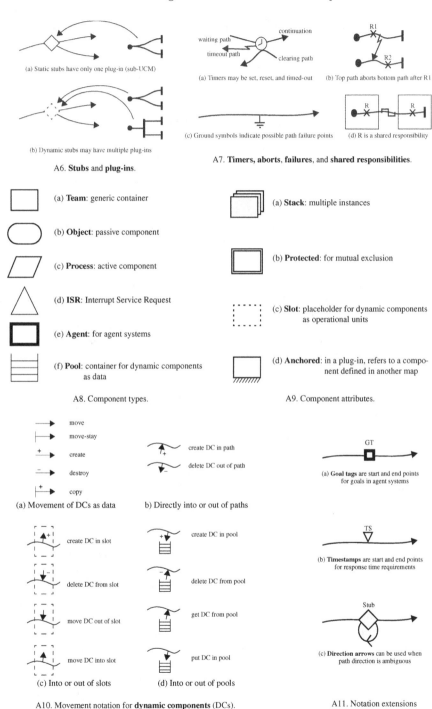

(a) Static stubs have only one plug-in (sub-UCM)

(b) Dynamic stubs may have multiple plug-ins

A6. **Stubs** and **plug-ins**.

(a) Timers may be set, reset, and timed-out

(b) Top path aborts bottom path after R1

(c) Ground symbols indicate possible path failure points

(d) R is a shared responsibility

A7. **Timers**, **aborts**, **failures**, and **shared responsibilities**.

(a) **Team**: generic container

(b) **Object**: passive component

(c) **Process**: active component

(d) **ISR**: Interrupt Service Request

(e) **Agent**: for agent systems

(f) **Pool**: container for dynamic components as data

A8. Component types.

(a) **Stack**: multiple instances

(b) **Protected**: for mutual exclusion

(c) **Slot**: placeholder for dynamic components as operational units

(d) **Anchored**: in a plug-in, refers to a component defined in another map

A9. Component attributes.

move
move-stay
create
destroy
copy

(a) Movement of DCs as data

create DC in path
delete DC out of path

b) Directly into or out of paths

create DC in slot
delete DC from slot
move DC out of slot
move DC into slot

(c) Into or out of slots

create DC in pool
delete DC from pool
get DC from pool
put DC in pool

(d) Into or out of pools

A10. Movement notation for **dynamic components** (DCs).

(a) **Goal tags** are start and end points for goals in agent systems

(b) **Timestamps** are start and end points for response time requirements

(c) **Direction arrows** can be used when path direction is ambiguous

A11. Notation extensions

An SDL Implementation Framework for Third Generation Mobile Communications System

Juha Sipilä and Vesa Luukkala

Nokia Research Center, Mobile Networks Laboratory,
P.O. Box 407 FIN-00045 NOKIA GROUP
{juha.p.sipila,vesa.luukkala}@nokia.com

Abstract. This paper presents an SDL implementation framework for the third generation mobile communication system protocols. The framework includes specific stylistic notations of SDL and extensive usage of ASN.1. Furthermore the framework requires that all protocols should define certain packages and combine certain functionalities into one process. This framework has allowed creation of a prototype implementation of third generation mobile system protocol stacks. This prototype implementation contains nearly all protocols from mobile station, radio access network and some protocols from the core network side. This implementation has and will be used in multiple prototype and validation system implementations within Nokia.

1 Introduction

The enormous growth in usage of mobile communications systems worldwide has created a need for new, better and more efficient services and thus higher bit rates. In addition to these quantities the future system should also be, unlike the current systems, globally standardized.

The current mobile communication systems are called second generation systems and their successors will logically be called third generation mobile communications systems. The most widely used second-generation system at the moment is GSM (Global system for Mobile Communications).

The most famous member of the third generation standard family is Universal Mobile Telecommunication System (UMTS). The work for UMTS standardization was completed in early 2000, however corrections are still being made. The standardization work was done by 3GPP (Third Generation Partnership Program). 3GPP consists of multiple national standardization bodies and representatives of the telecommunication industry from Europe, Japan and the USA.

The large size, concurrency and real-time nature of mobile communication systems present a set of difficulties when generating a good specification. Some of the problems are such that it is not possible to track them on paper, thus early implementations from the specification are required.

These early implementations set special requirements for the implementation language. The language should provide the possibility to do the implementation

R. Reed and J. Reed (Eds.): SDL 2001, LNCS 2078, pp. 288–299, 2001.

quickly, support distribution and concurrency, provide good tracing facilities and make possible to use formal methods for the most complicated parts of the system. On the other hand it is not absolutely necessary to have product level performance and memory consumption at this phase.

Desirable properties for a development environment in this context would be a notion of concurrency with proper support for simulation and tracing and well-defined semantics for the language used. With these requirements SDL [2] is a very good choice. SDL has been designed for telecommunication use and it hides all the irrelevant features of lower level programming languages (like pointers). Tool support for SDL is good as functional level debugging, tracing and simulation possibilities with the tools are excellent. SDL is a formal language which makes possible the use of verification and validation methods for selected parts of the system. SDL has also been used for a long time in telecommunication industry and it has been found out to be suitable even for product level implementation.

One problem with SDL is the presentation of data types. SDL as a language does not include the concept of PDU (Protocol Data Unit) and does not allow converting internal data representation to some transfer syntax. These problems can be solved by defining all the data types of the system with ASN.1 (Abstract Syntax Notation One) language. ASN.1 provides the ability to define a universal transfer syntax and thus makes it possible to distribute the different components of the system across diverse computer architectures.

Languages as such, even if they are good ones, are not the answer in themselves. Tools and frameworks are required to limit the degrees of freedom and make it possible for the implementor to focus on the essentials.

Telelogic SDT (Specification and Description Tool) [5] has been used for a long time in the telecommunication industry and it has stabilized its position as a mature SDL development environment. Due to its earlier use at Nokia Research Center (NRC) it was an obvious choice for an SDL tool within this project. SDT has support for ASN.1 language, but unfortunately it was found to be too limited for our needs and a special tool called ASN4SDT was developed to integrate NRC's ASN.1 tool set with the SDT tool.

From the beginning it was clear that the 3GPP specifications will be evolving and they would be in a state of constant change. This was one of the major cornerstones on which the system design was based.

To assure that the complete system stays modular and extendable it was essential to develop our own protocol implementation framework. Based on previous experiences, exact naming conventions for all the used SDL concepts and a basic process structure for protocols were defined. This process structure consists of certain packages and processes which must be defined for all protocols.

Section 2 gives a brief introduction to the context of this work, the UMTS radio network, Sect. 3 presents the high level architecture of the system and Sect. 4 presents the used framework. Section 5 gives an example of the principles outlined in Sect. 4 and finally Sect. 6 presents the conclusions.

2   UMTS Radio System

Mobile communication system consists of three main components:

- User Equipment
- Radio Network
- Core Network

The Radio network can be further separated into base station and base station controller. User equipment is the device that is connected to the system via radio interface and which is operated by the user. The base station, which is called Node B in UMTS, provides the radio interface for the user equipment. The base station controller (Radio Network Controller (RNC) in UMTS) controls and maintains the base stations and connects the Radio Network subsystem to the fixed part of the system. All the radio network subsystems together are called the Radio Access Network (RAN).

The core network consists of multiple computer units, which maintain the user information and connect the mobile communication system into the fixed networks like PSTN, X25 and Internet.

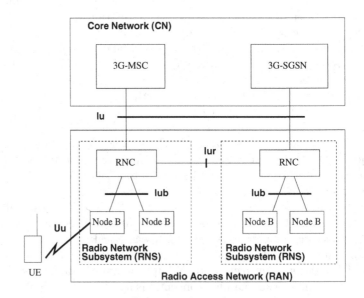

Fig. 1. UMTS architecture..

The radio interfaces (called Uu in UMTS) can be categorized based on their access technology. The newest of the access methods is called CDMA (Code Division Multiple Access). CDMA sets special requirements for the radio network. Because everybody is sending and receiving at the same frequency it is possible to transmit the data via multiple base stations. This increases the efficiency of the radio interface, but makes it harder to implement the radio access network. In addition to the classical base station - base station controller interface (called Iub in UMTS) a new interface is required for communication between base station controllers (called Iur in UMTS). This interface requires a new protocol and makes all the signaling procedures within the system more complicated because co-operation between multiple base stations must be taken into account.

The radio network as such is not useful unless it is connected into external networks. This functionality is taken care of by a group of computers grouped here under the name core network. The interface between the RAN and the core network is called Iu in UMTS. Due to growth of Internet and data traffic the interface is separated into packet switched and circuit switched parts, which are terminated into different access points within the core network. The architecture described in this section is pictured in Fig. 1.

3   3G SDL Library

The initial objective of the 3G SDL Library project was to implement prototype versions from all UMTS link (L2) and network layer (L3) radio related protocols with SDL for support of standardization. It was quickly discovered that the resulting system was suitable for more concrete purposes as well such as testing and trial systems.

A secondary objective for the project was to research and develop protocol implementation methodologies using SDL and ASN.1. This includes development of an implementation framework and also development and testing of ASN.1 and SDL/SDT tools. The project was designed to utilize and collect the knowledge of previous protocol implementation projects at NRC involving SDL and ASN.1.

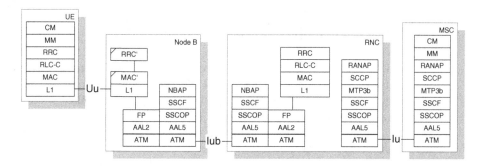

Fig. 2. Circuit switched control plane of UMTS system.

At the moment the project has produced models for nearly all UMTS L2 and L3 protocols. Physical layer (L1) related functionalities have not been implemented, but the interface between L1 and L2 has been specified, so the L1 can be added later or external L1 models can be used with minimal work. Figure 2 shows the control plane protocols of the UMTS circuit switched radio access network. In our model this translates to about 150000 lines of generated SDL-PR and 30000 lines of ASN.1 definitions which translate to about 2000000 lines of C code.

While the 3GPP standard specifications have been released and accepted, there are still numerous updates and corrections, so the functionality is constantly updated. It is notable that the constant changes of the specification (due to standardization) have been successfully accommodated into the model on a routine basis. The key to this success has been the developed framework and the usage of ASN.1.

4   Implementation Framework

4.1   General

The term framework means here a collection of rules for defining certain concepts in certain packages (using *package* as defined in [2]), augmented with the process structure, style guide and ASN.1 usage as described in this section.

The previous work involving SDL had problems in re-usability and extendibility that partly stemmed from undisciplined use of SDL, especially regarding the scope of data types. Thus one of the goals of the project was to research and develop protocol implementation methodologies using SDL and ASN.1.

The requirements for the project dictated the properties of the architecture: simplicity, extendibility, modularity and the possibility of ASN.1 usage. In anticipation of changing specifications and the need to communicate the concepts of the model easily, the basic architecture was chosen to be rather spartan and uniform. This is realized by constantly using SDL constructs that are presented as basic SDL in [1] (ignoring priority input). These are augmented with the following features: transition options, signallists, label/join and view/reveal. This restricts the allowed SDL features to a subset of SDL-92.

The decision of 3GPP to specify external PDU coding for some protocols with ASN.1 [3] together with extensive previous experience of ASN.1 tools at NRC made it feasible to start integrating existing NRC ASN.1 tools with SDT. The integration tool was named ASN4SDT. At the time it was known that SDT was also going to include ASN.1 support, but during the feasibility studies this support was found out to be inadequate for the needs of the project. It was also found out that at the time the mapping rules of Z.105 were not sufficient for efficient use of ASN.1 structures in SDL programs.

In addition to enabling distribution, ASN.1 enables simplification of testing and tracing by providing a standardized format for values of the data types. This allows easy translation of the encoded data to human readable form.

4.2   Process Architecture

The design goals called for extendibility and modularity, which led to extensive use of SDL packages. For each protocol implementation, a package including a protocol block type containing the process types of the protocol entities is defined. This allows multiple instantiation of the same entity in multiple locations.

The services that the protocol offers to higher level protocol are defined in external interfaces of a protocol, these are grouped in two packages. The external interface package defines the signals of the interface, but the data types of the signal's parameters are defined in a separate package. This allows a desirable separation of the relatively volatile data types and stable signaling and enables a hierarchy of data type visibility. Other auxiliary data types and procedures are also separated in packages.

The process architecture in the framework is based on the so-called Routing Encoding/Decoding (RED) architecture. The idea is to have a separate process that handles routing and data presentation conversions between internal and external presentations (encoding and decoding). This makes it possible to separate the behavioral functionality from the interface related functionality. In addition the RED-process may include some minor management related functionalities, such as creation of processes. A similar type of approach has also been presented in the protocol pattern context in [4], but the manager and routing processes are not combined in that work.

This solution has certain advantages. Keeping the routing table in one process is efficient and solves routing table consistency problems. From a debugging point of view it is useful to see the parameters before encoding and after decoding. This is possible because the parameters are passed to the RED-process as a signal and it is possible to see the signal parameters in a trace. It is crucial to ensure that all the transitions in the process are fast as otherwise the RED-process will be a bottleneck.

Another key feature is limiting the data type visibility. Here it is done by having four visibility levels: global, module, interface and local. Global definitions are visible across the system and the global data type module cannot use data types from other modules. Module-level data types are related to some logical concept or service and they should be visible to users of that concept. These are data types that clearly apply to more than one interface, but still are not needed by all entities. Module-level packages can include data types from global definitions and other module-level packages, the latter dependencies are regrettable but seemingly unavoidable. It is absolutely necessary that dependencies between modules are kept to a minimum. Interface-level data types contain all data types used in one interface and it may use data types defined at upper levels, but not at other interface levels. Local-level data types contain definitions for a single protocol implementation and again it may use data types from higher levels, but not from other local-level data types. Figure 3 shows the data type hierarchy.

4.3 Use of ASN.1

The 3G SDL library uses ASN.1 whenever possible. The 3GPP specification dictates the PDU structure in ASN.1 for some protocols, but at every interface it is possible to use ASN.1 to specify the parameter data types, which allows

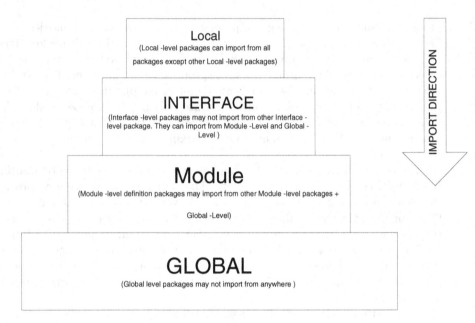

Fig. 3. The data type hierarchy.

distribution of the system at that interface. It was also found out to be feasible to use ASN.1 for data type specification within the protocol implementation.

As explained previously, a local tool ASN4SDT has been developed for integrating ASN.1 data types with SDT. ASN4SDT is based on NRC's ASN.1 tool set, which supports the required standardized PER and BER encoding rules. These tools also have support for use of ASN.1 value notation in the user interface (via 'ask' and 'print' functions).

ASN4SDT is based on the SDT code generation macros that define the data type translation from SDL types to C-types. ASN4SDT takes as input the ASN.1 definitions that have been compiled by other ASN.1 tools and outputs SDT Abstract Data Types (ADT). The resulting file is then included in the SDL model as a package. Each ASN.1 module translates to one file that contains all the ADTs generated from those ASN.1 definitions. By using code generation macros provided by SDT, ASN4SDT assures that the C-types that it generates from the ADTs are equivalent to the types generated directly by other ASN.1 tools. This ensures code-level compatibility across the existing NRC ASN.1 tool set and SDT.

4.4   Style Guide

The style guide contains naming rules for all SDL basic concepts. As in other languages these rules serve to clarify and partially document the system. However, in the case of a large system like the 3G SDL library, such naming rules are also required for tracing of the system execution, for maintenance and integration of the complete system.

The basic concept is the deterministic naming of elements. Most protocol names are named according to the following pattern:

<PROTOCOL NAME>_<NETWORK ENTITY>,

but the network entity identifier can be left out if it is not applicable. The goal is uniqueness of the name, but also traceability of the place of definition for the concept. All structural entities, like processes or non-PDU signals that belong to a protocol must have names that are prefixed with the *protocol entity prefix*. This is quite straightforward, but gates, channels and signal routes are handled slightly differently. Gates are named after the protocol or process entity for which the gate is intended, postfixed by _CTRL or _DATA where applicable. Names for signal routes and channels contain the names of the block types or processes that they connect. Unidirectional channels define the name in order of appearance, but in the case of bi-directional channels topographical rules based on the graphical representation of the blocks (for example, left to right) are used.

Naming conventions allow the introduction of new concepts into the framework. In this case the most central of them is the PDU (*Protocol Data Unit*). A PDU is used to pass data between protocol entities on the same level (peer-to-peer), but typically in different systems. Since PDUs are the only data structures that pass system boundaries, they must be encoded and tagged with some identification, before they can be sent to other systems. This requires precise definition of the external data representation (the actual bits in the air), which is achieved for example by using ASN.1 encoding rules, such as Packed Encoding Rules (PER) or Basic Encoding Rules (BER). PDUs are named in the protocol standards and this name, written in capital letters, is used as the name for a PDU signal.

Finally, naming conventions allow relatively easy development of specialized tools that gather otherwise unidentifiable information. Applications here include enhanced tracing analyzers, documentation tools and lint-like static checking tools. This is made possible by the existence of a textual representation for SDL.

5   Example

This section presents an example of the principles described in the previous sections. The RNC network element is represented by a system level block instance, where the protocol implementations of that network element are represented as sub block types (see Fig. 4).

The protocol implementations themselves are defined in packages as block types, so it is possible to reuse the same code by instantiating the block types in different locations.

Figure 5 shows a system consisting only of protocol RRC, named RNC_RRC.

1. The system includes packages that define the actual protocol implementation (RRC_RNC), but also packages for the interfaces that it uses to commu-

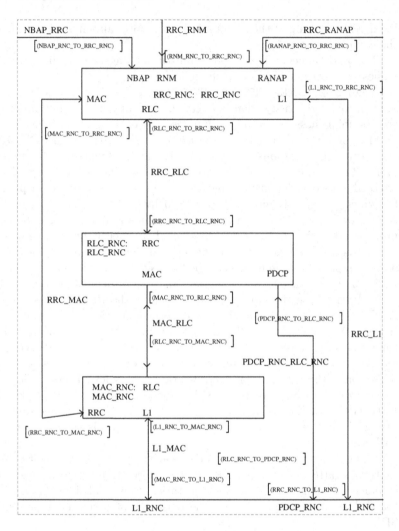

Fig. 4. Some protocol implementations within the block type for network element RNC.

nicate with protocol implementations that are below it (L1, MAC, RLC, NBAP). Furthermore there are packages for ASN4SDT specific definitions, SDT-related C data types, common data types and other global definitions (ASN4SDT, CTYPES, COMMON_IEs, GLOBALDEFS respectively).

2. The signal lists for the interface that this protocol offers for higher levels are defined in the RRC_RNC_EXTERNAL_INTERFACES package, while the lower level signal lists are defined in the external interface for that protocol.

3. The gates are named after the protocol entities that it connects, the names of the channels have the names of the protocols entities that the channel connects.

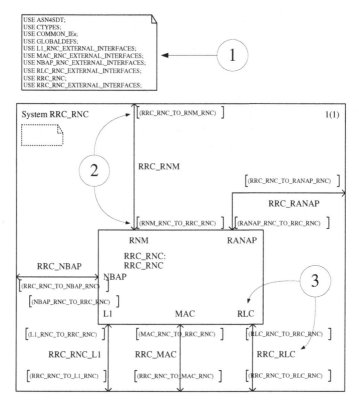

Fig. 5. A separate system for RRC protocol implementation.

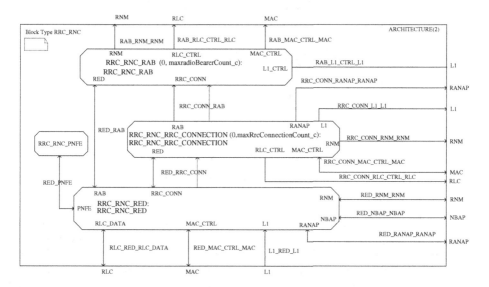

Fig. 6. Process types for RRC protocol implementation.

The same package and the same code is used in the system that constitutes the network element in Fig. 4. This allows the implementor to develop, test and debug a single protocol in isolation and within the network element context at will.

A typical process-level architecture of a protocol implementation is depicted in Fig. 6. Note that signal lists that are associated with channels and use clauses for ASN.1 data type packages have been removed for readability. A RED process is always present routing all of the actual communication, but typically there are also one or more connection processes present. A separate manager process can exist, but usually it is so simple that it is easily embedded in the RED process.

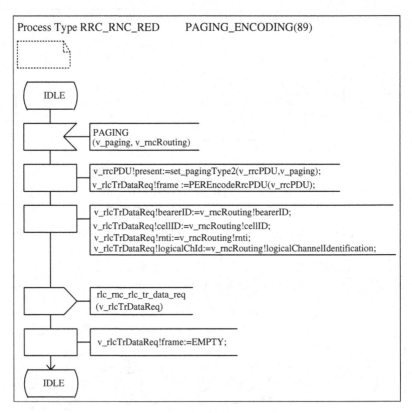

Fig. 7. Encoding of a PDU.

Finally, Fig. 7 contains a transition that shows typical handling of a PDU by a RED process. The PAGING PDU is received as a signal from RRC_RNC_PNFE, its parameters are encoded as a parameter of the primitive using the ASN4SDT provided RRC specific encoding function PEREncodeRrcPDU. Some routing information is filled in the primitive parameters and the primitive is sent further by using service primitives of the lower layer.

6   Conclusions

We have presented a framework that has been successfully used in creating a complex telecommunication system with relatively small resources. The implementation project could accommodate the continuous and occasionally major changes in specifications.

The modular structure of the framework has made integration easy, allowing distributed development of the protocols. The main components here are the agreed package structure and strict visibility rules for data types.

SDL is a very usable language in prototype protocol implementations and gives the implementor many possibilities to concentrate on the essentials. SDL, as any other language, should be used together with suitable and well understood methodologies and there should always be a corresponding implementation framework available. This framework should be based on experience and it should provide a clear 'cookbook' of instructions for the implementation. A good framework is an absolute must if the project has to deal with complex systems.

The ASN.1 language increases SDL's strength by adding the distribution possibility, both in the sense of peer-to-peer and process-to-process interfaces. A good ASN.1 tool set clarifies tracing and testing and enables enhancements in the code generation. As a tradeoff, use of homebrewed tools has disabled some functionality of the Telelogic tool set, most prominently, the automatic test generation and validation within the toolset.

References

1. J. Ellsberger, D. Hogrefe, and A. Sarma. SDL : Formal object-oriented language for communicating systems, 1997.
2. ITU. Specification and description language (SDL) ITU-T Recommendation Z.100, 1993.
3. ITU. SDL combined with ASN.1 (SDL/ASN.1) ITU-T Recommendation Z.105, ITU-T, 1995.
4. J. Pärssinen and M. Turunen. Patterns for protocol system architecture. In *PLoP 2000 Proceedings,* http://jerry.cs.uiuc.edu/plop/plop2k, 2000.
5. Telelogic AB. *SDT 3.5 User's guide, SDT 3.5 reference manual,* 1999.

OSPF Efficient LSA Refreshment Function in SDL*

Ostap Monkewich[1], Igor Sales[2], and Robert Probert[2]

[1] Nortel Networks Corporation, P.O. Box 3511 Station C
Ottawa ON K1Y 4H7 Canada
ostapm@nortelnetworks.com
[2] University of Ottawa, P.O. Box 450 Station A
Ottawa ON K1N 6N5 Canada
{isales,bob}@site.uottawa.ca

Abstract. An 11-router Internet Protocol network model based on the Open Shortest Path First (OSPF) routing protocol and Link State Advertisement (LSA) is expressed in the Specification and Description Language (SDL). The corresponding simulation data to verify the performance of a proposed more efficient OSPF LSA refreshment function is presented. Network traffic generated by the routing table refreshment activity using the new function is compared to the traffic generated in the Internet today when using the existing LSA refreshment function. The relative performance characteristics were found to depend on the number of LSA packets per router and the router startup sequence. Such dependencies when using protocol standards in natural language are not always visible until a number of implementations of the standard become available and tested in the field. SDL and tools provide an inexpensive but reliable way of verifying protocols under development in advance of implementation and final agreement.

1 Introduction

The Internet is an interconnection of computers called routers. Routers are responsible for routing of information packets through one or more interconnected Internet Protocol (IP) networks for delivery to specified destinations. If a router fails, the remaining routers must re-route the packets and ensure the correct delivery. This is a very complex task in a dynamic environment. For this purpose, the Internet Engineering Task Force (IETF) has defined a routing protocol called Open Shortest Path First (OSPF) protocol [1,2]. The operation of the OSPF protocol depends on link-state databases that are maintained by each router in the network and used in calculating shortest routes to the destination. Such databases need to be continually updated by having each router originate one or more LSA database entries containing information about each link connected

---

* Special acknowledgement is made to Dr. Ramachandra Munikoti of Nortel Networks Corporation for his guidance on the project and his contribution to the business value concepts and criteria used in this paper.

R. Reed and J. Reed (Eds.): SDL 2001, LNCS 2078, pp. 300–315, 2001.

to the router. The LSA exchanges lead to additional traffic on the Internet that could cause router outages if the traffic is not managed properly. Manufacturers strive to design OSPF routers that have average downtimes no greater than a few seconds per year.

1.1 Flooding Function Performance

An important property of an OSPF router is its ability to reconfigure dynamically when new routers are added or removed. If the topology of the network changes, each router must adapt quickly without losing information packets or disrupting the end user. To do this, OSPF routers communicate with one another over the network and generate additional traffic overhead. For example, routing tables maintained by every router are updated every 30 minutes by flooding the network with LSA packets from each router. According to Alex Zinin [3], Area Border Routers currently used may generate OSPF traffic on the network in bursts that may cause network exhaustion for several seconds. Depending on the router implementation, such bursts can cause routers to crash, bringing the area network down and requiring operator intervention. As a solution, Zinin proposed a method for dispersing this traffic over the entire 30-minute interval between floods. This proposal was contributed to IETF as an Internet Draft for a more efficient LSA refreshment function [3].

This paper presents the results of the work carried out to verify the operation of the new flooding function and to evaluate the SDL-based design process used in terms of a number of business values. An SDL [4] model of an 11-router OSPF network was developed and simulation data generated to verify the proposed refreshment function performance. Numerical data for the routing table refreshment traffic using the new function is presented based on SDL simulations and the data is compared to that for the currently used LSA refreshment function. Performance characteristics were found to depend on the number of LSAs per router and the router startup sequence.

1.2 Improving RFC Quality

Some important aspects of the refreshment function behaviour are not evident form the natural language specification and normally would require expensive implementation for their determination. SDL (in conjunction with commercial tools) provides an inexpensive approach for such evaluation during the standards development process. At the same time, the results obtained are reliable and have been supported by the individual experts in the field.

Traditionally, IETF Requests for Comment (RFCs) have been specified in ASCII format to permit reading of the content without special tools. However, verification of the RFCs cannot be carried out until a number of implementations enter into operation in the network. Implementation of new protocols are often delayed or not considered due to a lack of confidence in a proposal.

There is value in introducing a greater degree of precision and better documentation capability when specifying RFC content in IETF protocol work. Past experience indicates that there are identifiable problems with the current practice. The trend today is towards improving the quality of standards and specifications. It is appropriate to examine the possible benefits of formal methods in IETF.

2   The Project

This project was done with a good knowledge of SDL, commercial tools for SDL and communication networks in general, but without detailed knowledge of the OSPF/LSA concepts or operation. By the end of 4 months, much of the OSPF was expressed in SDL, complex routing operations were simulated and routing table refreshment traffic performance evaluated. Much of the rapid progress is attributed to the methodology and tools based on SDL, MSC [5] and UCM [6].

The project focused on developing an SDL model for a small, but reasonably realistic IP network using OSPF routers with the LSA refreshment function. Specific focus was on including the capability to switch to the recent proposal to make LSA refreshment more efficient as described in IETF Internet Draft *draft-ietf-ospf-refresh-guide.00.txt*. An important objective was to generate the programming code for the new LSA refreshment function from SDL, port the code to a Nortel Networks product platform and integrate it with the proprietary OSPF implementation for testing and evaluation. While the basis for the porting was established during this phase, the actual porting will be done during the next phase of the project.

To achieve these objectives, the following components were modeled in SDL:

1. the new LSA Refreshment Function and router functions as specified in IETF *draft-ietf-ospf-refresh-guide.00.txt*;
2. the original LSA refreshment function specified in RFC 2178;
3. the relevant parts of OSPF network topology necessary for the simulation;
4. three link types to interconnect the routers and link-router interfaces;
5. the interface to the Internet via the operating system Application Programming Interface (API) and the TCP/IP Socket.

From these components, C programming language code was generated from the combined SDL model and the routing table refreshment operations were performed using the 11-router executable network model connected to the Internet. The SDL methodology was then assessed against 5 selected business values and compared to the software design approaches based on manual programming procedures that are widely used in industry today.

At all stages, the simulation results were made available to the author of the IETF Internet Draft [3] for the Efficient LSA Refreshment Function who confirmed their validity. Numerical results were presented to the IETF OSPF Working Group meeting and the suggestions from the meeting were incorporated into the subsequent model refinements and simulations. The OSPF WG agreed that the results and the SDL model should be part of the Efficient LSA Refreshment in OSPF Informational RFC. Also, informal discussions with individual

IETF experts were held regarding common experiences with the magnitude of
effort required to implement OSPF functions. This was used in the evaluation
of the SDL-based methodology in terms of business values.

3   The Formal Process

Figure 1 illustrates the process that was used to produce the SDL model and the
numerical data. The initial understanding and capture of OSPF/LSA concepts
was done with the aid of hand-written and tool-written Use Case Maps (UCM).
UCM diagrams were used to capture key functional requirements as sequences of
actions and reactions that are not easy to capture directly in an Extended Finite
State Machine (EFSM) model. Rules for conversion of UCM representations to
MSC and SDL have been reported by a number of authors and were applied
here [7,8]. Production of SDL, Message Sequence Charts (MSC), C code, and
simulation/validation data was done using commercial tools [9].

4   Network Topology

The three network topologies that were used in the study are illustrated in Fig. 2.
Circles marked Ri represent routers that form an area network in which each
router may itself be connected to an area network. For example, R1 may also be

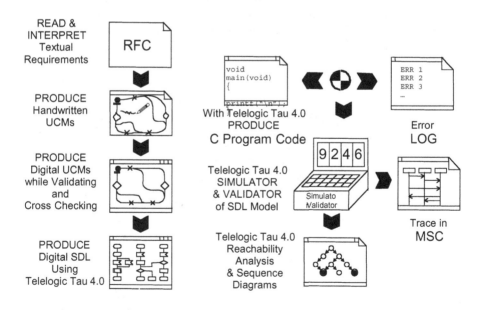

Fig. 1. The formal process flow including requirements capture, specification,
description and the use of commercial tools to perform simulation, validation,
analysis and C programming code generation.

connected to routers Ra and Rb (not shown) of an attached area network which would contribute to the routing table refreshment traffic. The arrows indicate bi-directional router links connected to router interfaces; the links may be of different type.

An LSA used in updating routing tables is a block of data produced by a router specifying the state of each link attached to it. Among other data, it contains link and router identifiers and details for shortest path calculations. Normally, every half hour, a router produces a fresh LSA. Using one or more update packets, the router sends over each link attached to it, the LSA it prepared as well as any LSAs received from the area network that might be attached to it. Currently, this is done regularly every 30 minutes by each router independently. In addition, an LSA may be sent on per-need basis, for example, when a new link or a router is attached. A new LSA is also sent when the previous LSA gets too old. That is, as LSAs get routed, their age is increased by an ageing algorithm. All routers in the network collect the LSAs, check them and use the data from valid LSAs to build identical routing tables.

Topology 1 and Topology 2 were used as component topologies to make up the overall 11-router network model in SDL. Simulation data was generated for all 3 topologies, however, Topology 3 was the main focus in the LSA refreshment function performance assessment.

Not all router functions are relevant to OSPF/LSA refreshment. A block diagram of the router functions that were modeled in SDL is shown in Fig. 3. The modeled components include: the OSPF Interface Function, the Flooding Function, the LSA refreshment functions and router port Interfaces.

The links interconnecting the router ports also perform packet processing functions. OSPF specifies 5 kinds of links listed in the boxed-in part of Fig. 4. Of

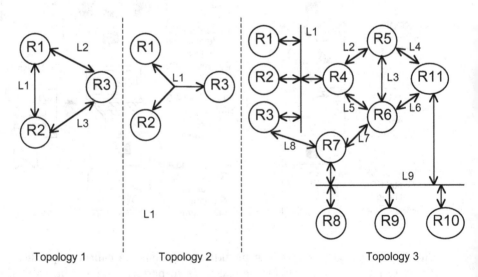

Topology 1   Topology 2   Topology 3

Fig. 2. The figure illustrates 3 topologies used

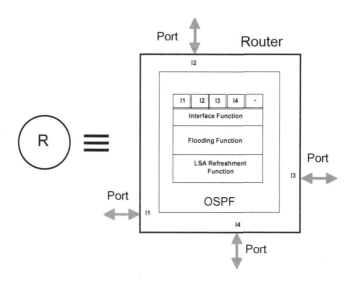

Fig. 3. Components of an OSPF router model which were expressed in SDL

the specified link types, only two types were modeled in SDL: Point-to-Point and Broadcast. The other link types were not considered to be essential to meeting the project objectives.

An additional link type, the Datagram link, was invented during the project to allow access by the SDL network model to other networks. To permit connection of the model to the Corporate Network and the Internet, a mechanism had to be found for integrating the SDL with the low-level operating system socket calls. Such mechanisms had to be modeled in SDL to act as links in the overall network model while preserving the network model topology.

Datagram links were made up of two separate links with different port numbers so that packets could be sent to the Internet on one link process and received by the same program on a different link process. The transfer of the update packets was done by encapsulating them inside User Datagram Protocol (UDP) packets. This is illustrated in Fig. 4. System sockets and the API were part of the Sun Microsystems Solaris functions for TCP/IP access to the Internet.

The links are illustrated in the SDL models in Fig. 5 through Fig. 8. The mesh topology of Fig. 5 is interconnected by Point-to-Point link types. On receipt of an update packet from a router, the Point-to-Point link forwards the packet to the two routers that are not the senders of the packet. The star topology of Fig. 7 uses the Broadcast link type. In this case, on receipt of an update packet from a router, the Broadcast link forwards copies of the packet to all routers in its router table.

Figure 6 shows a direct replacement of the Point-to-Point link type of Fig. 5 by two Datagram links, each of Point-to-Point type. Each Datagram link sends and receives UDP packets by means of system socket calls. In Fig. 8 the Broadcast link is replaced by 3 Datagram Broadcast link types that communicate with

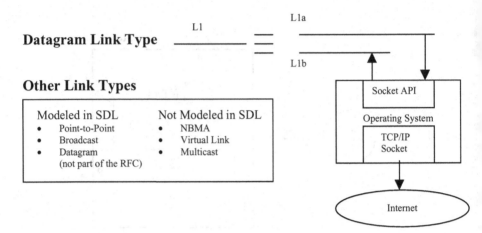

Fig. 4. Datagram link type to access the operating system API and TCP/IP socket. The table of other link types lists those link types which were modeled in SDL and those which were not.

each other. For example, when a packet is received on L1a it is sent to both, L1b and L1c.

5   The SDL

SDL models at process reference level for Topology 1 and Topology 2 network components are given in Fig. 5 through Fig. 8. Topology 3 is too complex to show in a single figure, however, with the aid of Fig. 2, the SDL model for Topology 3 can be easily inferred from the models for Topology 1 and Topology 2.

In addition to the 3 network topology models, the full SDL model includes a number of functions essential to the full OSPF/LSA refreshment operation. Figure 9 shows two variants of the delay calculation functions, one corresponding to the "Normal" flooding delay currently in use and the second corresponds to the "Zinin" delay algorithm that spreads the flooding operation over the Normal interval.

Other parts of the model that are not shown in the diagrams include components that perform configuration control of each topology, LSA ageing, LSA refreshment process management and the link details for update packet handling.

In addition, substantial part of the OSPF protocol was expressed in SDL, however, this part was not directly relevant to the operation of LSA refreshment. For example, routing calculations to find the shortest path through the network to deliver user data between end users is not a subject of this paper.

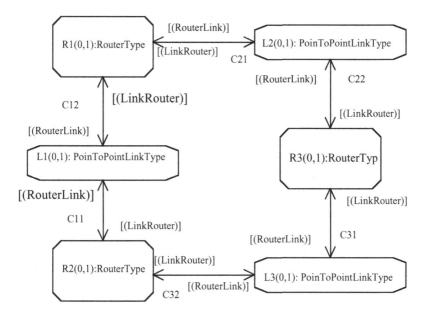

Fig. 5. Topology 1 – 3-router mesh topology in SDL with Point-to-Point links

6   Performance Data

The performance data quantifies the amount of traffic an OSPF IP network generates during the time intervals when routing tables are refreshed. At first, a simple case is presented to illustrate the flooding concept. A more complex case will be used to assess the performance of the proposed, more efficient flooding algorithm.

The graph in Fig. 10 corresponds to Topology 1 without any additional network areas attached to any of the routers. The graph compares the flooding traffic generated when executing the SDL model with Normal and with Zinin delays. The traffic was observed over a 7-hour period. Normal Delay refers to the LSA refreshment algorithm used in the Internet today. The delay for this case is calculated according to the SDL segment shown in Fig. 9(a). Zinin Delay refers to the new Efficient LSA Refreshment Function algorithm proposed by Alex Zinin. The delay for this case is calculated according to the SDL segment given in Fig. 9(b).

In the Normal Delay case, the network is simultaneously flooded at 30-minute intervals with update packets from all 3 routers. This means that 12 update packets (Vertical Axis), each carrying one LSA, are placed on the network every half hour (Horizontal Axis). In the Zinin Delay case, the 12 update packets appear in 3 groups of 4 packets and the groups are spread over the 30-minute period. The total number of update packets over the same interval is the same for both cases

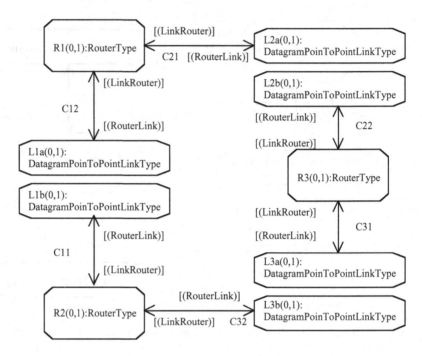

Fig. 6. Topology 1 – 3-router mesh topology in SDL with Datagram links to sockets

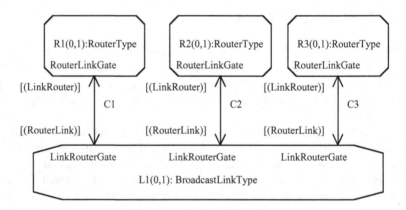

Fig. 7. Topology 2 – 3-router star topology in SDL with Broadcast links

since both must deliver the same amount of refreshment information. Zinin's approach floods more frequently but at lower traffic volume at each instant.

In Fig. 11, the simulation data corresponds to Topology 3. Here, the routers are interconnected by Point-to-Point and Broadcast link types to approximate the operation of a realistic network. The traffic pattern remains relatively simple because simultaneous flooding of the network by all routers is applied and only

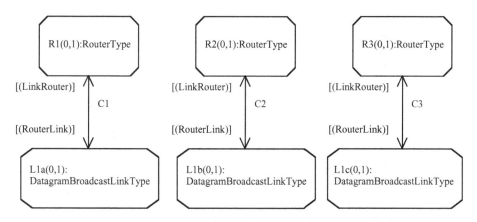

Fig. 8. Topology 2 – 3-router star topology in SDL with Datagram Broadcast links to sockets

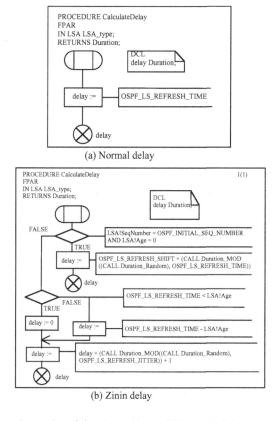

Fig. 9. SDL procedures for: (a) calculation of Normal delay used in the Internet today (b) calculation of Zinin delay to realize the efficient LSA refreshment function.

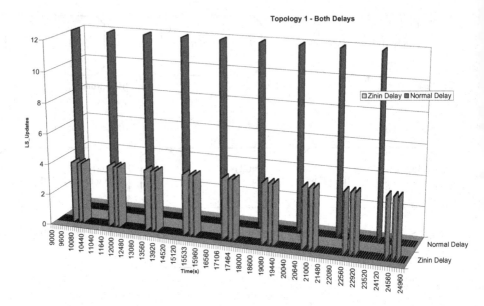

Fig. 10. The number of update packets carrying LSA refreshment are plotted with time. "Normal" designates the flooding method currently used on the Internet. "Zinin" designates the flooding method proposed by Zinin. In each case, the packet distribution is achieved by means of a delay algorithm. This graph corresponds to the simple case of Topology 1 of 3 routers and only 4 update packets per router, for a maximum peak traffic of 12 packets.

one LSA per update packet is used. Subsequent simulation data will illustrate the significance of multiple LSAs and independent router starts.

In practice, each router may produce a different number of LSAs depending on the number of links attached to it. Although each router sends all its LSAs spaced in time according to an internal algorithm, the routers may be started at arbitrary points in time during installation and are not synchronized with each other. To simulate this, fixed and random numbers of LSAs per router were used with 3 kinds of router start times: simultaneous, equally spaced and random. Many other combinations of topologies and parameters were used during the project, but are not reported here.

The graphs in Figs. 12, 13 and 14 present a good cross section of the results. At each instant in time, data points are plotted which are the sum of all LSAs sent by every router in the entire area network of 11-routers (including the LSAs from the area networks associated with each router). This sum is calculated by adding the LSAs on each link of the 11-router network at that instant in time. As required by the Zinin proposal, when a router has more than 10 LSAs to send, several update packets are sent in succession so that no update packet carries more than 10 LSAs. The number of LSAs plotted is all at separate time instants and therefore is not cumulative. The highest peaks on the graphs may be taken as maximum values.

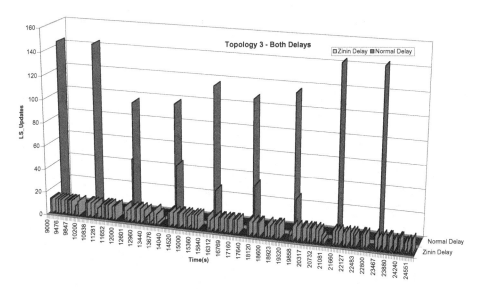

Fig. 11. This graph represents the simulation data for Topology 3 with simultaneous router starts and one LSA per update packet. The peak traffic of 143 update packets on the network may be concentrated at regular intervals or distributed evenly over time.

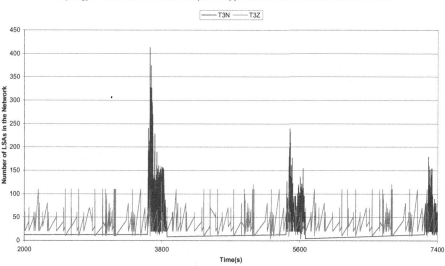

Fig. 12. This graph represents the total number of LSAs in the Topology 3 network model for the Normal (T3N) and Zinin (T3Z) flooding algorithms when all routers are started simultaneously and each router has a random number of LSAs to send.

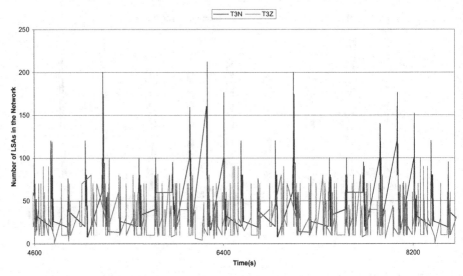

Fig. 13. This graph represents the data for the same parameters as those of Fig. 12 but the routers are started at equally spaced intervals of time.

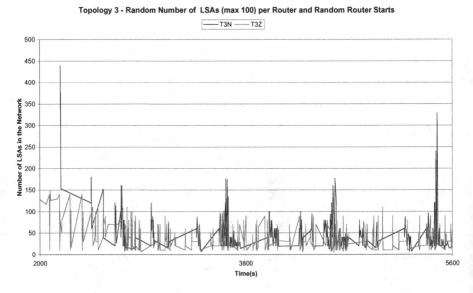

Fig. 14. This graph represents the data for the same parameters as those of Fig. 12 but the routers are started at random instances in time.

7   Process Assessment

The project presented an opportunity to assess the SDL-based design methodology in relation to a number of business values used in industry today. Table 1 summarizes the aspects of the methodology that were found to impact these values.

It is estimated that by using SDL and tools, the time to learn the complex OSPF protocol and to produce an executable implementation was about one half that quoted by some industry practitioners for the current process. Several IETF experts agreed that we had completed this task in a shorter time than they had thought possible.

The work was started with a good knowledge of SDL, MSC, UCM, C programming, Unix, Solaris and the commercial tools but with no prior experience with OSPF/LSA refreshment. The first simulations using the 11-router SDL network model were done only 4 months after the start of the project. The entire project, which included porting considerations, integration with the legacy implementation, interface to the Internet, numerous refinements and extensive simulations, took 8 months.

The size of the model turned out to be relatively small due to the high degree of module reusability considering the high complexity of OSPF/LSA refreshment operation. The resulting object code was only about 100 Kilobytes in size. Most of the modules that were part of the learning process were reused in the design.

As learning matured, the system was redesigned several times with very little loss of time, underscoring the ease and cost effectiveness of re-design. Early verification of the design could be done at the architectural and conceptual levels prior to commitment to detail. Several levels of management and the customer were able to work with the designer due to the user-friendly graphical interface of the commercial tools and the capability to animate the design on the screen. It made it easy to agree on customer's requirements and the deliverables.

No additional software, not intended to be part of the design itself, had to be produced. Commercial tools provided all the additional software requirements including the design environment and the tools for simulation, validation, analysis and generation of programming code from SDL. Interfacing the modules that were produced separately during the project was easy and the modules were reused many times throughout the design.

These and other components were identified as having had a clear impact on the project. The most important components are tabulated in Table 1 and the area of impact is indicated by an asterisk.

8   Conclusions

Zinin's proposal results in lower refreshment traffic at any instant compared to the refreshment traffic when using the current method. However, Zinin's approach requires routers to process refreshment traffic during the intervals when a router would route primarily data packets under the present method.

Table 1. Impact of SDL-based methodology factors on selected business values

BUSINESS VALUES	PROCESS COMPONENTS THAT DEMONSTRATED IMPACT ON BUSINESS VALUES													
	Reduced Learning Time	Understandability	Project Hand-over	Architecture Level Testing	Simulation Environment	Error-free Code	Traceability	Automated Test Generation	Intellectual Property Retention	Reuse	High Quality Standards	Error-free Test Cases	Scalability	Documentation
Speed to Market	*	*	*	*	*	*	*	*	*	*	*	*		*
High Availability						*					*			
High Performance				*		*					*	*	*	
Optimal Cost	*	*	*		*	*		*	*	*	*	*	*	*
Ease of Change		*	*				*	*		*			*	*

The quality of the RFCs can be improved by reducing the number of errors and ambiguities in the natural language specification. Accurate verification of protocol behaviour can be done well in advance of implementation.

The value in introducing a greater degree of precision through the use of SDL when specifying IETF protocol RFCs was demonstrated and it is expected that SDL will be more widely accepted in IETF.

IETF RFCs are adopted only after demonstrating interoperability between two or more implementations. Using SDL with commercial tools to specify RFCs can greatly enhance interoperability between independent implementations of the RFCs.

A reduced training interval and other aspects of the SDL-based methodology impact the business values of speed to market, high availability, high performance, optimal cost and ease of change and integration.

References

1. John T. Moy, *OSPF, Anatomy of an Internet Routing Protocol*, Addison-Wesley Longman Inc., 1998.
2. John T. Moy, *OSPF Version 2*, Internet Engineering Task Force Request For Comment 2328, April 1998.

3. Zinin, A., *Guidelines for Efficient LSA Refreshment in OSPF*, Internet Engineering Task Force Internet-Draft, *draft-ietf-ospf-refresh-guide.01.txt*, July, 2000.

4. ITU-T, *Languages for Telecommunication Applications – Specification and Description Language (SDL-2000)*. Geneva, ITU, 1999. ITU-T Recommendation Z.100 (11/99).

5. ITU-T, *Languages for Telecommunication Applications – Message Sequence Charts (MSC-2000)*. Geneva, ITU, 1999. ITU-T Recommendation Z.100 (11/99).

6. Buhr, R.J.A., *Use Case Maps as Architectural Entities for Complex Systems*, IEEE Transactions on Software Engineering, Special Issue on Scenario Management, Vol. 24, No. 12, December 1998.

7. A. Miga, D. Amyot, F. Bordeleau, D. Cameron, M. Woodside, *Deriving Message Sequence Charts from Use Case Maps Scenario Specifications*, this volume.

8. Igor S. Sales, Robert L. Probert, *From High-Level Behaviour to High-Level Design: Use Case Maps to Specification and Description Language*, SBRS 2000, Simpósio Brasileiro de Redes de Computadores, Belo Horizonte, Minas Garais, Brazi, May, 2000.

9. Telelogic Inc., *Telelogic Tau 4.0*, Telelogic AB, Malmo Sweden, February, 2000.

Using SDL in a Stateless Environment*

Vassilios Courzakis[1], Martin von Löwis[2], and Ralf Schröder[2]

[1] Siemens AG, ICM N IS E12, Siemensdamm 50, 13629 Berlin
Vassilios.Courzakis@bln1.siemens.de
[2] Humboldt-Universität zu Berlin, Rudower Chaussee 25, 12489 Berlin
{loewis,schroed2}@informatik.hu-berlin.de

Abstract. Telecommunication services are often implemented to maintain their state in persistent storage. The application logic program then
has no state variables of its own; it is seeded with a state depending
on call context. Superficially, this contradicts the notion of extended
state machines as they are defined by SDL-processes, where the state is
part of the state machine. This paper presents an approach to separate
state from program logic that is transparent to the SDL designer. This
approach has been implemented in the SITE SDL runtime system in
co-operation with Siemens, Berlin.

1 Introduction

Using SDL [1] to design and implement telecommunication services typically
requires maintenance of per-call state information over the lifetime of the call.
On a specific installation, many calls are processed both simultaneously (i.e. on
multiple processors) and in an overlapping fashion. Therefore, a natural architecture is to create an SDL process every time a call is initiated or detected, and
keep the per-call state in the variables of the process. Our experience shows that
this implementation approach allows elegant definition of the program logic.

However, the approach has a number of drawbacks when reliability and load-
sharing need to be considered. When translating SDL processes to target machine programs, process variables are typically translated to variables in a programming language for the target machine. Therefore, the per-call state will be
maintained in the operating system process executing the protocol logic.

When assigning SDL processes to operating system processes, it is often not
feasible to use a one-to-one mapping. Operating systems impose a limit on the
number of processes they can create, and this limit is typically much smaller
than the maximum number of concurrent calls that must be processed. As a
result, many SDL processes are executed in the context of a single operating
system process (potentially using operating system threads where possible). This
scenario is shown in Fig. 1. When implementing an SDL system in such a manner,
various aspects must be considered, and various strategies can be followed, see [2].

---

* We would like to thank our partners at Siemens, in particular Mr. Andreas Vogel,
for the numerous ideas and suggestions that lead to the technology described in this
paper.

R. Reed and J. Reed (Eds.): SDL 2001, LNCS 2078, pp. 316–327, 2001.

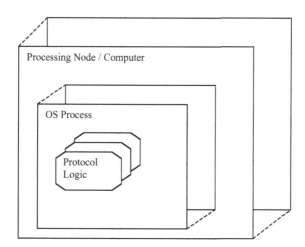

Fig. 1. Multiple SDL processes are executed in a single operating system process

To allow for load sharing, multiple processing nodes may be used to execute the same program logic. Likewise, multiple operating system processes may execute the program logic; such a scheme may be used to avoid the synchronisation that is common for multi-threaded runtime systems.

Unfortunately, the presented scheme for translating SDL processes into target system programs means that the association between SDL processes and operating system processes is fixed for the lifetime of the SDL process. The main reason is that the variables of each process are tied to the virtual address space of the executing process; moving them from one process to another is not supported by operating systems.

Likewise, reducing the number of available machines is difficult. Each operating system must be stopped before the machine can be switched off. In turn, each SDL process must stop before that, which means that each call must complete first. In applications with long-running calls, a significant amount of time may be needed to turn off the service on such a machine for maintenance.

Keeping the SDL process state in main memory also impacts the reliability and fault tolerance of the service. The potential failures range from hardware failures over operating system errors to errors in the SDL runtime system or the program logic. For many of these failures, not just the single SDL process will fail, but at least the entire operating system process, or the entire machine. That means that information on all calls that have been executed on that OS process or processing node is lost – even for calls that had no activity at the time of failure.

To provide fault tolerance, the state of a call must be stored persistently, or in a replicated way, or both. An SDL process whose state is stored persistently now needs to modify the persistent and replicated copy, instead of modifying the local OS virtual memory.

Computer systems typically provide a memory hierarchy, one layer of memory is a cache for the next layer of the hierarchy: for example, the main memory caches the hard disk, and the processor cache caches the main memory [3]. In these hierarchies, the different levels often vary in volatility and performance.

In implementations of persistent storage, throughput of the persistent storage is typically orders of magnitude lower than throughput of the main memory. This means that a trade-off between failure resistance and efficiency must be found. We have implemented different schemes for persistent state in an SDL process, using the following principles:

- The state of an SDL process needs only to be persistent during periods of inactivity (that is, in "stable" phases of the call): therefore, the state is only made persistent at the end of transitions (not necessarily after every transition);
- As a result, when a process is restored from persistent state, it will always restart at the beginning of a transition;
- Because the lifetimes of each call, and thus each process, are independent, saving of the system state always occurs on a per-SDL-process base, instead of saving the entire system simultaneously.

These principles have the consequence that failures occurring in the middle of a transition may result in loss of information. That means that recovery after a failure must consider the possibility that a certain transition was executed partially. For our applications, that is acceptable as protocols are typically designed to deal with duplicate information at some layer. In addition, after a restart, just continuing after the point of failure may be inappropriate, as that might generate the same failure again in case of an error in the program logic.

Readers should notice that a failure will affect at most the SDL processes that were actively executing transitions at the time of failure: processes that had no signals to process will have stored their state persistently.

In the following sections, we present a number of alternative approaches that we have considered. It should be pointed out that the resulting system is still experimental at the time of this writing, so no report on operational experience is available. All of these techniques have been implemented at least experimentally in the SDL Integrated Tool Environment, to determine whether a realisation is feasible.

1.1   The Integrated SDL Environment (SITE)

The SDL Integrated Tool Environment [4] supports the flexible integration of SDL into various application contexts. It consists of tools for the analysis of an SDL specification by based on an internal representation, and code generators that produce executable code for the target environment (see Fig. 2).

The C++ generator generates code that is still independent from the target environment; in particular, the code can be used to build both simulators and complete implementations of an SDL specification. The environment-specific semantics are brought into the C++ program by means of libraries, which are adjusted to the specifics of the underlying communication platform. These specifics include:

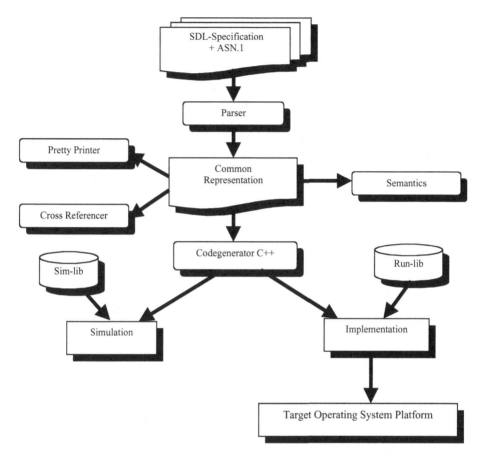

Fig. 2. Components of SITE

– the correlation between SDL processes and operating system processes;
– the transport mechanism(s) used to deliver SDL signals over the network;
– the scheduling strategy for choosing among ready SDL processes;
– deployment and run-time management details of the environment.

SITE currently supports SDL-96, and a number of SDL-2000 aspects. With respect to the storing process state into an external state, serialisation of potentially cyclic graphs is possible if object types (as introduced in SDL-2000) are used. Because SITE currently does not support object types, that aspect is not discussed in this paper. For many other aspects of SDL-2000 (such as composite states), we believe that they will have little, if any, impact on the strategies presented here.

2   Check-Pointing vs. Stateless Operation

In the following discussion, we will use the term "stateless" for a mode of operation where per-call information is available in main memory only for the small period of time where the actual processing of a message occurs; once the processing is complete, the state is deposited in permanent storage and removed from main memory. This is opposed to "check-pointing", where the state of the process is kept in main memory during the entire life of the process, and copied into permanent storage regularly.

The reader may have noticed that stateless operation is not required to achieve failure resistance. Instead, regular check-pointing of the current process state is sufficient: if a part of the installation (a machine or an operating system process) fails, the last persistent copy of the state of each process can be used to recover the process. Given the principles outlined above, processing of a transition proceeds in the check-pointing case as follows:

1. Based on the stimulus (input signal or timer), initiate the transition;
2. Run the transition code until it performs a `nextstate` terminator action;
3. Produce an external, persistent copy of the process state;
4. Record this copy as the last copy for this process.

Considering that producing a persistent copy of the process might be an expensive operation, we defined two alternative schemes for externalising a process. In the automatic scheme, the state is stored after each transition. In the application-controlled scheme, the application determines whether the state should be externalised at the end of the transition. To announce its decision to the SDL runtime system, it invokes `schedule_externalization`, an external procedure, which then records the applications' plan to invoke externalisation for the process, and does so once the process completes its transition.

For long-running processes, we expected that protocol designers would typically know in what state a long period of inactivity is anticipated, and request externalisation just before entering these states. However, we found that protocol designers will consider that aspect not part of their problem domain, and will initially refuse to introduce calls to this procedure in their program logic. Instead, the automatic scheme was used in most cases. Once detailed performance data are available, the decision to always save state may need reconsideration.

Completely stateless operation is necessary to support the load-balancing aspect in our systems. Once a process is stored into persistent state, its target language representation is removed from the operating system process. That not only has the advantage of providing failure resistance, but it also allows a reduction of the memory requirement in the target machines as idle SDL processes will not consume memory. For stateless operation, and assuming automatic saving of state after each transition, processing a signal for a process requires the following steps:

1. Determine the target process of the signal;
2. Find or create a target language object (such as a C++ or Java object) to hold the process state;
3. Find the process state in persistent memory;
4. Restore the process state into the programming language object;
5. Execute the transition;
6. Save the process state into the persistent storage, replacing the previous copy of the state;

This sequence requires that the persistent state of a process is present before the transition starts. To achieve this invariant, the start transition needs special treatment: either a persistent state of the process must be created before the start transition runs, or the `nextstate` action that completes the start must allocate a fresh chunk of persistent storage, instead of replacing a previous one.

3   Selecting the Process Representation

Depending on the underlying storage system used, different approaches can be taken to represent the process state. Considering the principles for storing processes, an external copy of a process will contain the following information:

- the value of implicit active expressions (`self`, `sender`, `parent`, `offspring`);
- values of process variables (including formal parameters);
- state of SDL process, and current position in process graph;
- procedures currently being invoked, and their variables;
- input queue of the process (including saved signals);
- the set value of running timers.

A number of items in this data set need further elaboration. While storing the `Pid` expressions `sender`, `parent`, and `offspring` is clearly necessary to restore the state, saving the value of self is questionable: When an inactive process receives a signal, its persistent storage must be located. To do so, the process identity must be used as a key. Therefore, the value of self is known before restoring the process, so it does not need to be part of the stored information set.

When storing the variables of a process, it is normally not sufficient to just store their value. SDL has the notion that variables may be unassigned: that is, they have no associated value. Since SDL-2000 defines that accessing such a variable will raise a predefined exception, knowledge must be preserved which variables are undefined[1].

To store the point of execution in the control flow, it is not sufficient to just store the `state` that the process is in. If a transition contains a procedure call, this procedure may enter a state, at which point the process might be externalised. To continue execution when the process is restored, the exact call in the

---

[1] Alternatively, a deviation from the SDL semantics (with no detection of access to undefined variables) may be acceptable to users.

transition must be identified as well, so that execution can continue after the call action once the procedure call returns. In a simplified scheme for externalising SDL processes, states in procedures could be ignored as points where externalisation occurs. As a result, saving the procedure stack will not be necessary.

When saving the timers of a process, the exact set of information stored for a timer was subject to debate. For a stateless model, it turns out that some OS-provided timeout event system must be used to initiate restoration of the SDL process. To support reliable operation, these operating system timers must be persistent and replicated themselves, should one of the machines fail. Therefore, the expiration time of the timer does not need to be part of the saved information set. On the other hand, most systems only provide a numeric identification for a running timer. That means that the actual set of parameter values must be stored with the timer, in addition to the system-dependent timer identification value.

Process variables will often contribute the largest part of the external process state. To store variables, some efficient and compact scheme is required, which yet supports the wide variety of SDL data types. As mentioned before, reference data (as possible with the SDL-2000 object types) are not considered here.

3.1   Flat Byte Stream Storage

To provide a flexible storage mechanism for SDL processes, one possible approach is to find some stream-based externalisation scheme. Such a scheme has the advantage of being modular: each piece of information needs to define its own marshalling scheme; aggregation of information occurs by concatenating pieces (potentially including structural information).

In our experience, the ASN.1 [5] Basic Encoding Rules [6] are well suited to provide a linearisation of arbitrary data in SDL – in particular when SDL is combined with ASN.1 [7]. In our applications, many data types are defined with ASN.1 by other parties, and we use these types unmodified. As a result, SDL developers rarely use "native" SDL type definitions (struct, string), except when using predefined types or generators that have no straightforward ASN.1 equivalent (array). In addition, the SITE environment provides generated encoding and decoding routines for ASN.1 data types.

Therefore, we have attempted to define the state of an SDL process in terms of an ASN.1 data type, which, when encoded according to the BER, provides a straightforward "stream format" for the persistent representation of the SDL process. This type has roughly the definition shown in Fig. 3

This data type definition uses a number of implicit assumptions about the SITE SDL code generator. For example, storing variables in a sequence-of without their name assumes that there is some explicit enumeration capability for variables in the run-time system, which always enumerates the variables in the same order. Likewise, storing the process type name as a character string means that it is possible to create instances of a process when only given its name. While these assumptions are all true for the SITE environment, this data type is by no means a general solution that directly ports to other SDL implementations. In particular, usage of the ANY type to store a variable's content assumes implementation-specific knowledge: values are always stored as the basic encoding of some ASN.1 type, even if the variable's sort is not defined in terms of ASN.1.

```
Process ::= SEQUENCE{
      typename    [1] IMPLICIT IA5String,
      self        [2] IMPLICIT PID,
      last-signal [3] IMPLICIT Signal OPTIONAL,
      parent      [4] IMPLICIT PID,
      offspring   [5] IMPLICIT PID,
      state       [6] IMPLICIT INTEGER,
      procedures  [7] IMPLICIT SEQUENCE OF
Procedure
                      OPTIONAL,
      variables   [8] IMPLICIT SEQUENCE OF
Variable,
      signals     [9] IMPLICIT SEQUENCE OF Signal
                      OPTIONAL,
      timers     [10] IMPLICIT SEQUENCE OF Timer
                      OPTIONAL
      }

   PID ::= … -- depends on addressing
infrastructure

   Procedure ::= SEQUENCE{
      typename    IA5String,
      state       INTEGER,
      return-label INTEGER,
      variables   SEQUENCE OF Variable
      }

   Variable ::= CHOICE{
      invalid [1] IMPLICIT NULL,
      value   [2] IMPLICIT ANY
      }

   Signal ::= SEQUENCE {
      typeid          INTEGER,
      sender          PID,
      parameters      SEQUENCE OF Variable
      }

   Timer ::= SEQUENCE {
      typename        IA5String,
      parameters      SEQUENCE OF Variable,
      }
```

Fig. 3. Sketch of ASN.1 data type for persistent SDL state

In addition, while it is possible to restore the state of a process using this data structure when the activity of a call continues, there is no support for versioning of the process logic. In an installation, it is sometimes required to perform updates of the process logic, without shutting the entire system down. To support upgrading, existing calls must continue to run with the old logic, since the layout of their external representation may have changed in the modified process logic (for example, due to addition of variables).

We found that this strategy of externalising an SDL process allows easy adaptation to different underlying storage APIs, as those APIs at a minimum offer to store a flat sequence of bytes. We also found that this representation is quite compact, requiring about 70 bytes for a trivial SDL process; depending on the number of variables. Using different encoding rules, such as. the Packed Encoding Rules (PER), further reduction might be possible.

During experimentation with that approach, we also found a number of drawbacks: Storage APIs sometimes require to give an upper bound of the maximum size that a byte array will consume in advance. It is difficult to estimate the maximum size of the BER-encoded process representation, as the number of procedure calls and the number of pending signals cannot be determined in advance. Likewise, variables that have variable-sized representations (such as sequences-of types) can be estimated only if a size constraint is given. In practice, the SDL designer will need to guess an upper boundary for the size of a process' state representation, and might then adjust the guess based on experimentation. If the estimate is too small, processes exceeding the limit might fail to externalise. If the estimate is too high, the storage management system might operate less efficiently.

3.2   Using Typed and Structured Storage

A number of environments for persistent and structured storage prefer an operation mode where each application defines an image layout in advance. The storage environment then offers an interface to associate main-memory address locations with persistent storage. To establish this mapping, the data type of each memory location must be known, so that the amount of persistent memory associated with the address is well known. The number of variables and their types then must not change over the runtime of the process.

Superficially, that approach significantly limits the expressiveness of SDL, since much more information about process variables must be known in advance (for example size constraints on string types), and since algorithms with potentially unbounded storage requirements (such as recursive procedure calls) cannot be adequately supported.

It turns out that expressiveness of the SDL is not as severely restricted if a few deviations from SDL semantics are admitted. First and foremost, it becomes necessary to annotate variables that are needed for persistent storage, while still allowing other variables that are not stored persistently. A number of annotation forms for SDL have been designed in the past. For this specific purpose, we chose a rather implicit annotation, by introducing a new operator for each data type, which takes a value of the type and returns it. For example, for `Integer`, this operator has the signature

```
SITEContextSave: Integer -> Integer; external;
```
It is used in the form
```
dcl counter Integer := SITEContextSave(0);
```
In this example, the variable counter is associated with a persistent value (i.e. one that is saved when the context of the process is saved). The variable also has

an initial value (0). Further assignments to the variable may change its value, but they won't change the property that the variable is persistent in the process context. It is only allowed to use the operator in a variable's initialisation (or a data type default initialisation).

This turns out to be a convenient notation, for a number of reasons:

- The storage systems sometimes require applications to provide initial values for the persistent state – with that notation, an initial value is always required;
- The operators are implicitly defined for every data type, so there is no need to declare them for the developer;
- It is possible to interpret a specification using these operators as if it was fully conforming to SDL, by providing definitions for the operators that just return their argument.

With that kind of annotation, it is possible for a protocol designer to specify which variables of the process carry long-living state, and which variables only hold values with "transition lifetime" (index variables, variables carrying signal parameters temporarily, etc.).

When a process using that annotation is executed in a stateless environment, the deviation from the SDL semantics becomes clear: variables not annotated in this way will not be stored persistently. As a result, they may lose the value they had in the previous transition. If the annotations are put in place carefully, this will appear unnoticed: if a variable is only used as a loop index, it will be initialised every time before the loop starts, so access to it in its uninitialised state cannot occur. It is possible to determine whether a variable might be used uninitialised in a transition, however we have not implemented such a test in SITE, yet.

3.3   Combining Flat and Structured Storage

To overcome the limitations of the structured approach, and to give back some of the expressiveness that extensive use of variables across transactions offers, a combined scheme is possible: Variables that are annotated as persistent are stored using the typed and structured API of the storage system; everything else is packed into a BER-encoded buffer for which we must guess a size limit.

With that approach, many variables will not be presented in the BER stream, but will be directly copied into persistent storage. That means that the variable-sized BER stream becomes smaller; in turn, it is more efficiently produced and more efficiently copied into the external storage system. Further refinements to that scheme might be necessary when operational experience about the actual data volume is obtained.

4   Operating SDL Stateless

To implement a stateless SDL environment, persistent storage is but one aspect of the runtime system. Other issues include:

- addressing of processes, especially in presence of processes that migrate between processing nodes;
- activation of processes on reception of a signal;
- keeping track of process instance sets, and the number of instances in it;
- distributing timers and time-out signals to processes;
- supporting dynamic deployment and upgrade of process logic.

A number of these aspects are highly dependent on the underlying hardware and operating system software, which are out of scope for this paper. Instead, we roughly outline how we deal with these issues: in some cases, we require certain functionality from the underlying communication platform. In other cases, we deliberately deviate from the SDL semantics, to allow for an efficient realisation of the complete system.

Since transfer of signals is implemented based on some communication infrastructure, addressing of processes has to take the addressing infrastructure into account. Specifically, the internal representation of a Pid value (and its external representation by means of the BER) must contain the "native" address value of the communication platform. In the case of stateless operation, this native address should *not* take the machine identification into account. Instead, the platform will deliver the message to an arbitrary machine (or rather, based on some policy such as load-balancing), at which point a restoration of the process state is initiated. Therefore, the platform must provide an association between a process address and an external storage location.

When receiving a signal, restoration will require executing some code that is specific to the SDL process. In particular, the decoding of BER-encoded data must occur; the knowledge of how to interpret the data is only present when the SDL process type is known. Reception of a signal will typically locate a routine in the C++ class implementing the SDL process first. That routine will then take over processing of the signal: it will restore the process state, execute the transition, and save the modified state back into external storage.

To reduce the amount of storage required to represent a process, a common heuristic is that processes typically have at most one pending signal in their input port. If the underlying communication platform guarantees to hold back further signals until the current signal is processed, and if the process does not contain any save-nodes or enabling conditions, there is no need to maintain a per-process input queue at all. Even though the SITE run-time system does not take this property into account, an empty input port will consume only a small number of bytes, which then can be considered in estimating the maximum size of process data.

Since processes "live" most of their life in external storage, it is inefficient to track the number of instances – especially if new instances of the same instance set are created on different processing nodes simultaneously. Not counting the number of processes means that create actions will not fail even if the maximum number of instances is reached. In our systems, we made the observation that process definitions typically fall into two categories:

- Some processes have exactly one initial instance, they are never created and they never stop. These we call *initial* processes. They typically serve initial requests, and create processes of the other category to deal with the request completion;
- All other processes have no initial instances. They are dynamically created, they stop when done, and there is no inherent limit on their number except for exhaustion of resources (such as memory).

If a system contains only these two kinds of process definitions, no monitoring of instance counts is necessary. If the system runs on multiple machines or operating system processes, each machine will have its own copy of initial processes. Instances of dynamic processes can then freely migrate from one copy of the system to another.

Support of timers basically operates similar to the processing of signals. The underlying platform must invoke an upcall routine whenever a timer expires. That will restore the process, including possible timer arguments, and allow to invoke a transition for the timer.

5   Future Work

This paper outlines the implementation strategies for providing a stateless execution model of SDL that we have been following so far. It shows that it is feasible to run SDL specifications even in an environment that does not fit the traditional model of SDL execution. Further work is necessary to validate all the principles demonstrated in this paper, and to obtain actual data on reliability and performance of our approach.

References

1. ITU-T. Recommendation Z.100 (11/99) - Specification and description language (SDL). Geneval, 1999.
2. Mitschele-Thiel. Systems Engineering With SDL: Developing Performance-Critical Communications Systems. JW Wiley, 2001.
3. Hennessy, Patterson. Computer Architecture: A Quantitative Approach. Morgan Kaufmann Publishers, 1996.
4. SDL Integrated Tool Environment of Humboldt University Berlin: http://www.informatik.hu-berlin.de/Themen/SITE
5. ITU-T. Recommendation X.680 (12/97) - Information technology - Abstract Syntax Notation One (ASN.1): Specification of basic notation. Geneva, 1997.
6. ITU-T. Recommendation X.690 (12/97) - Information technology - ASN.1 encoding rules - Specification of Basic Encoding Rules (BER), Canonical Encoding Rules (CER) and Distinguished Encoding Rules (DER). Geneva, 1997.
7. ITU-T. Recommendation Z.105 (11/99) - SDL combined with ASN.1 modules (SDL/ASN.1). Geneva, 1999.

An MSC Based Representation of *DiCons*

J.C.M. Baeten, H.M.A. van Beek, and S. Mauw

Department of Mathematics and Computing Science,
Eindhoven University of Technology,
P.O. Box 513, NL–5600 MB Eindhoven, The Netherlands.
{josb,harm,sjouke}@win.tue.nl

Abstract. We present a graphical MSC-based representation of the language *DiCons*, which is a formal language for the description of Internet applications.

1 Introduction

Building internet applications is not an easy task. Given the many problems involved it makes sense to investigate the use of formal methods since we think that formal methods can help to develop Internet applications more efficiently, and can help to improve the quality of applications.

Currently, a mix of different languages, at different levels, with a low degree of formality is used, e.g. Perl, C++ and Java. Recently, we have started a new line of research in order to remedy this. This has resulted in the first version of the language *DiCons* in [3] of which an extended abstract appeared as [2].

The most important feature of *DiCons* is that it is geared towards the highest level of abstraction, the communication level, and that aspects of lower levels are treated in separate parts of the language. The purpose of this paper is to give a graphical presentation of *DiCons* specifications.

The language *DiCons* focuses on a specific class of internet applications, a class we call Distributed Consensus (this explains the name of the language). This is the class of applications where several users strive to reach a common goal without having to meet face to face, nor will there be any synchronized communications between users. A central system, viz. an Internet application, must be used to collect and distribute all relevant information. Example applications are making an appointment, evaluating a paper, and selecting a winner.

Currently, we are working on the formal semantics of the language, which serves as the starting point for this paper. The papers [3,2] show the usefulness of the language in a number of examples which were first developed as MSC scenarios, and afterwards programmed in *DiCons*. MSC is the language Message Sequence Chart, that is used a lot in the telecommunications industry, as standardized in [11]. MSC has in common with *DiCons* that it is mainly concerned with the interaction between system components (here: a server and several clients) and that internal processing of information is less important.

A closer look reveals that drawing the MSCs does not leave out much information. Mostly, they contain just one scenario (or a couple of related scenarios).

R. Reed and J. Reed (Eds.): SDL 2001, LNCS 2078, pp. 328–347, 2001.

On the other hand, the examples in *DiCons* suggest that there is a main trace that admits variations occasionally. With the recent extensions of MSC [11] it could be possible to give a complete, or almost complete MSC specification of *DiCons* programs. This is the hypothesis that we investigate in this article.

There are several reasons why an MSC-like representation can have added value for a textual specification language like *DiCons*. Most important is that a visual interface can aid communication with a customer, who wants an application to be built. It is easier to understand by those not used to programming languages. Focusing on example scenarios can be very important in the initial design phase of a new application.

On the other hand, there are reasons why just using MSC as a trace description language is not enough. Most important, just describing traces can leave ambiguities, obscurities and misunderstanding about the working of an application.

In general, it is not advisable to give complete system specifications in MSC. However, it is interesting to note that *DiCons* is intended for restricted class of applications, and this makes it possible to define a complete MSC-like representation. Thus, in *DiCons* we only consider the behaviour of the server, depending on possible external stimuli and the internal state. This means we only have to define the complete behaviour of a single MSC instance. This appears to be a lot simpler than specifying the complete behaviour of several instances.

In this paper, we investigate giving an MSC-like representation of *DiCons* with comparable expressivity. A first try is to see whether MSC-2000 is powerful enough by itself, but it soon turns out that more is needed. For instance, *DiCons* involves several communication primitives that have to be represented in different ways. We see that *DiCons* has compound communications that go beyond the simple scheme of an asynchronous MSC communication. This requires extensions of MSC-2000 in order to raise the representation to the same level of abstraction. Our extensions are in the style of MSC-2000, for instance regarding the use of in-line expressions.

We are not in the business of proposing extensions of the language MSC. Rather, we look upon this work as a special application of MSC. In our experience, every (new) application domain of MSC will lead to a comparable adaptation of the language. This is due to the nature of the language MSC. On the one hand, MSC is so universal as to be applicable whenever there is a form of distribution and communication. On the other hand, the drive to express issues in the appropriate way and at the appropriate level will necessitate new features that have not been standardized by the ITU (yet). The present offering of MSC-2000 seems to have enough features already. There is a good basis of possibilities to express issues like modularization, data, time, and many more things, and it is not obvious that specific applications should lead to even more extensions of the language.

Rather, we see our work as defining a graphical layer on top of *DiCons* based on MSC, and not as an extension of MSC. This paper is exploratory: we do not give a formal graphical syntax and do not give a translation to the semantics.

On the other hand, we have tried to have a one-to-one correspondence with the semantical constructs.

Also, it is not our intention to use exactly the semantics of MSC. For example, we introduce several communication primitives that do not allow a simple reduction to existing primitives. Thus, the semantics of our MSC-like language will not arise by translation to the semantics of MSC, but rather by a translation to the abstract syntax of *DiCons* and from there to the semantics of *DiCons*. As a side remark, the semantics of MSC and the one of *DiCons* are not so different, since both are based on a translation to process algebra.

In the following section, we present a short introduction to the *DiCons* language and define our graphical representations for the language primitives. In the next section, we work out an example, the meeting scheduler. Finally, we present some conclusions.

2   *DiCons* Primitives and Their Graphical Representation

In this section we will discuss the considerations that led to the current design of the *DiCons* language and we describe the basic ingredients of *DiCons*.

In order not to have to face the complete problem of writing Internet applications in general, we restrict our problem setting in several ways. First of all, we focus on a class of applications which is amenable to formal verification with respect to behavioral properties. This means that the complexity of the application comes from the various interactions between users and a system, rather than from the data being exchanged and transformed. Implications for the design of the language are that the primitive constructs are *interactions*, which can be composed into complex behavioral expressions. Furthermore, it implies that the development of the language and its formal semantics must go hand in hand. Nevertheless, we will not discuss semantic issues in the current paper.

A further restriction follows from the assumption that although the users work together to achieve some common goal, there will be no means for the users to communicate directly with each other. We assume a single, central application that follows a strictly defined protocol in communication with the users.

The last consideration with respect to the design of *DiCons* is that we want to make use of standard Internet technology only. Therefore, we focus on communication primitives such as e-mail and Web forms. This means that a user can interact with the system with a standard Web browser, without the need for additional software such as plug-ins. Of course, it must be kept in mind that the constructs must be so general as to easily support more recent developments, such as ICQ or SMS messages. Currently, we only consider asynchronous communication between client and server.

2.1   Overview of Language Constructs

Keeping the above considerations with respect to the application domain and available technology in mind, we come to a description of the basic constructs of *DiCons*. Although *DiCons* is initially developed as a linear language we will only give a description of the graphical representation of *DiCons*. To this end we build on the work that is done on the development of the MSC standard. A graphical *DiCons* specification looks like an MSC, however, it *is not* an MSC. We will define some extensions to the MSC standard that are essential to reach the level of abstraction of the current linear *DiCons* language in a natural way. One can easily see whether a figure is a basic MSC or its *DiCons* version: keyword **msc** in the upper left corner is replaced with keyword **DiCons** in the *DiCons* version.

Apart from basic MSCs we will also make use of High-level MSCs (see [16]) to specify *DiCons* applications. In the *DiCons* version we extend HMSCs by adding constructs for declaring roles and variables.

We will first list the language ingredients and later discuss these in more detail, by defining their graphical syntax. We will not give a complete description of the syntax and semantics of (graphical) *DiCons*. The example in Sect. 3 will serve to show the flavor of the graphical *DiCons* syntax and the way in which the language can be used.

central application The central application is the main part of our *DiCons* language. All interactions take place via this application.

users and roles Since an application may involve different users, the application must be able to identify users. Moreover, since different users may want to use the system in the same way, it must be possible to group users into so-called *roles*.

interactions We have to identify the communication primitives, which we will call *interactions*. They form the basic building blocks of the behavioral descriptions. Interactions are abstract descriptions which are identified by their name and may carry input and output parameters.

behavior A number of interactions with the same user may be combined to form a *session*. Sessions and interactions can be composed into complex behavioral descriptions which define an *application*.

presentations The abstract interactions are presented to the user by concrete means of communication, such as e-mail and Web forms. This is called the *presentation* of an interaction.

data In order to transform (user) data and keep state information, we need a means to define and manipulate data (expressions, variables, data structures, etc.)

2.2   Central Application

Since all interactions take place between the central system and one of the users, the central system must be included in all graphical *DiCons* specifications. We can only give a decription of the behaviour of the central system. We cannot *force* users to interact in the way we intend, therefore we assume that they will do so in order to be able to give a useful specification. The application is represented by a gray-headed instance in a wide form not having an instance name. In Fig. 1 a graphical representation of the central application is given.

332   J.C.M. Baeten, H.M.A. van Beek, and S. Mauw

Fig. 1. The central application   **Fig. 2.** A user

2.3   Users and Roles

A user is a (possibly human) entity that can interact with the system. Users are grouped according to their role. Users with the same role are offered the same interaction behavior. In *DiCons* roles can be defined and variables can be declared which denote users with a given role.

In the graphical *DiCons* syntax we define roles by introducing them in the same way as we introduce variables (see Sect. 2.5). However, we use the **role** keyword instead of the **var** keyword. We represent a user by a regular instance, containing the name of the user in its head symbol (Fig. 2).

2.4   Interactions

The basic problem when defining the interaction primitives is to determine the right level of abstraction. In order to get a feeling of the level of abstraction which is optimally suitable, look at Fig. 3. In this drawing we sketch in MSC a typical scenario of an Internet application which is called the *Meeting Scheduler* (see [17]). This is an application which assists in scheduling a meeting by keeping track of all suitable dates and sending appropriate requests and convocations to the intended participants of the meeting.

Please note that this is not a graphical *DiCons* specification, but an ordinary MSC describing a possible scenario. We will give a more detailed graphical *DiCons* specification in Sect. 3.

The example shows that we have two roles, viz. initiator and participant. In this scenario, there is only one user with role initiator, while there are three users with role participant. The MSC shows that the initiator starts the system by providing it with meeting information. Next, the system sends an invitation to the participants who reply by stating which dates suit them. After collecting this information, the system informs the initiator about the options for scheduling the meeting and awaits the choice made by the initiator. Finally, the system informs the participants about the date and offers the users to have a look at the agenda. Only participant 2 is interested in the agenda.

This example nicely shows at which level of detail one wants to specify such an application. The arrows in the diagram represent the basic interaction primitives. First, look at the *invite* messages. Since the participants do not know that they will be invited for a meeting, the initiative of this interaction is at the server side. The way in which a server can actively inform a client is (for example) by sending an e-mail. This interaction only contains information transmitted

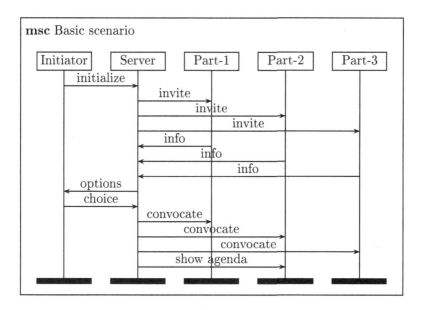

Fig. 3. An MSC Scenario of an Internet Application

from the server to the user. The messages *options* and *convocate* can also be implemented as e-mails.

Next, look at message *info*. This interaction is initiated by the user and is best implemented as a Web form supplied by the server, on request of the user and filled in by the user. The message *choice* also stands for a Web form being filled in.

The last message, *show agenda* contains information sent by the server to the user, on request of the user. This is simply the request and transmission of a non-interactive Web page.

Finally, we look at the first message, *initialize*. The initiator has to supply the system with various kinds of information, such as a list of proposed dates and a list of proposed participants. This will probably be implemented as a dialogue between the user and the system in the form of a series of Web forms. This is called a *session*.

We summarize the three basic interaction schemes in Fig. 4. Notice that the third scheme, the session, consists of a series of more primitive interactions. It starts with a client requesting a form and submitting it after having it filled in. This is the interaction which starts the session. Next, comes a series of zero or more submissions of Web forms. These are interactions which come in the middle of a session. And, finally, the session ends with the server sending a simple Web page after the last submission of the client. So a session is composed of three kinds of interactions.

In *DiCons* we have constructs for these five interaction primitives. We have used a naming scheme for the interaction primitives which is based on their

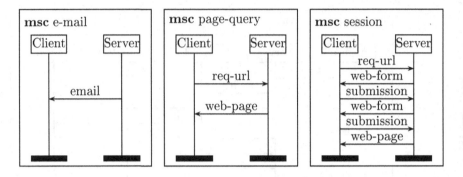

Fig. 4. Interaction Primitives

properties. First, we make a distinction based on the flow of information. If the information goes from the server to the client, we call this a *server push*, while if the information flows to the server, we call this a *server pull*. Notice that we reason from the viewpoint of the server in this respect. The direction of the arrow indicates whether the interaction involves a push or a pull.

The second distinction which we make is on which party takes the initiative for the interaction. Still reasoning from the viewpoint of the server we consider an *active* communication, which means that the server takes the initiative, a *reactive* communication, which means that the client takes the initiative, and a *session oriented* communication, which means that the communication is a response from the server to a prior submission of a Web form by the client. To graphically indicate the initiating party we place half a circle between its instance and the arrow representing the interaction. The filling of the circle indicates whether the interaction takes part in a session or not.

Finally, notice that we extend the interaction primitives with parameters to express which information is being transmitted. An output parameter denotes information sent by the server to the client, while an input parameter is a variable in the data space of the server which will contain the information sent by the client to the server. We make use of $\triangleright$ and $\triangleleft$ to graphically specify the direction in which data flows. Parameters left to a $\triangleright$ flow from left to right and parameters right to a $\triangleleft$ flow in the opposite direction.

The notation for our communication primitives is given below. We give a basic MSC to show in which order the different messages take place. Furthermore, we give the corresponding graphical *DiCons* syntax. In the figures given below, i_k $(0 \leq k \leq m)$ denotes an input parameter and o_k $(0 \leq k \leq n)$ an output parameter.

active server push The server takes the initiative to send information:

An *active push* takes place if the server sends a message to a client which is not directly the result of a request from that client. Such an interaction can only take place via an e-mail and not by sending a Web page, since this requires a client to request some URL.

We denote this interaction by a message from the server to the client (Fig. 5). The circle at the server side means that the server initiates the interaction. It is not filled since the interaction does not take part in a session. The direction of the arrow indicates that we have to do with a push.

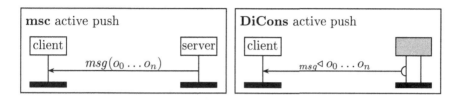

Fig. 5. Active push

reactive server push The server sends a Web page on request of the client: A *reactive push* takes place if the server sends a Web page, not containing a Web form, to a client which is the result of a normal request from that client, (that is, not generated by filling out a previously received Web form). Here, the circle is placed at the client side, meaning that the client initiates the interaction. Again, it is not filled since the interaction is not part of a session. Actually, the interaction may be seen as one that both starts and ends a session containing only this interaction.

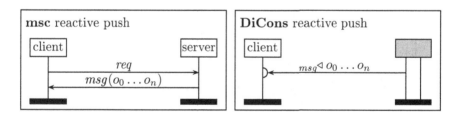

Fig. 6. Reactive push

reactive server pull This interaction takes place if a client sends a request to the server on which the server responds by sending a Web form. This form is filled in and submitted by the client. A reactive pull starts a session with one particular client. The client starts the interaction so the circle is at the client side. The upper half of the circle is not filled since no session existed prior to this interaction. Its lower half is filled, which means that after ending this interaction a session is open. Note that the direction of the arrow indicates a server pull.

The dashed line in the MSC in Fig. 7 is *not* part of this interaction primitive. It means that this interaction must be followed by an interaction which is started by sending a message from the server to the client.

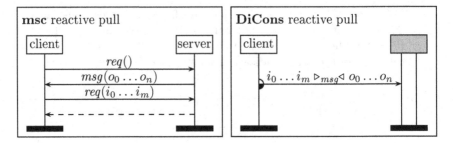

Fig. 7. Reactive pull

session-oriented server pull The server sends a Web form to the client as a response to a prior form submission by the client. After that, the client submits the filled in form. This interaction is repeated in the middle of a session.

The server initates this interaction and therefore the circle is at the server side (Fig. 8). It is completely filled, since a session existed at the beginning of the interaction and is still open at the end.

Again, the dashed lines represent mandatory messages preceding and following this interaction. Note that both a reactive pull and a session-oriented pull can precede this interaction and that a session-oriented pull or a session-oriented push can follow it.

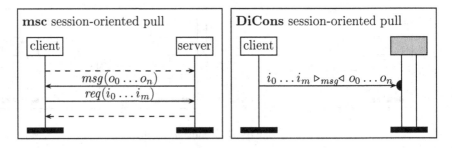

Fig. 8. Session-oriented pull

session-oriented server push The server sends a non-interactive Web page to the client in response to a prior form submission by the client. This interaction is the last interaction of a session.

If a server/client communication takes part in a session, the server can send a Web page, not containing a Web form, as a response on a submission of a Web form. Such a *session-oriented push* ends the session because the client can no longer fill out any forms. The server initiates this interaction so the circle is at the server side (Fig. 9). The upper half of the circle is filled, meaning that the interaction starts within a session. On the other hand, the

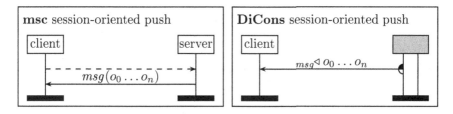

Fig. 9. Session-oriented push

lower half is not filled which denotes that the session is no longer open after ending this interaction.

The dashed line in the MSC indicates a message from the client to the server, which must precede this interaction.

Please notice that in our list of interaction primitives we did not mention the *active server pull*. The reason for this is simply that with standard Internet technology this interaction cannot be implemented. A Web server cannot take the initiative to obtain information from a client.

2.5   Behaviour

Now that we have defined the basic interaction primitives, we can discuss the means to compose them into sessions and applications. An application describes the protocol to be executed by the server. A number of standard programming language constructs are supported in *DiCons*. Since in most applications that we have studied users have to react before a given deadline, we have included a time-out construct in *DiCons*. A session is simply a program fragment with the requirement that execution starts with a session-start interaction and ends with a corresponding session-end interaction.

sequential composition We make use of the basic MSC based representation for sequential composition. We only specify the behaviour of the central application, so the order of the events on the instance axis of this application determines their causal connection.

conditional branching For the construct of a conditional branching we make use of an inline expression having the keyword **if** followed by the condition in its upper-left corner. See Fig. 10 for an example of the grapical syntax for expression *if b then X else Y fi*.

repetition We make use of two different statements for repetition. First of all we have the *for all $s \in S$ do X(s) od* statement. For all elements s in S, bodies $X(s)$ are sequentially executed in an arbitrary order, where S represents a finite set of data elements. Furthermore, we can specify a while loop by using the *while b do X od* statement. The while loop repeats statement X until test b proves false. In Fig. 11 the syntax for both loops is given. The X represents a collection of (inter)actions. If variable s occurs in X we indicate this by writing $X(s)$.

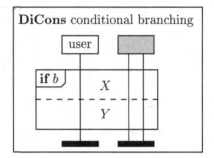

Fig. 10. Syntax for conditional branching

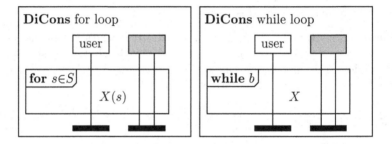

Fig. 11. Syntax for repetition

parallel composition We also have two operators for parallel composition at
our disposal: the *fork* and the *par* operator (Fig. 12).

Statement *fork* $u \in U$ *do X od* means that all users u ($u \in U$) can execute
(inter)actions X between user u and the central applications in parallel and
more than once. On the other hand, the *par* $u \in U$ *do X od* statement specifies
that all users u ($u \in U$) will execute (inter)actions X between user u and
the central applications in parallel but only once. So, in contrast to the *fork*
operation, after execution of X for all $u \in U$ the *par* operation ends.

A *fork* gives clients the possibility to start an interaction (for example via a
reactive push) while the *par* obliges a client to interact. This obligation only
makes sense if the initiative for the interaction is on the server-side, such as
an active server-push.

Note that both the instance head and the instance foot of u are placed inside
the operator's frame. This means that instance name u is bounded by the
operator and therefore it is not known outside the frame.

time-outs using conditional disrupts We introduce the   *until b do X od*
statement to specify conditional disrupts. This means that X is normally
executed until b becomes true. At that moment the statement ends, independent of the (inter)actions that are taking place at that moment. If X ends
before b becomes true the statement ends too. By placing a time check in b

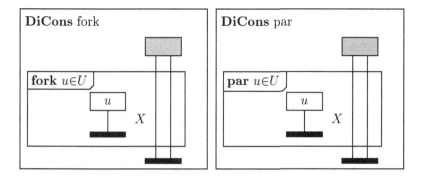

Fig. 12. Syntax for parallel composition

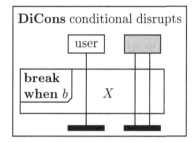

Fig. 13. Syntax for conditional disrupts

we can specify time-out interrupts. However, b may be an arbitrary boolean expression containing predicates on any part of the state space.

variable assignments and procedure calls In the textual syntax of *DiCons* we have a *data* part to introduce types, variables and functions. We can also introduce them in several ways in the graphical representation. First of all, we can introduce variables directly below the **DiCons** keyword (Fig. 14). These variables are available to all elements in the MSC, so the frame surrounding the instances defines the block in which these variables are known. As mentioned in Sect. 2.3, we can introduce roles in the same way by using keyword **role**.

Furthermore, we can introduce variables using an inline expression. The variables are only available within the box that is used for this variable introduction. If we want to assign an intial value to a variable we can do this at the place of declaration using the ":=" sign, however, this is optional.

Variables are owned by the central application only. This means that problems concerning individual variables in basic MSCs [10,9] do not arise in our graphical *DiCons* syntax.

There are three ways to change the value of a variable:

- Via a local action of the server, i.e. an assignment or a function call;
- Via an input parameter of an interaction;

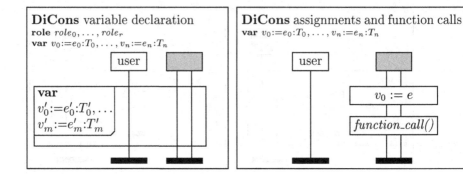

Fig. 14. Syntax for variable declarations, variable assignments and function calls

- Via the bind-construct in the header of a graphical *DiCons* specification.
High-level behavioural composition To be able to compose the behaviour description of a *DiCons* application in an hierarchical way we make use of the same notation as is used for the composition of MSCs by means of High-level MSCs (see [16]).

2.6   Presentations

Up to now, we only described the top level view of *DiCons* applications. This level is concerned with the composition of interactions into complete specifications of the behaviour of the applications. However, this is not the only level for which we would like to make a graphical representation. The other levels of the *DiCons* language concern the presentation of the interactions and the data definition. Examples of textual descriptions of these levels can be found in [3].

3   An Example: The Meeting Scheduler

The purpose of this section is to explain the use of graphical *DiCons* by means of an example. The example concerns the specification of a Meeting Scheduler, taken from [17]. Figure 3 in Sect. 2.4 already contains an informal example of a scenario of this application.

The purpose of a Meeting Scheduler is to support the process of scheduling a date for a meeting without the intended participants having to meet for selecting a date. A central server could easily take care of the administrative tasks and support the communication process. Of course there are several publicly available tools with capabilities similar to our Meeting Scheduler which offer much more functionality.

We present a top-down development of the case study. The highest level is represented in Fig. 15. This contains a drawing which resembles a High-level Message Sequence Chart (see [16]). It shows that the application consists of three phases, which are called *initialize*, *check*, and *select*. These three phases are elaborated in Figs. 16, 17 and 18. Roughly speaking, in the initialization phase the initiator starts the application and provides the server with the required information, in the check phase, the intended participants send in their selection of suitable dates, and in the select phase the initiator selects the meeting date. In these three phases, several global variables may be used. These are declared just below the header. Since we will not focus on the actual language used for defining these variables and other data-related objects, we will simply use an abstract mathematical notation. Users who interact with the system can have two roles, viz. initiator and invitee. Furthermore, we declare an initiator of type Initiator, a set of Invitees in which the list of invitees is stored, a set of optional dates, a deadline before which the intended participants must have replied, and for each intended invitee the list of dates that are not convenient to him.

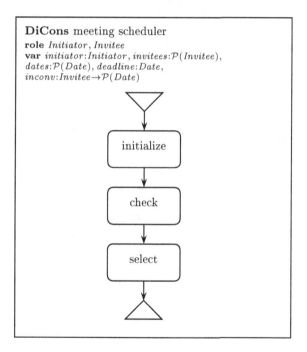

Fig. 15. High-level overview

Next, we look at the specification of the initialization phase in Fig. 16. This is a specification at the interaction level. There are three entities taking part in this description. The middle entity describes the system providing the required service. Its behavior is represented by a, so-called, fat instance axis. This is to clearly distinguish it from the other instances acting as clients. The left instance is named *initiator*. This is the user who initializes the application. Because any user is allowed to initialize the application, we add the **bind** construct in the header to indicate that the identity of the actual initiator is saved in the variable named *initiator*. From now on, this variable is instantiated. The rightmost entity has the name i. This is a variable local to the par-frame.

The first interaction, named *inv*, implies a flow of information from the initiator to the server which is stored in the variable with the name *invitees*. Of course the initiator must send data of the intended type (a set of invitees) in order to successfully complete this interaction. This can be enforced by using a web form and JavaScript, but we will not elaborate on these implementation related issues. Note that the interaction is a reactive server pull (see Sect. 2.4), specifying the beginning of a session. The second interaction within this session, a session-oriented pull, requires the initiator to send a list of possible dates, followed by an interaction which requires the initiator to send the deadline before which the intended participants must have replied.

After having received all this information, the server invites all users from the set of invitees to submit their selection of inconvenient dates. This can be implemented by providing them with a URL of some web page where each of them can select some of the proposed dates. The fact that all invitees are informed in this way is expressed in the upper left corner of the surrounding frame, which contains the *parallel* operator. Finally, the server ends the session by sending a confirmation to the initiator.

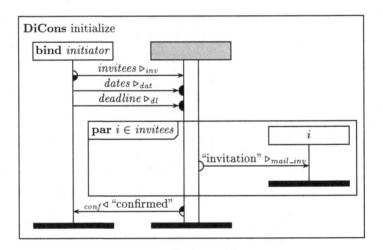

Fig. 16. Initialization phase

In the second phase (Fig. 17), all intended participants are allowed to submit their lists of inconvenient dates any number of times. This can be done until all participants have replied, or the given deadline for replying has passed. This is indicated by the **break**-frame. The contained behavior will be disrupted at the moment that the breaking condition becomes true. Please note that this frame in no way indicates any repetition itself. It just expresses that the enclosed behavior can be disrupted. The repetition is expressed in the **fork**-frame. This fork operator makes it possible for every invitee i from the set of invitees to start a session. It is allowed that these sessions run in parallel and that an invitee takes part in more than one session. It is even possible for one invitee to run several independent sessions in parallel. For each session within this **fork**-frame the server declares a local variable ds, which will contain the list of inconvenient dates as submitted by the invitee in this session. A **var**-frame is used to indicate the scope of this variable. An invitee starting such a session is prompted by the server with the list of optional dates and the list of inconvenient dates possibly provided by i in a previous session. In the same interaction i provides his (new) list of inconvenient dates ds. The assignment in the local action adds this list to the information maintained by the server, and the server ends the session by sending a confirmation to the invitee. Finally, the server adds the name of the invitee to the set of invitees *checked*. This is to keep track of which invitees have already submitted their information. This variable is used in the guard of the **break**-frame. If this guard becomes true, the fork will be disrupted and the server concludes that enough information has been collected in order for the initiator to select the most suitable date. This is indicated to the initiator in the final interaction of this drawing.

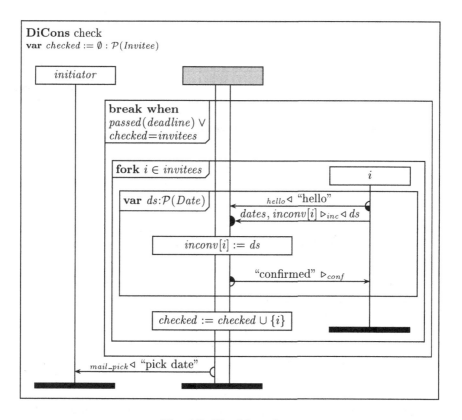

Fig. 17. Checking phase

In the last drawing (see Fig. 18) the initiator starts a session to select the date for the meeting and notify the participants of this date. In the first interaction the server provides the initiator with the list of inconvenient dates. Based on this information the initiator decides upon the final date and sends it to the server, where it is stored in local variable d. In a **par**-frame, the server sends an invitation or notification to all invitees, making a distinction between invitees that have indicated to be able to come on the selected date and those that cannot come. Finally, the server sends a confirmation to the initiator and the application ends.

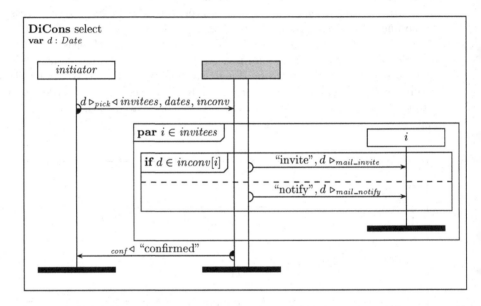

Fig. 18. Selection phase

4 Related Work

We introduced a graphical representation of a specification language for a specific class of Internet applications, viz. applications for distributed consensus. There are many different languages to specify Internet applications, but as far as we know, none of them is specifically designed to develop such applications.

Closest to our work is the development of the Web-language *Mawl* [1,13]. This is also a language that supports interaction between an application and a single user. Mawl provides the control flow of a single session, but does not provide control flow across several sessions. This is a distinguishing feature of *DiCons*: interactions involving several users are supported. On the other hand, Mawl does allow several sessions with a single user to exist in parallel, using an atomicity concept to execute sequences of actions as a single action.

Groupware [22] is a technology designed to facilitate the work of groups. This technology may be used to communicate, cooperate, coordinate, solve problems, compete, or negotiate. Groupware can be divided into two main classes: asynchronous and synchronous groupware. Synchronous groupware concerns an exchange of information, which is transmitted and presented to the users instantaneously by using computers. On the other hand, asynchronous groupware is based on sending messages which do not have to be read and replied to immediately. An example of asynchronous groupware that can be specified in *DiCons* is a calendar for scheduling a project.

Visual Obliq [4] is an environment for designing, programming and running distributed, multi-user GUI applications. Its interface builder outputs code in an interpreted language called *Obliq* [5]. Unlike *DiCons*, an Obliq application can be distributed over several so-called sites on a number of servers.

Collaborative Objects Coordination Architecture (COCA) [14] is a generic framework for developing collaborative systems. In COCA, participants are divided into different roles, having different rights as in *DiCons*. Li, Wang and Muntz [15] used this tool to build an online auction.

Further, there are languages that allow programming of browsing behaviour. These allow, for instance, the behaviour of a user who wants to download a file from one of several mirror sites to be programmed. For so-called *Service Combinators* see [6,12]. A further development is the so-called *ShopBot*, see [7].

However, none of the languages described above have a graphical representation.

5   Conclusions

We find that the nature of *DiCons* allows a complete MSC-like representation to be made in a straightforward manner. The examples show that although the specification does not become more compact, it does become clearer which agents communicate, and when they do so. This is exactly the strength of MSC.

In order to get a complete MSC representation, it was necessary to extend MSC with a number of new constructions. Maybe the presentation of these features is still a bit ad hoc, but the intention of this document is not to define a graphical layer of *DiCons*, but rather to investigate the feasibility of such an exercise.

In *DiCons*, we group separate communication actions into one aggregated interaction. In the semantics, this leads to questions concerning atomicity. In general, research is needed to consider atomicity with respect to possible race conditions. Our aggregated interactions are a special case of the notion of message refinement as discussed in [8].

Using MSC gives a whole range of possibilities in terms of precision and completeness, in our setting. On the one hand, there are trace-based descriptions of runs, requirements or test traces of a system, on the other hand, there is the complete description of the whole behaviour of the server. Further, MSC supports all levels in between, so that it is also possible to describe the composition of test runs, or test purposes. The testing of Internet applications is still not developed very much, and is usually based on the manual execution of ad hoc test sequences. Now that we have a formal notation for a class of Internet applications, that both supports the description of complete systems and of the behaviour of components, it might be possible to start testing in a more formal and structured manner, for instance using Autolink [20] or TorX [21]. In this respect, it is interesting to see what is the relation between our graphical language and the MSC-based graphical language for TTCN-3, see [19].

We have implemented a compiler which can be used to compile (textual) *DiCons* specifications into Java Servlets [18]. See our Web site[1] for some working examples. Except for generating a Servlet, the compiler checks a specification on its syntax and static semantics. As making graphical tools for *DiCons* is quite a big investment, we will not do a complete implementation before the language *DiCons* has become stable.

References

1. D. L. Atkins, T. Ball, G. Bruns, and K. Cox. Mawl: A domain-specific language for form-based services. *IEEE Transactions on Software Engineering*, 25(3):334–346, May/June 1999. Special Section: Domain-Specific Languages (DSL).
2. J. Baeten, H. van Beek, and S. Mauw. Specifying internet applications with *DiCons*. In *Proceedings of the 16th ACM Symposium on Applied Computing (SAC 2001)*, Mar. 2001.
3. H. v. Beek. Internet protocols for distributed consensus – the *DiCons* language. Master's thesis, Eindhoven University of Technology, Aug. 2000.
4. K. Bharat and M. H. Brown. Building distributed, multi-user applications by direct manipulation. In *Proceedings of the ACM Symposium on User Interface Software and Technology*, Groupware and 3D Tools, pages 71–81, 1994.
5. L. Cardelli. Obliq A language with distributed scope. SRC Research Report 122, Digital Equipment, June 1994.
6. L. Cardelli and R. Davies. Service combinators for web computing. *IEEE Transactions on Software Engineering*, 25(3):309–316, May/June 1999.
7. R. B. Doorenbos, O. Etzioni, and D. S. Weld. A scalable comparison-shopping agent for the world-wide web. In W. L. Johnson and B. Hayes-Roth, editors, *Proceedings of the First International Conference on Autonomous Agents (Agents'97)*, pages 39–48, Marina del Rey, CA, USA, 1997. ACM Press.
8. A. Engels. Message refinement: Describing multi-level protocols in MSC. In Y. Lahav, A. Wolisz, J. Fischer, and E. Holz, editors, *Proceedings of the 1st Workshop of the SDL Forum Society on SDL and MSC*, number 104 in Informatik-Berichte, pages 67–74, Berlin, Germany, June 1998. Humboldt-Universität zu Berlin.
9. A. Engels. Design decisions on data and guards in MSC2000. In S. Graf, C. Jard, and Y. Lahav, editors, *SAM2000. 2nd Workshop on SDL and MSC*, pages 33–46, Col de Porte, Grenoble, June 2000.
10. A. Engels, L. Feijs, and S. Mauw. MSC and data: Dynamic variables. In R. Dsoulli, G. von Bochmann, and Y. Lahav, editors, *SDL'99: The Next Millennium, Proceedings of the 9th SDL Forum*, pages 105–120, Montreal, Canada, June 1999. Elsevier.

---

[1] The *DiCons* Web site can be found at *http://dicons.eesi.tue.nl/*.

11. ITU-TS. *ITU-TS Recommendation Z.120: Message Sequence Chart (MSC2000)*. ITU-TS, Geneva, 2000.
12. T. Kistler and H. Marais. WebL — a programming language for the Web. *Computer Networks and ISDN Systems*, 30(1–7):259–270, Apr. 1998.
13. D. Ladd and J. Ramming. Programming the web: An application-oriented language for hypermedia service programming. In *Proc. 4th WWW Conf., WWW Consortium*, pages 567–586, 1995.
14. D. Li and R. R. Muntz. COCA: Collaborative objects coordination architecture. In *Proceedings of ACM CSCW'98 Conference on Computer-Supported Cooperative Work*, Infrastructures for Collaboration, pages 179–188, 1998.
15. D. Li, Z. Wang, and R. R. Muntz. Building web auctions from the perspective of collaboration. Technical report, UCLA Department of Computer Science, Sept. 1998.
16. S. Mauw and M. Reniers. High-level Message Sequence Charts. In A. Cavalli and A. Sarma, editors, *SDL'97: Time for Testing - SDL, MSC and Trends*, Proceedings of the Eighth SDL Forum, pages 291–306, Evry, France, September 1997.
17. S. Mauw, M. Reniers, and T. Willemse. Message Sequence Charts in the software engineering process. In *Handbook of Software Engineering and Knowledge Engineering, S.K. Chang, editor*. World Scientific, 2001. To appear.
18. K. Moss. *Java Servlets*. Computing McGraw-Hill, July 1998.
19. E. Rudolph, I. Schieferdecker, and J. Grabowski. HyperMSC – a graphical representation of TTCN. In *Proceedings of the 2nd Workshop of the SDL Forum Society on SDL and MSC (SAM'2000)*, Grenoble (France), June 2000.
20. M. Schimitt, A. Ek, J. Grabowski, D. Hogrefe, and B. Koch. Autolink – puting SDL–based test generation into practice. In A. Petrenko, editor, *Proceedings of the 11th International Workshop on Testing Comunicating Systems (IWTCS'98)*, pages 227–243. Kluwer Academic, 1998.
21. J. Tretmans and A. Belinfante. Automatic testing with formal methods. In *EuroSTAR'99: 7th European Int. Conference on Software Testing, Analysis & Review*, Barcelona, Spain, Nov. 1999. EuroStar Conferences, Galway, Ireland.
22. J. Udell. *Practical Internet Groupware*. O'Reilly & Associates, Inc., Oct. 1999.

Some Pathological Message Sequence Charts, and How to Detect Them

Loïc Hélouët

France Télécom R&D,
2 avenue Pierre Marzin, 22307 Lannion Cedex, France,
`loic.helouet@francetelecom.com`

Abstract. Some confusing Message Sequence Charts are identified, that can be considered as syntactically correct, but may lead to ambiguous interpretations. The first kind of MSC identified appears when parallel components of a parallel frame synchronize implicitly to continue an execution. The second case is called non-local choice, and appears when more than one instance is responsible for a choice. Non-local choice has already been studied before. This paper provides an extension of the definitions and corresponding detection algorithms. The third case is confluent MSCs, and appears when concurrency is expressed through a choice.

1   Introduction

Since the first standard appeared in 1992, Message Sequence Charts have gained a lot of expressive power. Many elements have been added to the original language: composition in MSC-96, additional time measurement possibilities, and variables in MSC-2000. All these improvements are obviously aiming at a better usability of Message Sequence Charts, and are mainly driven by expressed needs. However, adding a new element to a language may fully satisfy some users, and introduce at the same time confusion for all the others, or even semantics ambiguities.

The main elements of the language (instances, messages, timers) are usually well understood. Furthermore, as far as basic Message Sequence Charts are concerned, very few ambiguities can arise. It is not true when considering composition through parallel, choice, or loop operator, or through High-level Message Sequence Charts. Some graphical inconsistencies in MSC'96 were already pointed out by [7].

HMSCs, for example, allow for the definition of MSCs that are considered as syntactically correct, but the intuitive understanding of which are different from the behavior allowed by the semantics. In most cases, these HMSCs should be considered as pathological, and rejected. Fortunately, the cases introduced within this paper can be easily detected. This detection is based on global properties of the MSC.

As a correct syntax does not ensure the correct understanding of an MSC specification, we argue that a new document should be added to the appendices

R. Reed and J. Reed (Eds.): SDL 2001, LNCS 2078, pp. 348–364, 2001.

of recommendation Z.120, as a "methodological guideline", which should define what a valid MSC should be. The identification of at least two pathological cases and an extension of the definition of non local choice is a first contribution in this direction.

This paper is organized as follows: first we quickly recall the operational semantics of High-level Message Sequence Charts, as defined by [11,8]. Then, Sect. 3 shows the first ambiguous case, that arises when two parallel components synchronize without any communication. Section 4 recalls the definition of non-local choice, proposes an algorithm to detect it, and then discusses its pathological character. Section 5 identifies another pathological kind of MSC, and proposes an algorithm to detect it.

2   Operational Semantics

2.1   Basic Message Sequence Charts

Many semantics have been proposed for basic Message Sequence Charts (bMSC). The semantics retained for recommendation Z.120 is based on process algebra. However, we consider as natural to model bMSC as a finite, labeled partial order, as in [1,6].

A bMSC is a tuple $M = (E, \leq, A, I, \alpha)$ where:

- E is a finite set of events,
- $\leq$ is a partial order relation (antisymmetric, reflexive and transitive) called *causal order* on events,
- I is a set of names of instances that perform at least one action in M, and is called the set of *active instances* of M.
- A is a set of action names.
- $\alpha : E \longrightarrow A \times I$ is a labeling of events.

From now, we will note $\phi(e)$ the instance performing event e, ie the instance $i \in I$ such that $\alpha(e) = (a, i)$ for some $a \in A$. Slightly abusing the notation, we will note $\phi(E) = \{\phi(e) | e \in E\}$ the set of instances appearing in any set of events E. For any MSC M, we will note $min(M) = \{e \in E \mid \not\exists\, e' \neq e \text{ and } e' \leq e\}$ the set of minimal events of M. For any labeled order $M = (E_M, \leq_M, A_M, I_M, \alpha_M)$, for any set $E' \subseteq E_M$, we will denote by $M_{/E'}$ the restriction of M to events of E'. We will also denote by $M_\emptyset = <\emptyset, \emptyset, \emptyset, \emptyset, \emptyset >$ the empty MSC.

A bMSC defined with a partial order model has the same semantics as the process algebra definition of [11]. The main difference is that we use one single operational semantics rule :

$$\frac{e \in min(M), \alpha(e) = (a, i)}{M =< E_M, \leq_M, A_M, I_M, \alpha_M > \stackrel{a}{\longrightarrow} M_{E-\{e\}}}$$

2.2 High-Level Message Sequence Charts

High-level Message Sequence Charts allow for the definition of more elaborated behaviors. Their graphical syntax is given by means of graphs, the nodes of which are references to bMSCs, to HMSCs, or parallel composition of bMSCs and HMSCs references.

   One of the key points for understanding HMSC is the semantics of sequential composition. HMSC H_1 in Fig. 1 is a sequential composition of bMSCs M_1 and M_2. The semantics of the sequence of bMSCs defined in the standard is a weak sequential composition[1]. The result is an instance-by-instance concatenation, where, for each instance, the maximum event of the first bMSC is linked to the minimum event of the second bMSC. This gives to MSCs an interesting expressive power since communication messages can be accumulated between instances by concatenating basic patterns. Therefore, the HMSC in Fig. 2 should have the same operational semantics as the HMSC in Fig. 1. If the semantics of weak sequential composition gives HMSCs a huge expressive power, it is also the source of many misunderstandings. However, we think that weak sequence is essential for the design of asynchronous distributed systems, and that the ambiguous cases it may produce are an acceptable price to pay for it.

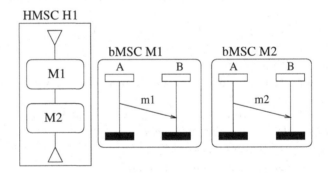

Fig. 1. HMSC H_1: sequence of bMSCs M_1 and M_2.

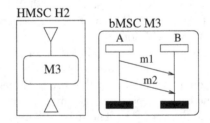

Fig. 2. HMSC H_2 equivalent to HMSC H_1 in Fig. 1.

---

[1] Weak sequential composition is close to the Pratt's local sequencing [10], where ϕ defines locality.

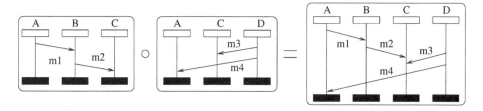

Fig. 3. Chain by chain concatenation of basic message sequence charts.

Let us define a sequencing operator $\circ$ on two MSCs $M_1 = (E_1, \leq_1, A_1, I_1, \alpha_1)$ and $M_2 = (E_2, \leq_2, A_2, I_2, \alpha_2)$: $M_1 \circ M_2 = <E, \leq_{M_1 \circ M_2}, I_1 \cup I_2, \phi>$, where:

- $E = E_1 \uplus E_2$ is the disjoint union of E_1 and E_2,
- $\forall e, e' \in E$, $e \leq_{M_1 \circ M_2} e'$ iff $e \leq_1 e'$ or $e \leq_2 e'$ or
  $\exists (e_1, e_2) \in E_1 \times E_2 : \phi_1(e_1) = \phi_2(e_2) \wedge e \leq_1 e_1 \wedge e_2 \leq_2 (e')$
- $A = A_1 \uplus A_2$, $I = I_1 \uplus I_2$,
- $\forall e \in E$, $\alpha(e) = \alpha_1(e)$ if $e \in E_1$ or $\alpha(e) = \alpha_2(e)$ if $e \in E_2$

More intuitively, sequential composition consists in ordering events e_1 in bMSC M_1 and e_2 in bMSC M_2 if they are situated on the same instance, and then calculating the transitive closure of the resulting order. An example of sequential composition is provided in Fig. 3. Due to the local sequencing, the emission of m_1 precedes the reception of m_4, and the reception of m_2 precedes the reception of m_3 in the resulting bMSC.

Basic Message Sequence Charts only allow for the definition of very simple scenarios. High-level Message Sequence Charts (HMSC) allow for the definition of more complex behavior, through parallel composition, choice, and sequence operators, and hierarchical construction. HMSC documents can be defined as a collection of graphs, as in [8]:

Definition 1. *A Message Sequence Chart document can be defined by a family of High-level Message Sequence Charts* $\mathcal{F} = \underset{i \in 1..N}{\{ H_i \}}$, *where each H_i is a High-level Message Sequence Chart.*

Definition 2. *A HMSC is a graph $H = (id, N, Ends, Start, \longrightarrow, \mathcal{M}, l)$, where:*

- *id is the name of the HMSC,*
- *N, Ends are disjoint finite sets of nodes (Ends is a set of end nodes).*
- *Start is a unique starting node,*
- *$\longrightarrow$ is the transition relation $(\subseteq (N \cup Ends \cup \{Start\})^2)$,*
- *$\mathcal{M}$ is a set of bMSCs, on disjoint sets of events. Each bMSC $M \in \mathcal{M}$ is a tuple $M = <E_M, \leq_M, I_M, \phi_M>$,*
- *l is a labeling function on nodes ($l : N \longrightarrow expr$), associating to each node a reference to a basic MSC, to the empty basic Message Sequence Chart $M_\emptyset$, to an HMSC, or a parallel composition of references. Nodes in Ends, choice*

and connector nodes will be labeled with $M_\emptyset$, and will be called empty nodes. To simplify notations and algorithms, we will consider the label associated to each node as a parallel composition of bMSC and HMSC:

$$l(n) = \underset{p \in 1..P}{\parallel} exp_p$$

with $exp_p = r_{M_p}$ for a reference to a basic MSC $M_p \in \mathcal{M}$ $exp_p = r_{H_p}$ for a reference to an HMSC H_p. Note that labeling of a node by a reference to a single bMSC or HMSC can be considered as the specific case $P = 1$.

We also require that no cyclic referencing such as in the example in Fig. 4 appears in MSC documents. This kind of situation will not be discussed here, but can also be considered as a pathological kind of MSC. It can be easily detected by constructing a graph connecting HMSC referencing each other, and then searching strongly connected components.

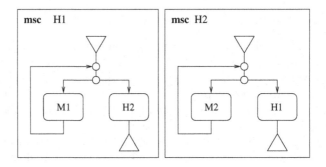

Fig. 4. Cyclic referencing

Definition 3. *A finite path of an HMSC H is a word $p = n_1..n_k \in (N \cup Ends \cup Start)^*$ such that $\forall i \in 1..k-1, (n_i, n_{i+1}) \in \longrightarrow$. An initial path is a path starting from Start.*

Definition 4. *A choice in an HMSC H is a node with more than one successor. A choice c defines an alternative between scenarios. Any loop-free path starting from c will be called a branch of the choice c.*

The semantics of HMSC [8] is given by means of regular expressions on bMSC, built from the operators $\mp$ (delayed choice), $\parallel$ (parallel composition), and $\circ$ (sequential composition). An HMSC defined as a graph can be easily transformed into a set of process algebra expressions with the same meaning. To each node $n \in (N \cup Ends \cup \{Start\})$, we associate an expression:

- if n is an end node, $n = \epsilon$
- if n is a choice node, $n = \underset{\{n_i | n \longrightarrow n_i\}}{\mp} ni$

- if n is a node with one single successor n', and $l(n) \neq r_{M_\emptyset}$, then $n = l(n) \circ n'$
- if n is a node with one single successor n', and $l(n) = r_{M_\emptyset}$, then $n = n'$

This definition by means of process algebra has the same expressive power as the hierarchical graphs. This is not really surprising, as HMSCs are just automata labeled with expressions. Hierarchical graphs are often more adapted to algorithmic considerations, and process algebra facilitates discussions on semantics points. During the rest of the paper, we will use both representations.

Let us now recall some operational semantics rules for HMSCs defined in [8]. Note that we slightly adapted the rules to fit our partial order representation. Furthermore, we only recall rules that will be needed in the next sections.

First, a permission relation $\cdots \overset{a}{\to}$ is defined. Using our partial ordering definition for MSCs, this permission relation becomes :

$$\frac{\phi(a) \notin \phi(E_x)}{x \cdots \overset{a}{\to} x} \qquad \frac{x \cdots \overset{a}{\to} x', y \cdot \!/\!\cdot \overset{a}{\to}}{x \mp y \cdots \overset{a}{\to} x'} \qquad \frac{x \cdots \overset{a}{\to} x', y \cdots \overset{a}{\to} y'}{x \circ y \cdots \overset{a}{\to} x' \circ y'}$$

More intuitively, $expr \cdots \overset{a}{\to}$ if all events of $expr$ and action a are independent, and can be executed concurrently. We can now give some rules that will be needed in the rest of the paper. For a complete semantics, interested readers are referred to [8].

$$\frac{x \cdots \overset{a}{\to} x', y \overset{a}{\to} y'}{x \circ y \overset{a}{\to} x' \circ y'} \qquad \frac{x \overset{a}{\to} x'}{x \| y \overset{a}{\to} x' \| y} \qquad \frac{y \overset{a}{\to} y'}{x \| y \overset{a}{\to} x \| y'}$$

$$\frac{x \overset{a}{\to} x', y \overset{a}{\not\to} y'}{x \mp y \overset{a}{\to} x'} \qquad \frac{y \overset{a}{\to} y', x \overset{a}{\not\to} x'}{x \mp y \overset{a}{\to} y'} \qquad \frac{x \overset{a}{\to} x', y \overset{a}{\to} y'}{x \mp y \overset{a}{\to} x' \mp y'}$$

3   Implicit Synchronization

MSC are supposed to be very intuitive and visual. However, the interpretation of some constructions may be in contradiction with the semantics. In most cases, this divergence between the effective behavior and the intended behavior is just due to a bad knowledge of the language semantics, or to some abusive use of its constructs. The first pathological MSC kind defined in this paper is *implicit synchronization*.

Consider the HMSC of Fig. 5. From the previous translation rules, we can define the same HMSC with the following expressions:

$Pathology1 = Start_{P1}$ $H1 = Start_{H1}$ $H2 = Start_{H2}$
$Start_{P1} = np1$ $Start_{H1} = n1h1$ $Start_{H2} = n1h2$
$n1p1 = n2p1$ $n1h1 = n2h1$ $n1h2 = n2h2$
$n2p1 = (r_{H1} \| r_{H2}) \circ n1p1$ $n2h1 = n3h1 \mp n4h1$ $n2h2 = n3h2 \mp n4h2$
                             $n3h1 = r_{M1} \circ n1h1$ $n3h2 = r_{M3} \circ n1h2$
                             $n4h1 = r_{M2} \circ n5h1$ $n4h2 = r_{M4} \circ n5h2$
                             $n5h1 = \epsilon$ $n5h2 = \epsilon$

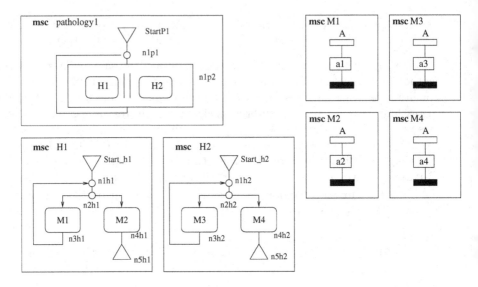

Fig. 5. Implicit synchronization

According to the operational semantics rules, as $H1$ and $H2$ are composed in parallel, $a1$ and $a2$ can be executed concurrently with $a3$ or $a4$, and conversely. This may mean that instance A is composed of more than one entity, that may evolve concurrently. So, events $a3$ and $a4$ cannot prevent $a1$ and $a2$ from being executed. However, $np1$ rewrites to $r_{H2} \circ np1$ after executing action $a2$. As $H2 \cdot \overset{a1}{\not\rightarrow}$ and $H2 \cdot \overset{a2}{\not\rightarrow}$, actions $a1$ and $a2$ cannot be executed from $r_{H2} \circ np1$, unless $a4$ is executed, allowing to rewrite $r_{H2} \circ np1$ into $np1$. The automata corresponding to this semantics is provided in Fig. 6. The language defined this way is $\left((a1+a3)^*.((a2.a3^*.a4)+(a4.a1^*.a2))\right)^*$. A synchronization between the two parallel components is defined. One may wonder if this behavior was really intended, and if HMSC of Fig. 7, in which the two parallel components behave in parallel without synchronizing does not fit better the expected behavior.

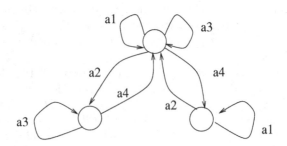

Fig. 6. Operational semantics for HMSC in Fig. 5

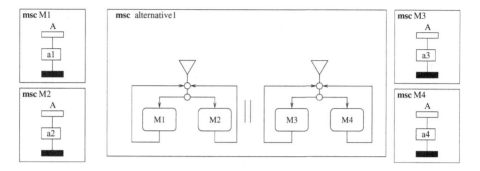

Fig. 7. Alternative intended behavior

Of course, in the general case, it is false that $(x\|y)^*$ and $(x^*\|y^*)$ have the same operational semantics. Hence, the end of the parallel frame acts as a synchronizing point, which may not be obvious for the designer. According to the semantics, one may suppose that there is a fork and join between the two operands of the parallel composition, that is abstracted in the MSC representation.

This implicit synchronization is not dramatic, as HMSCs of Figs. 5 and 7 are both correct. However, the fact that an HMSC may express more than what was really intended points out the need for simulation tools exhibiting possible behaviors of a specification. Furthermore, it should be clear that events composed in a parallel frame can be interleaved, but are not always independent, which explains the synchronization when exiting the frame.

One may note that there is not explicit synchronization construct in Message Sequence Charts. Adding this new feature to the language should not be too difficult, as synchronization reduces concurrency, and consequently the set of behaviors described by an MSC. However, the discussion of a synchronization construct is beyond the scope of this paper.

4   Non-local Choices

Implicit synchronization is not the only kind of MSC where control is hidden. Another situation potentially leading to erroneous interpretation is called *non-local choice*. The generally admitted meaning of non-local choice [2] is when more than one instance can decide to perform a scenario or another at a choice node. The intended behavior is that the first instance able to perform the choice chooses a behavior. The next instances reaching the same occurrence of this choice have to conform to the chosen scenario. This results in a behavior in which an instance "knows" what to do at a choice node without any communication. An example of non local HMSC is given in Fig. 8. In this example, if instance A chooses to send message $m1$ then instance B must conform to scenario $M1$ and receive $m1$. Conversely, if instance B chooses to send message $m2$ then instance A must conform to scenario $M2$ and receive $m2$. To implement only these two scenarios, a designer would have to use a consensus or synchronization mechanism, which is not explicitly represented.

A formal definition of non-local choice was previously given in [2]. This definition does not take parallel composition into account, and assumes that any instance should perform a communication in each bMSC referenced by a successor of the choice node. This assumption limits the search for non-local choice to the set of immediate successors of a choice node of an HMSC. However, [4] shows that when weak sequential composition of bMSCs with a disjoint set of instances is considered, non-local choice is not a local property. Therefore, a global definition of non-local choice must be provided. Consider HMSC in Fig. 9: choices seem to be local, but the decision to perform a scenario can be taken by A or C. This shows that non-locality is a global property, which has to be computed on the HMSC structure. The definition has thus to be extended, and should take into account parallel composition and hierarchy.

Definition 5. *Let* $expr = \underset{i \in 1..N}{\mp} n_i$ *be an expression defining a choice. We will say that expr is a non local choice if there are two events* e_1 *and* e_2 *such that* $expr \xrightarrow{e_1}$, $expr \xrightarrow{e_2}$, *and* $\phi(e_1) \neq \phi(e_2)$.

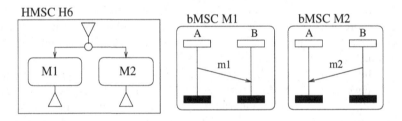

Fig. 8. non local HMSC

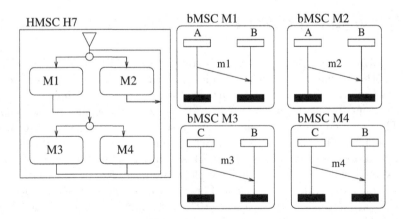

Fig. 9. An example HMSC that may seem local

We have shown that the locality of a choice is a global property. So, it may be very difficult to detect it, even with good knowledge of the semantics. As the graphical and process algebra representations of HMSCs are dual, there must be a definition of non local choice holding for graphs. We now show that non locality can be expressed as a global property on the paths of a HMSC document. This immediately provides us with an algorithm for non local choices detection.

Definition 6. *Let H_i be a HMSC, and x be a node of H_i A maximal path starting from x of H_i is either :*

- *a finite path of the form $w = x.n_1...n_k$ where $n_k \in Ends_i$,*
- *a finite path of the form $w = x.v$ such that it is impossible to leave the sequence $(v)^\omega$ (any path of the HMSC H starting with $x.v$ is a prefix of $x.v.v^\omega$).*

Algorithm: Calculus of the maximal paths of HMSC H_i starting from node x
Maximal_path$(H_i, x)=$
$P = \{x\}$ /* set of path of H */
$AP = \emptyset$ /* set of acyclic path */
while $P \neq \emptyset$ do
$\quad AP = AP \cup \{w.n | w = n_1...n_k \in P , n_k \longrightarrow_i n , n \in End_i\}$
$\quad\quad \cup \{w = n_1...n_k.n...nk \in P | n_k \longrightarrow_i n\}$
$\quad$ /* paths for which adding node n creates a cycle */
$\quad P' = \{w.n | w \in P, w = n_1...n_k,$
$\quad\quad\quad n_k \longrightarrow_i n, and w \neq n_1...n_k.n...nk, n \notin End_i\}$
$\quad P = P'$
end while
$MAP = \emptyset$
/* remove paths that stop on a cycle but are not prefixes of
    infinite paths of the form $v.(v')^\omega$ */
for all $w \in AP$ do
$\quad$ if $\not\exists w' \neq w \in AP$ such that $w' = w.v$ then
$\quad\quad MAP = MAP \cup \{w\}$
$\quad$ end if
end for
return(MAP)

Consider, for example the HMSC of Fig. 10. The set of maximal acyclic path of this HMSC is :
$MAP(H) = \{$ $start.n1.n2.n4.n5;$ $start.n1.n2.n3.n2.n4.n5;$
$\quad\quad\quad start.n1.n2.n6.n7.n8.n7;$ $start.n1.n2.n3.n2.n6.n7.n8.n7\}$
$n1.n2.n3$ is not a maximal path, as it is always possible to leave cycle $(n2.n3)^*$, for example by choosing $n4$. Note that the maximal paths are computed without considering what a node references. Maximal path will be used to detect minimal events on each branch of a choice. Once maximal paths have been found, scenarios attached to each path are computed.

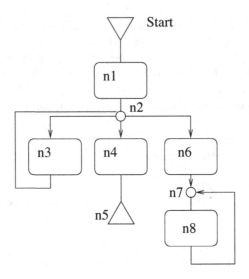

Fig. 10. Example of HMSC

Definition 7. *The path order family $POF(w)$ associated to a path $w = n_1..n_k$ is a set of partial orders built from the sequences of bMSCs along this path. It is defined recursively:*

- *let n_1 be a node and v be a path of a HMSC. $POF(n_1.v) = \{o_1 \circ o_2 | o_1 \in POF(n_1), o_2 \in POF(v)\}$*
- *For any node n, $POF(n) =$*
  - *if $l(n) = \epsilon$ then $M_\emptyset$*
  - *if $l(n) = r_M$ then $\{M\}$*
  - *if $l(n) = \|r_{H_i}$ then*
    $$\bigcup_{\substack{\pi \\ permutation \\ on\ 1..N}} \{o_{\pi(1)} \circ \cdots \circ o_{\pi(N)} | \exists w_i \in MAP(H_i, Start_i), o_i \in POF(w_i)\}$$

Theorem 1. *Let c be a choice node in a HMSC H, and $expr_c$ the process algebra expression associated to this choice. $expr_c \xrightarrow{e}$ if and only if $\exists w \in MAP(H, c), \exists o \in POF(w)$, and $e_1 \in min(o)$.*

proof: To complete the proof, we need some lemmas:

Lemma 1. let $expr_n$ be the expression associated to a node n.

If $expr_n = x \circ y$ and $y = \mp exp_i$ then $expr_n$ can be rewritten into $expr'_n = \mp y \circ exp_i$,

If $expr = x$ and $x = y$, then $expr_n$ can be rewritten into $expr'_n = y$
   We will say that $expr'_n$ is obtained by rewriting x.

For any event e, we have $expr_n \xrightarrow{e} \Longleftrightarrow expr'_n \xrightarrow{e}$.
proof: From the definitions. $\Box$

Lemma 2. Let c be a choice node. If there is a sequence of rewritings $w = n1.n2.\ldots.nk$ of $expr_c$, such that any pair (n_i, n_{i+1}) is unique in w then w is a maximal path starting from c.
proof: From the definitions. $\Box$

Lemma 3. Let $expr_n$ be the expression associated to a node n. $expr_n \cdots\xrightarrow{e}$ if and only if
$\exists o =< E_o, \leq_o, A_o, I_o, \alpha_o >\in POF(n)$, such that $\phi(E_o) \neq \phi(e)$.
proof: By induction on the depth of HMSC references. $\Box$

From Lemma 1, and lemma 2, we know that $expr_c \xrightarrow{e}$ if and only if:

- for all w, sequence of rewritings of $expr_c$ rewriting $expr_c$ into $expr'_c$
- $w \in MAP(H, c)$, and
- $expr'_c \xrightarrow{e}$

As $expr'_c \xrightarrow{e}$, then there exists one sequence $w = n1.\ldots.n_k$ and $i \in 1..k$ such that $n_i \xrightarrow{e}$, and for all $x < i$:

- there is an expression $n_x = expr \circ n_{x+1}$ with $expr \cdots\xrightarrow{e_1}$, or
- $n_x = \underset{j\in1..N}{\mp} n_j$ with $n_{x+1} = n_j$ for some j, or $n_x = n_x + 1$

So, w is of the form $w = v.n_i.v'$, and using lemma 3, we can say that $\exists o_v \in POF(v)$, $\phi(e) \notin \phi(E_o)$, and $\exists o_i \in POF(n_i)$ such that $e \in min(o_i)$.

Consequently, $expr_c \xrightarrow{e}$ if and only if $\exists w \in MAP(H, c)$, $\exists o \in POF(w)$, and $e \in min(o)$. $\Box$

From this definition, the search for non local choice from graphs is straightforward.

Algorithm: Calculus of the orders associated to a path w
```
Compute_orders(w)=
Ow = ∅; i = 1
while  i <= |w|  do
    if  l(w[i]) ≠ M∅  then
        if  l(w[i]) = rM  then
            O'w = {o ∘ M|o ∈ Ow}
        else if  l(w[i]) =  ||   exprj  then
                          j∈1..P
            /* compute the POF associated to each expression */
            for all  j ∈ 1..P  do
                if  exprj = rM  then
                    TABO[j] = {M}
```

$$\textbf{else if }\ expr_j = H_j\ \textbf{ then}$$
$$TABO[j] = \bigcup_{p \in MAP(H_j)} Compute\_Orders(p)$$
end if
end for
$O_H = \emptyset$
for all π permutation on $1..P$ **do**
$\qquad O_H = O_H \cup \{o_1 \circ \dots o_P | o_i \in TABO[\pi(i)]\}$
end for
$O'_w = \{o \circ o' | o \in O_w, o' \in O_H\}$
end if
$O_w = O'_w$
end if
$i = i + 1$
end while
return(O_w)

Algorithm: Non local choice detection
Non_Local(c)=
$MAP = $ Compute_MAP(c)
$INST = \emptyset$
for all $p \in MAP$ **do**
$\qquad O_c = $ Compute_Orders(p)
$\qquad$ **for all** $o \in O_c$ **do**
$\qquad\qquad INST = INST \cup \phi(min(o))$
$\qquad\qquad$ /* set of instances participating to */
$\qquad\qquad$ /* the decision at a choice node */
$\qquad$ **end for**
end for
if $|INST| > 1$ **then**
$\qquad$ **return**(true) /* c is a non local choice */
else
$\qquad$ **return**(false)
end if

Non local choices do not create HMSC that are very ambiguous, they just define multiple scenarios. However, they may require additional synchronization mechanisms when an implementation allowing only the defined behaviors is planned. They can therefore be considered as too abstract for some purposes. Hence, we think that non-local choice is a property that may cause misunderstanding between the requirement capture and the specification phase of a system development when it is not detected.

5   Confluence

We now define another pathological kind of Message Sequence Charts, called confluent MSC. An MSC is said to be confluent if a parallel composition is expressed by means of a choice node. Consider, for example, HMSC of Fig. 11.

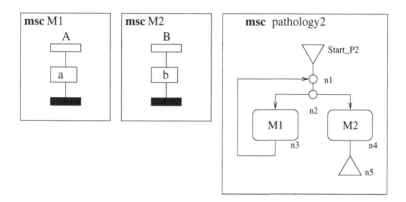

Fig. 11. Confuence

$$Pathology2 = Start_{P2} \quad n3 = r_{M1} \circ n1$$
$$Start_{P2} = n1 \qquad\qquad n4 = r_{M2} \circ n5$$
$$n1 = n2 \qquad\qquad\quad n5 = \epsilon$$
$$n2 = n3 \mp n4$$

According to the operational semantics rules, $n2 \xrightarrow{a}$ and $n2 \xrightarrow{b}$. Furthermore, as $n3 \overset{b}{\cdots\rightarrow}$ the execution of b may result from unbounded unfoldings of $n3 = r_{M1} \circ n1$. Therefore, executing b does not indicate how many a's should be executed, and the language defined by this MSC is $a^*.b.a^*$. This behavior can also be exhibited by the HMSC of Fig. 12, through a parallel composition. However, we can suppose that the intended behavior was to define b as an exit event for the loop. So, the desired behavior would have defined the language $a^*.b$. Obviously, a system where the exit condition of a loop is set by a process that never communicates with processes activated in the loop is highly pathological.

This kind of specification is called confluent, and was already identified in [9], and proved undecidable. However, this undecidability is a consequence of a weaker ordering relation than what is usually used for MSC ([9] considers that receptions of messages coming from different senders are not ordered even if they are sequential on an instance axis). We now give a formal definition for confluence, and show that it can be detected using a simple algorithm. Up to now, the definition and algorithm only concerns HMSC referencing bMSC, but can probably be extended to include parallel frames and hierarchy.

Definition 8. *Let H be an HMSC referencing bMSCs, and n be a choice node of H. n is said confluent if and only if there is a cycle c = n.n₁....nₖ, and a path*

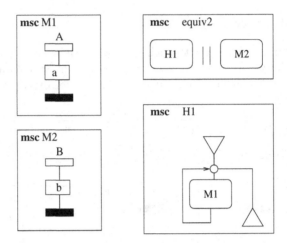

Fig. 12. HMSC with the same behavior as in Fig. 11

$p = n.x_1...x_k$ *starting from* n, *and there is a minimal event* e *of* O_p *such that* $\phi(e)$ *is not an active instance of* O_c

Note that a confluent Message Sequence Chart is necessarily non-local. We do not define the cycle computation, that can be performed using the well known Tarjan algorithm [13].

Algorithm: Confluence detection for a choice node s of a HMSC H

Confluent(H, n)=

CY = compute_cycles(n) /* set of cyclic paths containing n */

MAP = compute_map(H,n)

                          /*maximal acyclic paths starting from n*/

for all $c \in CY$ **do**

    for all $p \in MAP$ **do**

        if $\exists i \in \phi(min(Op))$ such that $i \notin \phi(O_c)$ **then**

            choice node n is confluent.

        end if

    end for

end for

Confluent Message Sequence Charts can be considered as a bad use of choice in a specification. Obviously, they may lead to misunderstanding, and therefore should be avoided. Yet, confluence is again a global property, that a correct syntax cannot prevent. One may remark that the correction of confluence is easy for the example of this section (transform HMSC of Fig. 11 into HMSC in Fig. 12). However, it may be more difficult or even impossible to find another HMSC exhibiting the same behavior.

6   Conclusion

This article has shown three kind of High-level Message Sequence Charts that may cause misunderstanding. The first case concerns implicit synchronization in loops, and can be considered as a bad specification or not, depending on the designer's awareness of the HMSC semantics. The second case is the well known non-local choice. Again, non-local specifications should be considered as incomplete rather than erroneous. However, when HMSCs are designed as a premise for the production of a specification, non-locality points out parts of the specification where additional information is needed. Therefore, a special attention should be paid to these parts of the specification where erroneous behaviors may be introduced after the requirement phase. The last situation described is confluence, and clearly appears as a bad specification: choices are used as parallel composition. Note that all these cases are expressed with MSC-96. We can also expect new constructs of MSC-2000 to introduce more interpretation problems, due to inheritance, or variables, that will be revealed when the language is put into practice.

Up to now, there is no documentation providing guidelines for correct message sequence Charts construction. We think that such a document could reference cases of bad usage of MSCs, that the grammar cannot reject. The responsibility for avoiding such MSCs would then be left to the designer, or to automated detection algorithms implemented in MSC tools when possible.

References

1. Alur R., Holzmann G. , Peled D., *An analyzer for Message Sequence Charts*, Tools and Algorithms for the Construction and Analysis of Systems (TACAS'96), LNCS no 1055 , pp 35-48, Passau, Germany, 1996.
2. Ben-Abdallah H., Leue S., *Syntactic Detection of Process Divergence and non-Local Choice in Message Sequence Charts*, Proceedings of TACAS'97, Lecture Notes in Computer Science, Vol. 1217, Brinksma.E editor, Springer-Verlag publisher, pp. 259-274, 1997.
3. Graubmann P. , Rudolph E. , Grabowski J., *Towards a Petri Net Based Semantics Definition for Message Sequence Charts*, In: SDL'93 - Using Objects (Editors: O. Faergemand, A. Sarma), North-Holland, October 1993.
4. Hélouët L., Jard C., *Conditions for synthesis from Message Sequence Charts*, 5th international workshop on Formal Methods for Industrial Critical Systems (FMICS), Berlin, April 2000.
5. ITU-T, *Message Sequence Chart (MSC-2000)*, ITU-T Recommendation Z.120 (11/99), 2000.
6. Katoen J.P., Lambert L., *Pomsets for message sequence charts*, proceedings of *SAM98:1st conference on SDL and MSC*, pp. 291-300, 1998.
7. Loidl S., Rudolph E., Hinkel U., *MSC'96 and Beyond - a Critical Look*, Proceedings of the Eight SDL Forum, SDL'97: Time for Testing - SDL MSC and Trends, A. Cavalli and A. Sarma, editors, Evry, France, 23-26 September, 1997.
8. Mauw S., Reniers M. , *High-level Message Sequence Charts*, Proceedings of the Eight SDL Forum, SDL'97: Time for Testing - SDL MSC and Trends, pp 291-306, A. Cavalli and A. Sarma, editors, Evry, France, 23-26 September, 1997.

9. Muscholl A., Peled D., *Message sequence graphs and decision problems on Mazurkiewicz traces*, Proc. of MFCS'99, Lecture Notes in Computer Science 1672, pp. 81-91, 1999.

10. Pratt.V , *Modeling Concurrency with Partial Orders*, International journal of Parallel Programming, Vol. 15, No. 1, pp. 33-71, 1986.

11. Reniers M., Mauw S., *An algebraic semantics for basic message sequence charts*, The Computer Journal", Vol. 37, No. 4, pp. 269-277, 1994.

12. Reniers M., *Message Sequence Charts: Syntax and Semantics*, PhD Thesis, Heindhoven University of Technology, 1998.

13. Tarjan.R, *Depth-first search and linear graph algorithms*, SIAM Journal of Computing, 1(2), 1992.

An Execution Semantics for MSC-2000*

Bengt Jonsson[1] and Gerardo Padilla[2]

[1] Dept. of Computer Systems, P.O. Box 325, S-751 05 Uppsala, Sweden
bengt@docs.uu.se
[2] Telelogic AB, Uppsala, gpadilla@docs.uu.se

Abstract. Message Sequence Charts (MSCs) is a visual notation for expressing requirements on communicating systems. Their expressive power has traditionally been somewhat limited, and additional information is usually needed by tools that manipulate them: for example, to derive test suites. The new standard MSC-2000, developed by ITU-T, extends earlier versions by constructs for data and high-level control, so that it may be possible to derive test sequences directly from MSC requirements, without the need for additional information. Motivated by this, we present an execution semantics for a significant part of the MSC-2000 standard. The semantics has the form of an Abstract Execution Machine, which can either accept or generate sequences of events that are consistent with a given MSC. In the former case, the Abstract Execution Machine can be used as a test oracle, in the latter as a test sequence generator.

1 Introduction

Message Sequence Charts (MSCs) is a graphical notation for description and specification of interaction between entities of a communicating system. MSCs may be used for requirement specification, interface specification, simulation and validation, test case specification, and documentation. The main use of MSCs has been as abstract description of the communication behavior of communicating systems, while omitting other aspects of a system's behavior. Used in this way, MSCs do not, by themselves, contain enough information for the generation of test sequences, or for attempting to synthesize a system design. Tools for test generation, such as Autolink or TestComposer [17], and TGV [6] derive test sequences from the combination of a test purpose specification in MSC and a design model in SDL.

One of the motivations for our work is to investigate whether and how MSCs can be used to describe requirements more completely, in a way that allows test sequences to be generated without involving other information about the system (such as in an SDL model). MSCs must then be able to express many aspects of a system that can be represented in, e.g., an SDL model. The new standard

---

* Supported in part by Telelogic AB, and by NUTEK through the ASTEC Competence Center. This work was carried out while Gerardo Padilla was at Telelogic AB, Uppsala

R. Reed and J. Reed (Eds.): SDL 2001, LNCS 2078, pp. 365–378, 2001.

MSC-2000 [11], developed by ITU-T, extends the earlier standard MSC-96 [10] by constructs for data and high-level control, so that this may be possible.

In many approaches to test generation from MSCs (such as [7,18]), an MSC test purpose is viewed as specifying a set of acceptable sequences of events of the implementation. In the approaches of Autolink, TestComposer, or TGV, the test purpose is often incomplete (missing parameters and data values), and the missing information is supplied by matching it with an SDL model. There are also other approaches to testing, see [4,14,15], where a requirement is translated to a test oracle, which monitors test execution and reports deviations from the original requirements.

Thus, part of the test generation in the above approaches consists in viewing an MSC as an acceptor or generator of sequences of events. It is not difficult to understand what sequences are represented by a simple basic MSC. However, this is less obvious for MSCs following the newer standard MSC-2000, which contains advanced control structures and treatment of data. When trying to understand how MSC-2000 can be used for testing, we found a need for an understandable and implementable description of what sequences of events are represented by an MSC. This motivates our development of a semantics for a part of MSC-2000, which views an MSC as a (structured) abstract machine, whose executions generate or accept sequences of events in the same way as runs of a finite automaton generate or accept strings.

In this paper, we present the major aspects of our execution semantics for MSC-2000. Our aim has been to define a semantics, which in a straight-forward way explains how an MSC "executes", defined in terms of its graphical structure. The semantics derives from each MSC a state machine, which maintains the "current state" of the MSC. Based on the current state of the MSC, transitions can be performed, which are equipped with labels that explain what action is currently performed, and which result in new states of the MSC. The possible transitions are defined through rules for how an MSC may transform between global states while performing events (such as transmitting or receiving messages). We have attempted to make these rules close to an intuition obtained from the graphical presentation of MSCs. Our rules thus provide an operational semantics for the MSC, which can be presented and understood in rather simple terms and be implemented quite directly.

As a contribution, this paper thus presents an operational semantics for MSCs, which is presented in a direct manner by showing how an MSC may "execute" by changing its global state, without translation to an immediate formalism. Each construct of MSC-2000 that we consider is defined by one or two rules. In particular, we present a simple operational semantics for high-level control constructs (such as inline expressions) and data aspects of the MSC-2000 standard [5]. A main problem here is that inline expressions impose control structures that are global to several instances, which otherwise execute asynchronously. In the execution semantics, this "global control" is represented by global execution steps where participating instances are synchronized.

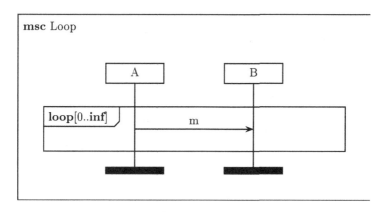

Fig. 1. An MSC with a Loop Inline Expression.

Our semantics naturally implies a certain amount of interpretation of the ITU-T standard. For instance, we use the weak form of sequential composition (in the terminology of [11]). This means that the MSC shown in Fig. 1 allows executions in which many messages can be sent by A before any of them are received by B. Our interpretation is that the standard adopts this view. However, the semantics we propose in this paper can easily be adapted to a stronger interpretation of sequential composition, or to a stronger synchronization between instances in the presence of control constructs.

Related Work There are several approaches to defining the semantics of MSCs in terms of other formalisms. For basic MSCs, semantics have been given in terms of Petri Nets [8,9], automata [12], process algebra [13], etc. In comparison with Petri Net semantics, our work gives a more direct semantics without translation to an intermediate formalism. Ladkin and Leue [12] make a more restrictive interpretation of communication: disallowing buffered communication.

Several approaches defined the semantics compositionally, by building a process algebra with a repertoire of composition operators, in which MSCs are defined. An advantage of this approach is its compositionality, which is not present in our, more direct, semantics. Reniers [16] presents a carefully worked out compositional semantics of MSC-96. Data is not considered.

Works that are closer in spirit to ours include the work on *Live Sequence Charts* by Damm and Harel [3]. They present an extension of MSCs which is different from the standards MSC-96 and MSC-2000. Their semantic definition is similar in spirit to the one defined in this paper, but differs, for example, by adding explicit "program pointers" that range over control locations in the instances. Alur and Yannakakis [1] present a simple semantics for Hierarchical basic MSCs.

Outline In the next section, we informally present the ideas behind our execution semantics. In Sect. 3, we present the main features of MSC-2000 that we considered in a form suitable for the definition of semantics. A semantics for these features is proposed in Sect. 4. Section 5 contains conclusions and directions for future work.

2   Informal Presentation of Ideas in the Semantics

In this section, we present the ideas underlying the execution semantics for MSCs. Consider first a simple Basic Message Sequence Chart, such as the one in Fig. 2. This MSC defines some possible scenarios of interaction between the three

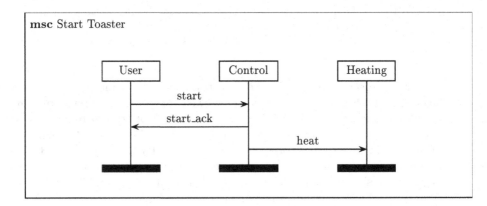

Fig. 2. A Basic MSC.

instances User, Control, and Heating. Our aim is to derive a transition system (that is, a finite-state or infinite-state machine) which is able to perform the six events (three sends and three receives) in exactly those orders that are allowed by the MSC. A straight-forward approach is to let the state of this transition system for each instance record the sequence of remaining events. For example, for the *User* instance, the state will initially consist of the sequence

$$send(\text{start})\quad rec(\text{start\_ack})\qquad .$$

In order to enforce the ordering between corresponding send and receive events, we need a mechanism by which a reception becomes enabled only after the corresponding send event has been performed. In the above example, the reception of start_ack, denoted $rec(\text{start\_ack})$ will become enabled only after the corresponding sending of start_ack, denoted $send(\text{start\_ack})$, by instance *Control*, has been performed. Based on these ideas, rules can be formulated for how the transition system can perform events, and how its internal state will be changed accordingly.

Let us now extend the MSCs with control structures, such as inline expressions. Figure 3 shows an MSC with inline expressions. After the transmission

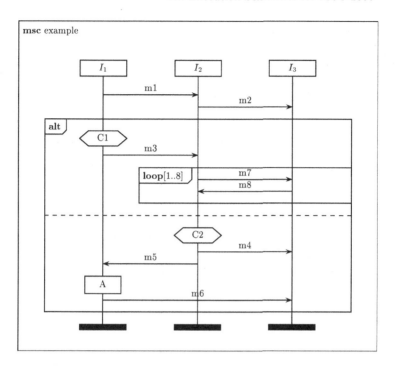

Fig. 3. An MSC with Inline Expressions.

of messages $m1$ and $m2$, the MSC contains an alternative inline expression, in which the execution can follow one of two possibilities. The first alternative starts by checking condition $C1$ in instance I_1, whereafter I_1 sends $m3$. The second alternative starts by instance $m2$ checking condition $C2$ and then sending $m4$.

In order to derive a transition system which gives the semantics of this MSC, let us consider how to extend the ideas presented for the previous basic MSC. The transition system will still maintain a local state for each instance, and let message receptions become enabled by corresponding transmissions. The execution of the MSC in Fig. 3 may follow either of two paths, depending on which alternative is chosen in the outer inline expression. The crucial point here is that all three instances must choose the same alternative even if they do not arrive simultaneously to the inline expression in their execution. In Fig. 3, the choice between the two alternatives means that if instance I_1 passes condition $C1$ and sends $m3$, then instance I_2 must not attempt to pass $C2$ after sending $m2$, but should rather wait for message $m3$. Thus, instance I_1's entry into the first alternative has global consequences in that it "forces" the other instances to also choose the first alternative.

The preceding discussion shows that control structures must be captured by a mechanism which is global to all instances. In our semantics, we treat inline expressions as "global" procedures, which may span over several instances. In the containing MSC, an inline expression is represented by a name in the control structure of each participating instance. Before any instance can "enter" into an inline expression, the inline expression name must be "expanded" and replaced by its defining inline expression. This expansion takes place simultaneously in all participating instance: the entry into a section of an inline expression is a global event. Note that at expansion time, some instances may not yet have reached the inline expression in their execution, but the expansion forces them to make a consistent choice when they reach it later in their execution.

Let us finally consider data, which is introduced in the MSC-2000 standard. As an illustration of the data definition mechanisms, consider the MSC in Fig. 4. In this MSC, instance I_1 has a local variable x, instance I_2 has a local variable y, and instance I_3 has a local variable z. In its first action, instance I_1 assigns an arbitrary value, represented by a wildcard, to x. This value is then transmitted as a parameter of message $m1$ to instance I_2. Instance I_2 may then use the value of x in local manipulations, such as assignments to its variable y, and as a parameter in subsequent message transmissions. We say that instance I_2 *inherits* the binding to x from the reception of the first message, where x is bound to a specific value.

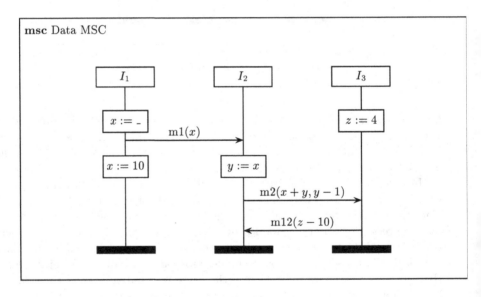

Fig. 4. An MSC with Data.

In our semantics, we extend the state of each instance by an environment which maintains the values of local variables. Thus, instance I_1 has a local environment which maintains the current value of x, and similarly for instances I_2 and I_3. In order to model the fact that the transmission of the first message also transmits a binding of the local variable x from I_1 to I_2, we let the local environment of I_2 also contain a binding to x after reception of the message.

In our semantics, when the transmission of message m1 is performed, then the corresponding message reception becomes enabled *and* extended with an extra update action which assigns appropriate values to inherited variables (x in this case). When the message is received, this extra update action will be performed, thus adding the inherited variables to the local environment of the receiving instance.

3   Definition of Message Sequence Charts

In this section, we define MSCs in a form which is suitable for our execution semantics. The MSC syntax presented here corresponds closely to the representation of its state in the execution semantics.

Basic Concepts We assume a basic vocabulary, containing *instance names*, *message names*, and *condition names*. We assume *data types*, and a set of typed data *variables*. A *wildcard* is a special symbol to denote a don't care value. An *expression* is formed from variables, wildcards, and operators in the data language. A *binding* corresponds to an assignment, written $x := y + 3$[1]. It is assumed that expressions, bindings, etc. are type-consistent in the usual sense.

We will consider *actions* that consist of bindings, such as the four actions in Fig. 4, and *guarding conditions* which contain boolean expressions as guards. By adding suitable control variables, we could treat setting conditions as actions that assign appropriate values to the control variables, and treat guarding conditions containing condition names as conditions whose guards are boolean expressions over the control variables.

Each arrow representing a message is in our definition represented by a unique *transmission name*. We use Y to range over transmission names. The transmission of the message in the sending instance is represented by a *sending reference* of form $Y!$. The receiving end of the message is in the receiving instance represented by a *receiving reference* of form $Y?$. The label *mlabelofY* of Y is an expression of form $m(exp_1, \ldots, exp_k)$ where m is a message name, and $exp_1, \ldots, exp_k$ is a (possibly empty) tuple of expressions.

Each occurrence of an inline expression is represented by a unique *chart name*, ranged over by X. A chart name is analogous to a procedure call, which occurs once in each participating instance. Each chart name has a definition (corresponding to the body of the procedure), which is expanded (simultaneously in all concerned instances) when execution reaches the chart name.

Sequence Charts A *thread* is a sequence of actions, conditions, transmission references, and chart names, or constructs of form $\mathbf{par}(\pi_1, \cdots, \pi_k)$, where $\pi_1, \ldots, \pi_k$ are threads. We use π, possibly with subscripts, to range over threads. A *chart* C is a composition of threads of form

$$I_1 : \pi_1 \; \| \; \cdots \| \; I_m : \pi_m$$

---

[1] The MSC-2000 standard also allows other syntaxes for assignments, which we do not consider here

where $I_1, \ldots, I_m$ are distinct instance names and $\pi_1, \cdots, \pi_m$ are threads. We refer to $\{I_1, \cdots, I_m\}$ as the *instances* of C, denoted $IsofC$. The chart must satisfy certain consistency conditions, concerning ordering between occurrences of chart names, transmission references, etc. We do not present them here, since they are not needed for understanding the remainder of the paper.

Inline Expressions An *inline expression* consists of an operator applied to a tuple of charts. If C and $C_1, \cdots, C_k$ are charts with $IsofC_1 = \cdots = IsofC_k$, and if l and u are natural numbers or ∞ with $0 \leq l \leq u \leq \infty$ and $l < \infty$, then

- **loop**$([l..u], C)$ is a *loop* inline expression, which intuitively denotes at least l and at most u repetitions of the chart C.
- **alt**$(C_1, \ldots, C_k)$ is an *alternative* inline expression, which denotes a nondeterministic choice of one of the charts $C_1, \cdots, C_k$, and
- **par**$(C_1, \ldots, C_k)$ is a *parallel* inline expression, denoting the parallel independent execution of the charts $C_1, \cdots, C_k$.

Each chart name has a definition as an inline expression. We write $eval(X) = $ *inline expression* to denote that the chart name X is defined as the inline expression *inline expression*. It is defined:

$$\begin{aligned} Isof\mathbf{loop}([l..u], C) &= IsofC \\ Isof\mathbf{alt}(C_1, \ldots, C_k) &= IsofC_1 \\ Isof\mathbf{par}(C_1, \ldots, C_k) &= IsofC_1 \\ IsofX &= Isof\,eval(X) \end{aligned}$$

Remark: There are other forms of inline expressions, which can be expressed in terms of the above ones. Here are two examples.

- A coregion with events e_1 , $\cdots$, e_m can be defined as the expression **par**$(e_1, \cdots, e_m)$.
- An optional inline expression can be written as a loop inline expression which is repeated 0 or 1 times, so we omit it.

Example The MSC in Fig. 3 is, according to our definitions, represented as the chart

$$I_1 : Y_1! \; X_1 \parallel I_2 : Y_1? \; Y_2! \; X_1 \parallel I_3 : Y_2? \; X_1$$

where

$$\begin{aligned} eval(X_1) &= \mathbf{alt}(C_2, C_3) \\ C_2 &= I_1 : C1 \; Y_3! \parallel I_2 : Y_3? \; X_2 \parallel I_3 : X_2 \\ C_3 &= I_1 : Y_4? \; A \; Y_6! \parallel I_2 : C2 \; Y_4! \; Y_5! \parallel I_3 : Y_4? \; Y_6? \\ eval(X_2) &= \mathbf{loop}([1..8], C_4) \\ C_4 &= I_2 : Y_7! \; Y_8? \parallel I_3 : Y_7? \; Y_8! \\ mlabelofY_i &= mi \qquad \text{for } i = 1, \ldots, 8 \end{aligned}$$

Data Variables can be of two types:

- A *static variable* is used to parameterize an MSC, and is declared in the head of the MSC. A static variable cannot be modified after the instantiation of the MSC, and the scope of the variable is the entire MSC body. We do not here treat static variables.
- A *dynamic variable* belongs to an instance, and must be declared in the MSC. A dynamic variable can be modified by the owning instance using the binding (i.e., assignment) mechanism.

Dynamic variables are manipulated by conditions and actions. A transmitted message can also carry an implicit binding of the variables that appear in its label. For instance, if one instance transmits $m(x)$, where an actual value v occurs for x, then the receiving instance can also use the variable x in subsequent bindings and message transmissions, thus "inheriting" the binding to x in the received message.

4   Execution Semantics

Our semantics defines the execution of an MSC as a sequence of computation steps between successive *configurations* of the MSC. Each configuration corresponds to the "current state" of the MSC. Possible Computation steps are defined by a *labeled transition relation*, which consists of a set of triples, each of which is denoted $\gamma \xrightarrow{l} \gamma'$ where γ and γ' are configurations and l is a *label* which represents the "observable view" of the transition. Intuitively, $\gamma \xrightarrow{l} \gamma'$ denotes that in the configuration γ, the label l can be observed, whereby the configuration evolves into γ'. Labels are either of form $I : e$, meaning that the instance I performs the event e, or are the "empty" label ϵ. Depending on how the transition semantics is intended to be used, the labels may be chosen differently, since they do not affect the structure of the semantics.

A *receive event* is a term of form $rec(m(v_1, \cdots, v_k))$, where m is a message name and $v_1, \cdots, v_k$ is a tuple of values. An *extended thread* is a sequence of actions, conditions, receive events, transmission references, and chart names, or constructs of form $\mathbf{par}(\pi_1, \cdots, \pi_k)$, where $\pi_1, \cdots, \pi_k$ are extended threads.

A *local environment* of an instance I is a (partial) mapping from the dynamic and inherited variables of I to values. Given an environment σ and an expression exp, let $\sigma(exp)$ denote the *set* of possible values of exp in the environment σ^2. We use $\alpha(\sigma)$ to denote the effect of action α on the environment σ. We use $\sigma \models c$ to denote that the guarding condition c is satisfied in the environment σ. Given a tuple $exp_1, \cdots, exp_k$ of expressions and a corresponding tuple $v_1, \cdots, v_k$ of values, define $bindings(exp_1, \cdots, exp_k, v_1, \cdots, v_k)$ as the binding $x_{j_1} := v_{j_1}, \ldots, x_{j_l} := v_{j_l}$ which assigns to the variables among $exp_1, \cdots, exp_k$ the corresponding values in $v_1, \cdots, v_k$.

---

[2] Different possible values arise from different values of wildcards in exp.

A *local configuration* of an instance I is a pair (π, σ), where π is an extended thread and σ is a local environment of I. A *(global) configuration* is of form

$$I_1 : (\pi_1, \sigma_1) \parallel \cdots \parallel I_m : (\pi_m, \sigma_m)$$

where (π_j, σ_j) is a local configuration of instance I_j. We use γ to range over global configurations.

We use the notation $\pi[X]$ to denote that X occurs in the thread π, and use $\pi[\pi'/X]$ to denote the result of replacing the *first* occurrence of X in π by π'.

Action: Let α be an action

$$\cdots \parallel I_j : (\alpha\, \pi_j, \sigma_j) \parallel \cdots \xrightarrow{I_j : \alpha} \cdots \parallel I_j : (\pi_j, \alpha(\sigma_j)) \parallel \cdots$$

Guarding Condition: Let c be a guarding condition.

$$\frac{\sigma_j \models c}{\cdots \parallel I_j : (c\, \pi_j, \sigma_j) \parallel \cdot \xrightarrow{I_j : c} \cdots \parallel I_j : (\pi_j, \sigma_j) \parallel \cdots}$$

Parallel Composition 1:

$$\frac{\cdots \parallel I_i : (\pi_{i_j}\, \pi, \sigma_{i_j}) \parallel \cdots \xrightarrow{I_i : e} \cdots \parallel I_i : (\pi'_{i_j}\, \pi, \sigma'_{i_j}) \parallel \cdots}{\cdots \parallel I_i : (\mathbf{par}(\cdots, \pi_{i_j}, \cdots)\, \pi, \sigma_{i_j}) \parallel \cdots \xrightarrow{I_i : e} \cdots \parallel I_i : (\mathbf{par}(\cdots, \pi'_{i_j}, \cdots)\, \pi, \sigma'_{i_j}) \parallel \cdots}$$

Parallel Composition 2:

$$\cdots \parallel I_i : (\mathbf{par}(\epsilon, \ldots, \epsilon)\, \pi, \sigma_i) \parallel \cdots \xrightarrow{\epsilon} \cdots \parallel I_i : (\pi, \sigma'_i) \parallel \cdots$$

Message Transmission: Let Y be a transmission name with $mlabelofY = m(\overline{exp})$, and let α be the action $bindings\,(\overline{exp}, \overline{v})$ minus bindings to the local variables of I_j.

$$\frac{v_1 \in \sigma_j(exp_1), \ldots, v_k \in \sigma_j(exp_k)}{\begin{array}{l}\cdots \parallel I_i : (Y!\pi_i, \sigma_i) \parallel \cdots \parallel I_j : (\pi_j[Y?], \sigma_j) \parallel \cdots \\ \cdots \parallel I_i : (\pi_i, \sigma_i) \parallel \cdots \parallel I_j : (\pi_j[rec(m(\overline{v}))\, \alpha], \sigma_j) \parallel \cdots\end{array}} \xrightarrow{I_i : send(m(\overline{v}))}$$

Message Reception:

$$\cdots \parallel I_j : (rec(m(v_1, \cdots, v_k))\, \pi_j, \sigma_j) \parallel \cdots \xrightarrow{I_j : rec(m(v_1, \cdots, v_k))} \cdots \parallel I_j : (\pi_j, \alpha(\sigma_j)) \parallel \cdots$$

Fig. 5. Rules for Labeled Transitions

Transition Relation. The relation $\longrightarrow$ for MSCs without inline expressions is defined by the rules in Fig. 5.

Let us explain the intuition of these rules. The rule **Action** states that if a local action α occurs first in the thread of an instance, then that action can be performed, whereby the effect of the action is applied to the local environment of the instance[3]. The rule **Guarding Condition** states that a local condition c

---

[3] For simplicity, we consider only deterministic actions, such as assignments without wildcards.

can be passed whenever it is satisfied in the local environment of the concerned instance.

According to **Parallel Composition** 1, a parallel composition such as $\mathbf{par}(\cdots, \pi_{i_j}, \cdots)$ of "subthreads" may execute whatever any one of its subthreads π_{i_j} may execute. The rule **Parallel Composition** 2 states that a parallel composition $\mathbf{par}(\pi_{i_1}, \ldots, \pi_{i_k})$ of "subthreads" may terminate when all its subthreads have terminated. We do not associate any particular label on the transition, just the empty label, represented by ϵ.

The rule **Message Transmission** describes how a sending reference $Y!$ with $mlabelofY = m(exp_1, \cdots, exp_k)$ is "unfolded" into a message transmission. A sending reference $Y!$ can always be executed. When doing so, then values $v_1, \cdots, v_k$ for the parameters of the message are obtained by evaluating $exp_1, \cdots, exp_k$ in the local environment. The transmission of the message is modeled by "enabling" the corresponding (unique) receiving reference $Y?$. This is done by replacing $Y?$ by the receive event $rec(m(exp_1, \cdots, exp_k))$ together with an action which updates the local environment of the receiving instance by values of inherited variables[4]. The rule **Message Reception** can now be seen simply as an instance of the rule **Action**, since it merely performs the receive event when it is first in the local thread.

Rules for Inline Expressions As described previously, inline expressions are represented by *chart names*, that work analogously as procedure identifiers, except that a chart name may be referenced in several instances. This means that control structures are global, and that the unfolding of chart names must be made in the same way in all concerned instances. In our semantics, each unfolding of chart names is done in a transition which unfolds the chart name simultaneously for all concerned instances.

In the following, let C be a chart $I_1 : \pi_1 \parallel \cdots \parallel I_m : \pi_m$. For $i = 1, \ldots, k$, let C_i be a chart $I_{j_1} : \pi^i_{j_1} \parallel \cdots \parallel I_{j_l} : \pi^i_{j_l}$ such that

$$IsofC_1 = \cdots = IsofC_k = \{I_{j_1}, \ldots, I_{j_l}\} \subseteq IsofC \quad,$$

and such that the thread π_j in C contains at least one occurrence of X if and only if $j \in \{j_1, \ldots, j_l\}$[5].

The rules for unfolding or expanding chart names into inline expressions are given in Fig. 6. Let us give an intuitive explanation of these rules. If X is defined as a parallel inline expression, then it can be expanded by replacing the name X, in the thread of each instance I_j of X, by the parallel composition of the appropriate threads $\pi^1_j, \cdots, \pi^k_j$ from each C_i. If X is defined as an alternative inline expression, then it can be expanded by first choosing one of its arguments, say C_p, and then replacing the name X, in the thread of each instance I_j, by the appropriate thread π^p_j in C_p. If X is defined as a loop inline expression, then it can be expanded either by replacing the name X in each of its threads by the

---

[4] this is how the transmission of bindings is modeled in our semantics

[5] an instance may contain several occurrences of a chart name as a result of unfolding a loop

Parallel inline expression:

$$eval(X) = \mathbf{par}(C_1, \cdots, C_k)$$
$$\pi'_j \quad = \text{if } j \in IsofC_1 \text{ then } \pi_j[\mathbf{par}(\pi_j^1, \cdots, \pi_j^k)/X] \text{ else } \pi_j$$

$$\overline{I_1 : (\pi_1, \sigma_1) \parallel \cdots \parallel I_m : (\pi_m, \sigma_m) \xrightarrow{\epsilon} I_1 : (\pi'_1, \sigma_1) \parallel \cdots \parallel I_m : (\pi'_m, \sigma_m)}$$

Alternative inline expression:

$$eval(X) = \mathbf{alt}(C_1, \cdots, C_k)$$
$$1 \le p \le k$$
$$\pi'_j \quad = \text{if } j \in IsofC_1 \text{ then } \pi_j[\pi_j^p/X] \text{ else } \pi_j$$

$$\overline{I_1 : (\pi_1, \sigma_1) \parallel \cdots \parallel I_m : (\pi_m, \sigma_m) \xrightarrow{\epsilon} I_1 : (\pi'_1, \sigma_1) \parallel \cdots \parallel I_m : (\pi'_m, \sigma_m)}$$

Loop inline expressions: the "continue in loop" case. Let X' be a fresh chart name with $eval(X') = \mathbf{loop}([l \ominus 1..u - 1], C_1)$, where $\ominus$ is the monus operator.

$$eval(X) = \mathbf{loop}([l..u], C_1)$$
$$u \ge 1$$
$$\pi'_j \quad = \text{if } j \in IsofC_1 \text{ then } \pi_j[\pi_j^1 \ X'/X] \text{ else } \pi_j$$

$$\overline{I_1 : (\pi_1, \sigma_1) \parallel \cdots \parallel I_m : (\pi_m, \sigma_m) \xrightarrow{\epsilon} I_1 : (\pi'_1, \sigma_1) \parallel \cdots \parallel I_m : (\pi'_m, \sigma_m)}$$

Loop inline expressions: the "exit loop" case.

$$eval(X) = \mathbf{loop}([0..u], C_1)$$
$$\pi'_j \quad = \text{if } j \in IsofC_1 \text{ then } \pi_j[\epsilon/X] \text{ else } \pi_j$$

$$\overline{I_1 : (\pi_1, \sigma_1) \parallel \cdots \parallel I_m : (\pi_m, \sigma_m) \xrightarrow{\epsilon} I_1 : (\pi'_1, \sigma_1) \parallel \cdots \parallel I_m : (\pi'_m, \sigma_m)}$$

Fig. 6. Rules for Unfolding of Inline Expressions.

appropriate thread π_j^1 in C_1 followed by an "unrolled" version of the loop, or by the empty thread. Both of these cases have natural conditions in terms of the loop boundaries.

The above rules do not impose restrictions on when chart names can be expanded. It is required that a chart name be expanded only "on demand", i.e., if the following transition concerns a construct which appears due to the chart name expansion. A sequence of chart expansions and a following transition with a non-empty label can be merged into one transition with the non-empty label.

Example To illustrate how the execution semantics works, let us consider how it generates executions of the MSC in Fig. 3, which is defined as the chart

$$I_1 : Y_1! \ X_1 \parallel I_2 : Y_1? \ Y_2! \ X_1 \parallel I_3 : Y_2? \ X_1$$

Since this example contains no data, we can omit the local environments and consider this chart to be the initial configuration in an execution.

Initially, only one transition is enabled, which is the unfolding of $Y_1!$ in instance 1. This transition is labeled by $send(m1)$ and results in the configuration

$$I_1 : X_1 \parallel I_2 : rec(m1) \ Y_2! \ X_1 \parallel I_3 : Y_2? \ X_1 \quad .$$

In this chart, either the chart name X_1 can be unfolded or the message $m1$ may be received. If the former is carried out, and the first alternative in the alternative inline expression is chosen, whereafter the condition $C1$ is passed (assuming that it is enabled), we arrive at the configuration

$$I_1 : Y_3! \parallel I_2 : rec(m1) \; Y_2! \; Y_3? \; X_2 \parallel I_3 : Y_2? \; X_2 \qquad .$$

Here, we can choose to transmit message $m3$, obtaining the configuration

$$I_1 : \; \parallel I_2 : rec(m1) \; Y_2! \; rec(m) \; X_2 \parallel I_3 : Y_2? \; X_2 \qquad .$$

At this point, the only enabled transition is the reception of $m1$. After receiving $m1$, then sending and receiving $m2$, and finally receiving $m3$, the configuration becomes

$$I_1 : \; \parallel I_2 : X_2 \parallel I_3 : X_2 \qquad .$$

The chart name X_2 can now be expanded and its first event (sending m7) performed, resulting in

$$I_1 : \; \parallel I_2 : Y_8? \; X_2' \parallel I_3 : rec(m7) \; Y_8! \; X_2' \qquad .$$

where $eval(X_2') = \mathbf{loop}([0..7], C_4)$. In a short execution, it is now possible to exit the loop after receiving $m7$ and finally transmitting and receiving $m8$.

High-Level MSCs An additional control mechanism, present also in MSC-96 is the high-level MSC. Essentially, an HMSC is a graph of MSCs. The semantics of High-Level MSCs can be represented by "expansion" of MSCs in a manner analogous to Inline Expressions. We omit the details in this paper.

5   Discussion and Conclusions

In the paper, we have outlined an execution semantics for the main constructs of the standard MSC-2000. The semantics defines the meaning of and MSC in terms of a state machine, whose execution follows the structure of the MSC. We have not yet implemented the semantics, but we expect such an implementation to be rather straight-forward, based on the semantics given here. Our main motivation is to use such an implementation for test generation, and for generating test oracles.

In our opinion, our semantics also has independent interest by virtue of its simple structure. The meaning of each construct is represented by a few rules. In our experience, these rules have been very valuable to several people for understanding the meaning of constructs in MSC-2000. Of course our semantics has implied some amount of interpretation of the standard, but we expect that alternative interpretations can be accommodated by modest changes to the rules in this semantics.

Acknowledgments We are grateful to Anna Eriksson at Telelogic AB for discussions and support during the work, and to the reviewers for helpful comments. The MSCs in this paper were "drawn" using the MSC LaTeX package [2].

References

1. R. Alur and M. Yannakakis. Model checking of message sequence charts. In *CONCUR 99*, volume 1664 of *Lecture Notes in Computer Science*, 1999.
2. V. Bos and S. Mauw. A LaTeX macro package for message sequence charts, April 1999. http://www.win.tue.nl/~sjouke/mscpackage.html.
3. W. Damm and D. Harel. LSCs: Breathing life into message sequence charts. In P. Ciancarini, A. Fantechi, and R. Gorrieri, editors, *Proc. 3rd Int. Conf. on Formal Methods for Open Object-Based Distributed Systems*, pages 293–312. Kluwer Academic Publishers, 1999.
4. D. Drusinsky. The remporal rover and the ATG rover. In K. Havelund, editor, *SPIN Model Checking and Software Verification, Proc. 7th SPIN Workshop*, volume 1885 of *Lecture Notes in Computer Science*, pages 323–330, Stanford, California, 2000. Springer Verlag.
5. A. Engels. Design decisions on data and guareds in MSC2000. In *Proc. 2st Workshop of the SDL Forum Society on SDL and MSC - SAM'2000*, Grenoble, France, June 2000.
6. J.-C. Fernandez, C. Jard, T. Jéron, and C. Viho. An experiment in automatic generation of test suites for protocols with verification technology. *Science of Computer Programming*, 29, 1997.
7. J. Grabowski. The generation of TTCN test cases from MSCs. Technical Report IAM-93-010, University of Berne, Institute for Informatics, April 1993.
8. J. Grabowski and P. Graubmann an E. Rudolph. Towards a petri net based semantics definition for message sequence charts. In O. Færgemand and A. Sarma, editors, *SDL'93 - Using Objects - Proc. 6th SDL Forum*, Darmstadt, 1993. Elsevier.
9. S. Heymer. A semantics for MSC based on petri-net components. In *Proc. 2st Workshop of the SDL Forum Society on SDL and MSC - SAM'2000*, Grenoble, France, June 2000.
10. ITU-T. Recommendation Z.120, Message Sequence Chart. Geneva, April 1996.
11. ITU-T. Recommendation Z.120, Message Sequence Charts. Geneva, Nov. 1999.
12. P.B. Ladkin and S. Leue. What do message sequence charts mean? In *FORTE 93*. North-Holland, 1993.
13. S. Mauw and M.A. Reniers. Operational semantics for MSC'96. In A. Cavalli and D. Vincent, editors, *SDL'97 - Time for Testing - SDL, MSC and Trends*, pages 135–152. Elsevier, Sept. 1997.
14. T.O. O'Malley, D.J. Richardson, and L.K. Dillon. Efficient specification-based test oracles for critical systems. In *Proc. 1996 California Software Symposium*, April 1996.
15. D.K. Peters and D.L. Parnas. Using test oracles generated from program documentation. *IEEE Transactions on Software Engineering*, 24(3):161–173, March 1998.
16. M.A. Reniers. *Message Sequence Chart: Syntax and Semantics*. PhD thesis, Eindhoven University of Technology, June 1999.
17. M. Schmitt, M. Ebner, and J. Grabowski. Test generation with autolink and testcomposer. In *Proc. 2nd Workshop of the SDL Forum Society on SDL and MSC - SAM'2000*, June 2000.
18. M. Schmitt, A. Ek, J. Grabowski, D. Hogrefe, and B. Koch. Autolink - putting sdl-based test generation into practice. In *11th Int. Workshop on Testing of Communicating Systems (IWTCS'98)*, Tomsk, Russia, Sept. 1998.

Comparing TorX, Autolink, TGV and UIO Test Algorithms

N. Goga

Department of Mathematics and Computing Science,
Eindhoven University of Technology,
P.O. Box 513, NL–5600 MB Eindhoven, The Netherlands
`goga@tue.nl`

Abstract. This paper presents a comparison of four algorithms for test derivation: TorX, TGV, Autolink and UIO algorithms. The algorithms are classified according to the detection power of their conformance relations. Because Autolink does not have an explicit conformance relation, a conformance relation is reconstructed for it. The experimental results obtained by applying TorX, Autolink, UIO and TGV to the Conference Protocol case study are consistent with the theoretical results of this paper.

1   Introduction

There are several algorithms for test generation using different techniques. From a user point of view (having to select one of them for use) it is interesting to know what the possibilities and limits of these algorithms are.

Possibly the most used algorithm in this area is Autolink (part of the commercial tool TAU/Telelogic [15]) which uses SDL–92 specifications [3] and MSC–92 test purposes [4] for deriving tests. Another technique for test generation is realized in the UIO and UIOv methods ([2]) on which different tools are based (for example the Phact tool used inside of Philips [6]). We can also enumerate other famous algorithms such as TGV developed at IRISA/INRIA Rennes and Verimag Grenoble ([9]) and TorX ([16]), which is the test generation algorithm of the CdR (*Côte-de- Resyste*) project: a project in the area of automatic test derivation formed by Dutch research groups from industry and universities.

One pioneering effort in the direction of comparing the test generations algorithms was done inside the CdR project: the first year of the project was devoted to realizing a benchmarking between the performances of the four algorithms mentioned above ([4,5,9]).

Because the comparison was made by practical experiments, we found it necessary to complete it by doing research for classifying the four algorithms from a theoretical point of view. For researchers acquainted with the different approaches to automated test derivation, the results of the research presented here may seen straightforward. However, due to the fact that the different schools use different notations, we think that the mere act of expressing these methods

R. Reed and J. Reed (Eds.): SDL 2001, LNCS 2078, pp. 379–402, 2001.

in the same framework is already worthwhile. Another interesting finding of our research is that Autolink does not have an explicit conformance relation. So our work, in which a conformance relation is reconstructed for this algorithm, consolidates the conformance foundation of it. Moreover, our theoretical classification of the four algorithms is consistent with the classification that resulted from the practical experiment. The theory presented in this paper is quite general because it treats a set of test generation algorithms that are well known and come from a large range of domains: academia, commercial, industry.

In our comparison, we judge the error detection power of the algorithms as a function of the conformance relation which they implement. Because two algorithms (TGV and TorX) from four use the *ioco* theory as their formal foundations, we expressed the conformance relations of all algorithms in terms of *ioco* theory (considering that it is the most general one for all of them).

The algorithms are judged in the limit case when they exhaustively generate a large number of tests (as an approximation of an infinite set). In the limit, they can detect only the erroneous implementations which their conformance relation can detect.

The comparison is presented in the paper in the following way:

- Section 2 gives a summary of the *ioco* theory; in this summary TorX and its properties are also described;
- the next sections are dedicated to one algorithm each: Section 3 for TGV, Section 4 for UIO (UIOv) algorithms and Section 5 for Autolink;
- Section 6 describes the conclusions of our research.

Acknowledgements: We thank Sjouke Mauw and Loe Feijs for their stimulating discussions and useful comments regarding the content of this paper.

2   The Ioco Theory for Test Derivation (TorX)

The TorX test generation algorithm is at the heart of the TorX architecture. The algorithm has a sound theoretical base, known as the *ioco* theory. Below, we will give a brief summary of this theory. For a full description of the *ioco* theory see [16].

In this theory the behaviours of the implementation system (physical, real object) are tested by using the specification system (mathematical model of the system). The behaviours of these systems are modelled by labelled transition systems. A labelled transition system is defined as follows.

Definition 1. *A labelled transition system is a quadruple $\langle S, L, \rightarrow, s_0 \rangle$, where S is a (countable) non empty set of states; L is a (countable) non empty set of observable actions; $\rightarrow \subseteq S \times (L \cup \{\tau\}) \times S$ is a set of transitions; $s_0 \in S$ is the initial state.*

The universe of labelled transition systems over L is denoted by $\mathcal{LTS}(L)$. A labelled transition system is represented in a standard way as a graph or by a process–algebraic behaviour expression.

The special action $\tau \notin L$ denotes an unobservable action. A trace σ is a sequence of observable actions ($\sigma \in L^*$) and $\Rightarrow$ means the observable transition between states ($s \stackrel{\sigma}{\Rightarrow} s'$ indicates that s' can be reached from state s after performing the actions from trace σ). The empty trace is denoted by ϵ. In some cases the transition system will not be distinguished from its initial state (or the state on which it is). Furthermore we will use $s \stackrel{a}{\rightarrow}$ (or $s \stackrel{\sigma}{\Rightarrow}$) to denote $\exists s' : s \stackrel{a}{\rightarrow} s'$ (or $\exists s' : s \stackrel{\sigma}{\Rightarrow} s'$).

Definition 2. *Consider a labelled transition system $p = \langle S, L, \rightarrow, s_0 \rangle$ and let $s \in S$, $\sigma \in L^*$ and $A \subseteq L$. Then: 1) $traces(s) =_{def} \{\sigma \in L^* \mid s \stackrel{\sigma}{\Rightarrow}\}$ (the set of traces from s); 2) $init(s) =_{def} \{\mu \in L \cup \{\tau\} \mid s \stackrel{\mu}{\rightarrow}\}$ (the set of initial actions of a state); 3) s **after** $\sigma =_{def} \{s' \in S \mid s \stackrel{\sigma}{\Rightarrow} s'\}$ (the set of reachable states after $\sigma \in L^*$).*

A failure trace is a trace in which both actions and refusals, represented by a set of refused actions, occur. For this, the transition relation $\rightarrow$ is extended with refusal transitions (self–loop transitions labelled with a set of actions $A \subset L$, expressing that all actions in A can be refused) and $\Rightarrow$ is extended analogously to $\stackrel{\varphi}{\Rightarrow}$, with $\varphi \in (L \cup \mathcal{P}(L))^*$ (φ is trace which leads to a state of the system in which all the actions from a set $A \subset L$ can be refused).

Definition 3. *Let $p \in \mathcal{LTS}(L)$; then we define the failure traces of p as follows. Then: $Ftraces(p) =_{def} \{\varphi \in (L \cup \mathcal{P}(L))^* \mid p \stackrel{\varphi}{\Rightarrow}\}$.*

A special type of transition systems, the input–output transition systems, are used. In these systems the set of actions can be partitioned in a set of input actions L_I and a set of output actions L_U. The universe of such systems is denoted by $\mathcal{IOTS}(L_I, L_U)$. Because $\mathcal{IOTS}(L_I, L_U) = \mathcal{LTS}(L_I \cup L_U)$, these input–output transition systems are also labelled transition systems.

For modelling the absence of outputs in a state (a quiescent state) a special action $null$ (δ in the notation of [16], $null \notin L$) is introduced and the transformation of the automaton in a *suspension automaton* is used. Formally a state s is quiescent (denoted as $null(s)$) if $\forall a \in L_U \cup \{\tau\}$, $\not\exists s' : s \stackrel{a}{\rightarrow} s'$.

Definition 4. *Let $p \in \mathcal{LTS}(L)$ ($L = L_I \cup L_U$). Then the* suspension automaton *of p is the labelled transition system $\langle S_{null}, L_{null}, \rightarrow_{null}, q_0 \rangle \in \mathcal{LTS}(L_{null})$, where $S_{null} = \mathcal{P}(S) \setminus \{\emptyset\}$; $L_{null} =_{def} L \cup \{null\}$; $\rightarrow_{null} =_{def} \{q \stackrel{a}{\longrightarrow}_{null} q' \mid q, q' \in S_{null}, a \in L, q' = \cup_{s\in q}\{s' \in S \mid s \stackrel{a}{\rightarrow} s'\}, \exists s \in q, s' \in q', s \stackrel{a}{\rightarrow} s'\} \cup \{q \stackrel{null}{\longrightarrow}_{null} q' \mid q' = \{s \in q \mid null(s)\}, q \neq \emptyset, q' \neq \emptyset\}$; $q_0 =_{def} \{s_0\}$.*

The suspension traces of $p \in \mathcal{LTS}(L)$ (p is a normal automaton) are: $Straces(p) =_{def} Ftraces(p) \cap (L \cup \{L_U\})^*$ (for L_U occuring in a suspension trace, we write $null$).

Informally, an implementation i is a correct implementation with respect to the specification s and implementation relation **ioco**$_\mathcal{F}$ if for every trace $\sigma \in \mathcal{F}$ each output or absence of outputs, that the implementation i can perform after having performed sequence σ, is specified by s.

Definition 5. Let $i \in \mathcal{IOTS}(L_I, L_U)$, $s \in \mathcal{LTS}(L_I \cup L_U)$ (i and p are normal automatons), and $\mathcal{F} \subseteq L^*_{null}$, then: i ioco$_{\mathcal{F}}$ s $=_{def}$ $\forall \sigma \in \mathcal{F}$: out(i **after** σ) $\subseteq$ out(s **after** σ), where out(p **after** σ) $=_{def}$ init(p **after** σ) $\cap$ $(L_U \cup \{null\})$.

If $\mathcal{F} = traces(s)$ then $ioco_{\mathcal{F}}$ is called $ioconf$ and if $\mathcal{F} = Straces(s)$ then $ioco_{\mathcal{F}}$ is called $ioco$. The correctness of an implementation with respect to an implementation is checked by executing test cases (that specify a behaviour of the implementation under test). A test case is seen as a finite labelled transition system that contains the terminal states Pass and Fail. An intermediate state of the test case should contain either one input or a set of outputs. The set of outputs is extended with the output θ which means the observation of a refusal (detection of the absence of actions). A test suite is a set of test cases. When executing a test case against an implementation the test case can give a Pass verdict if the implementation satisfies the behaviour specified by the test cases or a Fail verdict if the implementation does not satisfy the behaviour.

The conformance relation used between an implementation i and a specification s is $ioco_{\mathcal{F}}$. In the ideal case, the implementation should pass the test suite (completeness) if and only if the implementation conforms. In practice because the test suite can be very large, completeness is relaxed to the detection of non-conformance (soundness). Exhaustiveness of a test suite means that the test suite does not just assure conformance but it can also reject conforming implementation. For deriving tests, the following specification of an algorithm is presented in [16]:

The specification of the test derivation algorithm
Let S be the suspension automaton of a specification and let $F \subseteq traces(S)$; then a test case $t \in \mathcal{TEST}$ (L_I, L_U) is obtained by a finite number of recursive applications of one of the following three nondeterministic choices:

1. (*terminate the test case*) $t = $ Pass;
2. (*supply an input for the implementation*) Take $a \in L_I$ such that $F_a \neq \emptyset$
   $t = a; t_a$
   where $F_a = \{\sigma \mid a\sigma \in F\}$, $S \xrightarrow{a}_{null} S_a$ and t_a is obtained by recursively applying the algorithm for S_a and F_a;
3. (*check the next output of the implementation*)
   $t = \sum\{x;\text{Fail} \mid x \in L_U \cup \{\theta\}, \overline{x} \notin \text{out}(S), \epsilon \in F\}$
   $\quad + \sum\{x;\text{Pass} \mid x \in L_U \cup \{\theta\}, \overline{x} \notin \text{out}(S), \epsilon \notin F\}$
   $\quad + \sum\{x; t_x \mid, x \in L_U \cup \theta, \overline{x} \in \text{out}(S)\}$
   where $\overline{x}$ is a notation for x in which the $null$ action is replaced by θ action and vice versa, $F_x = \{\sigma \mid \overline{x}\sigma \in F\}$, $S \xrightarrow{\overline{x}}_{null} S_x$ and t_x is obtained by recursively applying the algorithm for S_x and F_x. The summation symbol $\sum$ means: choice as usual in process algebra.

In the implementation of the algorithm initially F equals $traces(S)$.

The algorithm has three *Choices*. In every moment it can choose to supply an input a from the set of inputs L_I or to observe all the outputs $(L_U \cup \{\theta\})$ or to finish. When it finishes, because this does not mean that the algorithm detected an error, it finishes with a Pass verdict. After supplying an input, the input becomes part of the test case and the algorithm is applied recursively for building the test case. When it checks the outputs, if the current output is present in $out(S)$, that output will became also part of the test case and the algorithm will be applied recursively. If the output is not present in $out(S)$ the algorithm finishes in almost all the cases with a Fail verdict (if the empty trace is considered an element of $\mathcal{F}$). If the empty trace is not in $\mathcal{F}$ then the verdict will be Pass.

This algorithm satisfies the following properties (for a proof see [16]):

Theorem 1. *1. A test case obtained with this algorithm is finite and sound with respect to ioco$_{\mathcal{F}}$.*

2. The set of all possible test cases that can be obtained with the algorithm is exhaustive.

3   TGV

TGV is a tool for automatic test derivation [12] developed at IRISA/INRIA Rennes and Verimag Grenoble [9]. TGV is interconnected with several simulators: the SDL–92 Simulator of ObjectGeode from Verilog [3] and the Lotos Simulator from CADP [8]. The inputs of TGV are a state graph produced by a simulator and an automaton formalizing the behavioural part of a test purpose. The algorithm was used on–the–fly and batch–oriented (the test case was produced in the standard TTCN format – [1]) for testing different IUT (industrial and military applications [12]). The test purposes can be generated automatically (for example using TVEDA – see [10]) or by hand. For a precise description of TGV see [12,7].

In the early version of TGV, the tool was based on the *ioconf* conformance relation [7]. A formal proof that TGV is sound and exhaustive with respect to this conformance relation it was given in [7]. The new versions of TGV extended the tool to the *ioco* conformance relation (see [12]) by adding loops labelled with a particular output *null* to quiescent states of the specification. TGV keeps its properties (soundness and exhaustiveness) with respect to the *ioco* conformance relation [12].

Because TorX and TGV are both exhaustive with respect to *ioco* this means that theorically in the limit case, when they both exhaustively generate a big (infinite) number of tests, TGV and TorX have the same detection power (*ioco* detection power).

4   UIO and UIOv Algorithms (Phact)

One of the early methods used by the algorithms for test derivation was the UIO [2] – Unique Input/Output sequence – technique. Later this method was developed in a new methodology (called UIOv – [19]) more powerful in the detection of the IUTs which do not conform to a specification. The UIO methodology is still used today. One example is Phact [6,14] – a tool developed at the Philips Research Laboratories in Eindhoven for deriving test cases in a TTCN format and executing them. The heart of the tool is the Conformance Kit [18] developed by KPN. Phact was also used in the benchmarking experiment with the Conference Protocol [4].

Although [19] shows that the UIO relation is not complete in all cases almost all the algorithms from industry (such as the Conformance Kit) implement the UIO method, due to the complexity of the UIOv method. Nevertheless, in our comparison between the four algorithms, there are some words about the UIOv algorithms in this section. As for the other cases, the detection power of Phact is judged by looking at the detection power of its conformance relation: UIO. This will be realized by going through the following steps: Subsection 4.1 will give a survey of the UIO and UIOv theory; Subsection 4.2 will present the translation of the UIO and UIOv concepts to *ioco* concepts and Subsection 4.3 will give the classification of the UIO, UIOv and *ioconf* conformance relations.

4.1   A Survey of the UIO and UIOv Theory

The UIO and UIOv methods are applied to deterministic FSM models (finite–state–machine), which are commonly referred to as a Mealy machine [13].

Definition 6. *An FSM is a quintuple $M = \langle S, I, O, \mathrm{NS}, Z \rangle$ where S is a finite set of states, denoted by $\{s_0, ..., s_n\}$; I is a finite set of inputs, denoted by $\{i_1, ..., i_m\}$; O is a finite set of outputs, denoted by $\{o_1, ..., o_r\}$ $(O \cap I = \emptyset)$; NS (next state) is the transfer function: $S \times I \rightarrow S$; Z is the output function: $S \times I \rightarrow O$.*

Here it is understood that NS and Z are total functions.

The FSM has:

1. an initial state $s_0 \in S$;
2. a reset transition $r \in I$ (which assures that from every state it is possible to return in the initial state – for every state $s \in S$ we have that $NS(s, r) = s_0$);
3. a $null \in O$ output produced by some inputs.

Every transition of an FSM is labelled with a pair (input/output) of signals.

The reflexive, transitive closure of the transfer function is defined as in the following:

Definition 7. *Let $M = \langle S, I, O, \mathrm{NS}, Z \rangle$ be an FSM. Let $s \in S$ be a state, ϵ the null sequence, $\sigma \in I^*$ a sequence of inputs and $i \in I$ an input. Then the closure $\mathrm{NS}^* : S \times I^* \rightarrow S$ of the transfer function is $\mathrm{NS}^*(s, \epsilon) = s$; $\mathrm{NS}^*(s, \sigma i) = \mathrm{NS}(\mathrm{NS}^*(s, \sigma), i)$.*

The language of the FSM is $\mathcal{L}(M) = \{w \in (I \times O)^* \mid w = i_1/o_1...i_k/o_k, \forall w' = i_1...i_j, j < k \in \mathbb{N} : Z(s_0, i_1) = o_1 \wedge Z(NS^*(s_0, w'), i_j) = o_j\}$. An FSM is minimal if there does not exist another FSM which has the same language and which has a smaller number of states.

An FSM is connected if from every state s of the FSM it is possible to reach every other state t via a sequence of transferring inputs.

Definition 8. *An FSM $M = \langle S, I, O, NS, Z \rangle$ is connected iff: $\forall s, t \in S, \exists \sigma \in I^* : NS^*(s, \sigma) = t$.*

A transition of an FSM is a quadruple $\langle q, i, o, t \rangle$ with $q, t \in S$ and $i \in I$ and $o \in O$ such that $NS(q, i) = t$ and $Z(q, i) = o$. We write $q \xrightarrow{i/o} t$ for this transition.

An *UInOut* – Unique Input Output – for a state s is an I/O behaviour not exhibited by any other state. It was shown that an FSM which is connected and minimal has an *UInOut*-sequence for every state (see [17]). In general there are more *UInOuts* for a given state. In such a case we can choose one of them. In the remainder of this paper by *UInOut(s)* we mean the *UInOut* for a state s.

Example Let us consider the FSM of the specification from Figure 1. For this FSM, the set of inputs is $I = \{a, d\}$ and the set of outputs is $O = \{b, c, e\}$. Examples of *UInOut* sequences of the states of the specification are given in Figure 2 (a). The state 1 has a distinguishing input/output pair a/b which is its *UInOut(1)*; the *UInOut* for the states 2 and 3 are formed by concatenating two transitions: *UInOut(2)* is $a/c.a/b$ and *UInOut(3)* is $d/e.a/b$.

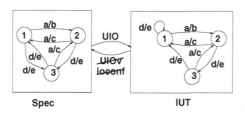

Fig. 1. The difference between UIO, UIOv and *ioco* conformance relations

For applying the UIO and UIOv methods it is necessary that:

1. the FSM of the specification is connected,
2. it is minimal and
3. the number of the states of the implementation (FSM_i) is less than or equal to the number of the states of the specification (FSM_s).

For the UIO and UIOv algorithms the approach taken for checking the correctness of an implementation FSM_i to its specification FSM_s is to test that every transition of the specification is correctly implemented by FSM_i. The procedure for testing a specified transition from state q to state t (of the specification FSM_s) with input/output pair i/o consists of three steps:

1. FSM_i is brought to state q;
2. input i is applied to FSM_i and the output produced is checked for whether it is o;
3. the state FSM_i reached after the application of input i is checked for whether it is t.

These three steps are only an abstract procedure. There are several ways, for example, to perform the check of step (3). We turn this into a more concrete procedure (organized on three sequences to be executed in succession), which we call **TestCase**. Regarding the parameters of this procedure, we will give some explanations below.

Step (3) of the procedure **TestCase** can be implemented in a different way in an UIO algorithm compared with an UIOv algorithm. Therefore we will introduce the type of the algorithm as a parameter of the procedure (**TestCase(UIO,...)** or **TestCase(UIOv,...)**). For building a test, the knowledge of the *UInOuts* of the states is used. So the set of pairs $\langle$state, *UInOut*(state)$\rangle$ (*SUIO*) will be another parameter of the procedure. Another observation is that, although we use FSM_i inside of the procedure, the building of the test does not depend on it; so we will simply remove FSM_i from the list of parameters.

$$\textbf{TestCase(UIO/UIOv, } \textit{SUIO, } \textbf{FSM}_s, q \xrightarrow{i/o} t)$$

1. a sequence ρ which brings FSM_i from its current state s_0 to state q (we call this a transferring sequence);
2. input i to check the output o;
3. a characterizing sequence to verify that FSM_i reached t (we explain this below).

Step (3) of the procedure makes the difference between the UIO and UIOv algorithms. The UIO algorithm implements it by simply checking that the implementation exhibits the *UInOut* of that state (t). Using the UIO algorithm, some faults may go undetected, however [19]. The problem of the UIO algorithm is that although each UIO sequence is unique in the specification, this uniqueness need not hold in a faulty implementation (see the example from the end of this subsection). The solution proposed by the UIOv algorithm is:

1. to check that the implementation exhibits the *UInOut* behaviour of that state (t);
2. it does not exhibit the *UInOut* behaviour of any other state k'.

Because UIO is simpler than UIOv, usually the practical tools implement the UIO method (Phact uses the UIO method).

Once it is fixed, the *UInOut* sequences of the states of the specification, every implementation that passes all the tests that are generated by an UIO (UIOv) algorithm is considered correct, all others are considered incorrect. In other words, the tests generated by the algorithm decide correctness. The UIO and UIOv methods are implicit correctness conformance relations.

The authors of UIOv [19] argue that their method can detect any missing and erroneous state and I/O of an IUT. So it can be said that the UIOv conformance relation is complete (in the sense that it can detect all erroneous IUTs), under the assumptions on which it is applied:

A) the specification FSM is connected,
B) the specification FSM is minimal,

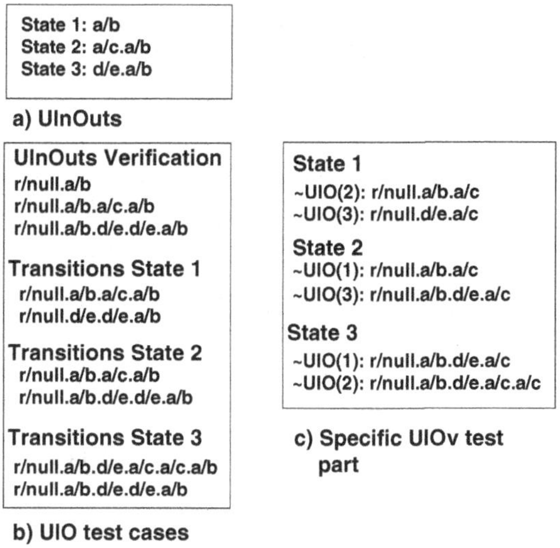

```
State 1: a/b
State 2: a/c.a/b
State 3: d/e.a/b
```

a) UInOuts

```
UInOuts Verification
r/null.a/b
r/null.a/b.a/c.a/b
r/null.a/b.d/e.d/e.a/b

Transitions State 1
r/null.a/b.a/c.a/b
r/null.d/e.d/e.a/b

Transitions State 2
r/null.a/b.a/c.a/b
r/null.a/b.d/e.d/e.a/b

Transitions State 3
r/null.a/b.d/e.a/c.a/c.a/b
r/null.a/b.d/e.d/e.a/b
```

b) UIO test cases

```
State 1
~UIO(2): r/null.a/b.a/c
~UIO(3): r/null.d/e.a/c

State 2
~UIO(1): r/null.a/b.a/c
~UIO(3): r/null.a/b.d/e.a/c

State 3
~UIO(1): r/null.a/b.d/e.a/c
~UIO(2): r/null.a/b.d/e.a/c.a/c.a/c
```

c) Specific UIOv test part

Fig. 2. Test cases derived for UIO (b) and UIOv (b, c) conformance relations

C) the number of the states of the implementation is less than or equal to the number of the states of the specification.

Example The specification from Figure 1 has three states and the implementation also has three states. The FSM of the specification is connected and minimal. So the requirements for applying the UIO (UIOv) method are fulfilled.

There is a subtle difference between the IUT and the specification: transition d/e of the state 1 goes in the specification to the state 3 and in the IUT to the same state 1. The $UInOut(3)$ (Figure 2 (a)) does not appear only in state 3 of the implementation but it appears also in state 1. So applying an UIO algorithm on the set of $UInOut$ from Figure 2 (a) will let the faulty IUT go undetected, but with an UIOv algorithm it will be detected. So in this example the IUT does conform with the specification according to an UIO relation but does not conform according an UIOv relation.

The test derivations using the UIO and UIOv methods are given in Figure 2 (b) and (c) (r is the reset signal which is present in every state; for simplification it was not represented in the figure). The verdict is not shown, it is implicit: if the IUT does not produce the expected output the verdict will be **fail**; if the IUT correctly executes all the tests cases the verdict will be **pass**. The tests from the first subpart of Figure 2 (b) (the $UInOuts$ $Verification$ part) check the $UInOut$ of every state in the implementation. The next part from Figure 2 – (b) checks every transition of the specification. We will exemplify the building of the tests from part (b) by showing it for the test: $r/null.a/b.d/e.a/c.a/c.a/b$. This test is built from four parts:

1. the sequence $r/null$ brings the FSM in the initial state;
2. the sequence $a/b.d/e$ brings the FSM in the state 3;

388   N. Goga

3. the sequence a/c is the transition which is checked;
4. now a correct implementation is supposed to be in state 2; the sequence $a/c.a/b$ is $UInOut(2)$.

The specific part of the UIOv algorithm (Figure 2 (c)) tests that every $UInOut$ is not exhibited by another state of the implementation. From the set of tests from part (c), let us take the test $r/null.a/b.d/e.a/c.a/c$. This test is formed by:

1. the reset sequence $r/null$ which brings the FSM in the initial state;
2. the sequence $a/b.d/e$ which brings the FSM in the state 3;
3. the sequence $a/c.a/c$ which is formed by the sequence of inputs $a/.a/$ (from $UInOut(2)$) and by the sequence of outputs $/c./c$ produced by the specification when $a/.a/$ is applied in the state 3.

The set of tests produced by the UIOv algorithm is composed by the union of the set of tests from Figure 2 – (b) and from Figure 2 – (c).

4.2   The Translation of the UIO and UIOv Concepts to the *Ioco* Concepts

For making the translation of the UIO and UIOv concepts to the *ioco* concepts, first we should represent an FSM as a labelled transition system. The conversion can be made easily by adding intermediate states to the automaton after each input.

Definition 9. *The labelled transition system* $M' = \langle S', L', \longrightarrow, s_0 \rangle$ *of an FSM* $(M = \langle S, I, O, NS, Z \rangle$ *is* $S' = S \cup (S \times I)$; $L' = I \cup O$; $\longrightarrow \subseteq S' \times L' \times S'$, $\longrightarrow = \{s \xrightarrow{i} \langle s, i \rangle, \langle s, i \rangle \xrightarrow{Z(s,i)} NS(s,i) \mid s \in S, i \in I\}$; s_0 *is the initial state.*

By $\langle s, i \rangle \in (S \times I)$ we denote an intermediate state of a labelled transition system of an FSM, which separates the input i from the output $Z(s, i)$.

Example Let us consider the FSM specification from Figure 1. The labelled transition system of this specification is given in Figure 3.

The correspondence between an FSM and its labelled transition system is very simple (because the labelled transition system of an FSM has only extra states which separate an input from an output). We will exemplify this for some elements of the FSMs.

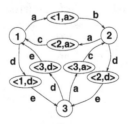

Fig. 3. The labelled transition system of the specification

A sequence $\sigma = i_1/o_1...i_n/o_n$ $(n > 0, i_j \in I, o_j \in O, j = 1, ..., n)$ produced by an FSM is translated into a correspondent trace σ' for the labelled transition system of the FSM, as $\sigma' = i_1 o_1 ... i_n o_n$. In the same way, we can obtain a correspondent set of traces for the set of tests produced by the procedure **TestCase**. The same thing can be also done for the *UInOut* of the states of an FSM.

The procedure **TestCase** can be applied on the labelled transition system of an FSM in the following way:

1. the transferring sequence is a transferring trace;
2. input i to check the output o;
3. the characterizing sequence is a characterizing trace.

Therefore in the remainder of this paper, FSM should be taken to mean as the labelled transition system of an FSM, and S should be taken to mean the set of states which are not intermediate.

As we explained in Subsection 4.1 the UIO and UIOv algorithms represent implicit conformance relations. So we will give a formal definition of the UIO and UIOv conformance relations. But, before doing this, we will introduce some terminology.

A test produced by the procedure **TestCase** is a trace. So a test suite is a set of traces. An FSM **passes** a test suite if the test suite is contained in the set of traces of the FSM.

Definition 10. *Let w be an FSM and T a test suite. Then: w passes $T \Leftrightarrow T \subseteq$ traces(w)*

The Unique Input Output Sequence of a state is

Definition 11. *Let FSM_s be an FSM of a specification and k a state of FSM_s. A trace σ is an $\mathrm{UInOut}(k)$ of $k \Leftrightarrow k \stackrel{\sigma}{\Longrightarrow} \wedge \forall s', s' \neq k : s' \stackrel{\sigma}{\not\Longrightarrow}$.*

The tests suites produced by **TestCase** are

Definition 12. *Let FSM_s be an FSM of a specification and FSM_i be an FSM of an implementation. Let I and S be the inputs and states set of FSM_s. Let SUIO be the set formed by the pairs of the states from S and their UInOut (SUIO $= \{\langle s, \mathrm{UInOut}(s)\rangle \mid s \in S\}$). Then: 1) $T_{\mathrm{UIO}} = \cup_{i\in I, q, t\in S}$ **TestCase(UIO, SUIO, FSM_s, $q \stackrel{i/Z(q,i)}{\longrightarrow} t$)** is the test suite formed by the set of test cases produced by the procedure **TestCase** using an UIO algorithm on the set SUIO for the step (3) of the procedure; 2) $T_{\mathrm{UIOv}} = \cup_{i\in I, q, t\in S}$ **TestCase(UIOv, SUIO, FSM_s, $q \stackrel{i/Z(q,i)}{\longrightarrow} t$)** is the test suite formed by the set of test cases produced by the procedure **TestCase** using an UIOv algorithm on the set SUIO for the step (3) of the procedure.*

Now we will formally define the UIO and UIOv conformance relations.

Definition 13. *Let FSM_s be an FSM of a specification and FSM_i be an FSM of an implementation. Let T_{UIO} and T_{UIOv} be the tests suites produced by the*

procedure **TestCase** *(using an UIO or an UIOv method on step (3)). Then: 1)*
FSM_i *UIO* FSM_s $\Leftrightarrow$ FSM_i **passes** T_{UIO}; 2) FSM_i *UIOv* FSM_s $\Leftrightarrow$ FSM_i **passes**
T_{UIOv}.

The central ideas of expressing the UIO relation in *ioco* terms are the following:

1. the implementation conform with the specification if it passes T_{UIO};
2. the implementation passes T_{UIO} if every trace from T_{UIO} is contained in the
   set of traces of the implementation and this is true if
3. for every trace in T_{UIO} the implementation produces after every input prefix
   of that trace (a prefix which ends in an input) the same output as the
   specification does
   $(\forall i_1 o_1...i_n o_n \in T_{UIO}, \forall i_1 o_1...i_j, j < n \in \mathbb{N} : out(FSM_i \text{ after } i_1 o_1...i_j) = out(FSM_s \text{ after } i_1 o_1...$
   $i_j)$ – the specification and the implementation are FSM, so they can produce
   only one output after an input; for this reason we put = in place of $\subseteq$). The
   same holds for the UIOv relation.

In the steps enumerated above we used for a trace the notion of the set of
its prefixes which ends into an input.

Definition 14. *Let L be an label set. Then:*

1. *(a) let $t \in L^*$, then ϵ and t are **prefix** of t;*
   *(b) let $x, y \in L^*$ and $a \in L$, then x **prefix** $y \Rightarrow x$ **prefix** ya.*
2. *let $S \subseteq L^*$ be a set of traces; then the prefix closure operator* PC: $S \longrightarrow L^*$
   is: $PC(S) = \{t \in L^* \mid \exists t' \in S : t \text{ **prefix** } t'\}$.
3. *let T be a set of traces; then the set of its prefixes P_T which ends into an
   input is $P_T = \{t \in PC(T) \mid \text{**last**}(t) \in I\}$ where the function* **last** *gives the
   last element of a trace.*

Now we have all the ingredients defined for expressing the UIO (UIOv) relation in *ioco* terms. But this will be done in the following subsection.

4.3   Classification of UIO, UIOv and *Ioconf* Conformance Relation

Lemma 1 gives a translation of the UIO and UIOv conformance relations in *ioco*
terms. Theorem 2 and Theorem 3 use the results of the lemma to relate the
ioconf, UIO and UIOv conformance relations.

In Lemma 1 we build two sets of traces F_{UIO} and F_{UIOv} such that: (FSM_i
UIO FSM_s iff FSM_i $ioco_{F_{UIO}}$ FSM_s) and (FSM_i UIOv FSM_s iff FSM_i $ioco_{F_{UIOv}}$
FSM_s). As might be expected, the set of traces F_{UIO} is the set of the prefixes
which ends into an input of the set T_{UIO} ($F_{UIO} = P_{T_{UIO}}$); the same holds for
F_{UIOv}.

Example For the specification from Figure 1 the test suite T_{UIO} is represented in Figure 2 (b) and is the set of traces $\{a/b, a/b.a/c.a/b, a/b.d/e.d/e.a/b, d$
$/e.d/e.a/b, a/b.d/e.a/c.a/c.a/b\}$. In this set we did not put the $r/null$ sequence,

because it is implicitly understood that before running each trace, FSM_i and FSM_s should be in the initial state so that a new test case will be run. Now the set of its inputs prefixes is $F_{UIO} = \{a\} \cup \{a/b.a/c.a, a/b.a, a\} \cup \{a/b.d/e.d/e.a, a/b$ $.d/e.d, a/b.d, a\} \cup \{d/e.d/e.a, d/e.d, d\} \cup \{a/b.d/e.a/c.a/c.a, a/b.d/e.a/c.a, a/b.d/e$ $.a, a/b.d, a\} = \{a, a/b.a/c.a, a/b.a, a/b.d/e.d/e.a, a/b.d/e.d, a/b.d, d/e.d/e.a, d/e.$ $d, d, a/b.d/e.a/c.a/c.a, a/b.d/e.a/c.a, da/b.d/e.a\}$.
For the UIOv relation $F_{UIOv} = F_{UIO} \cup \{d/e.a\}$.

Lemma 1. *Let* FSM_i *be an implementation and let* FSM_s *be a specification. Then there exist two sets of traces* F_{UIO} *and* $F_{UIOv} \in L^*$ *such that:*

1. FSM_i UIO $FSM_s \Leftrightarrow FSM_i \; ioco_{F_{UIO}} \; FSM_s$;
2. FSM_i UIOv $FSM_s \Leftrightarrow FSM_i \; ioco_{F_{UIOv}} \; FSM_s$.

Proof. In conformance with Definition 13 and Definition 10

1. FSM_i UIO $FSM_s \Leftrightarrow FSM_i$ **passes** $T_{UIO} \Leftrightarrow T_{UIO} \subseteq traces(FSM_i)$;
2. FSM_i UIOv $FSM_s \Leftrightarrow FSM_i$ **passes** $T_{UIOv} \Leftrightarrow T_{UIOv} \subseteq traces(FSM_i)$.

Let $\sigma \in T_{UIO}$ (or T_{UIOv}) be a trace. Let $P_{\{\sigma\}}$ be the set of its input prefixes. The trace σ is contained in $traces(FSM_s)$ and FSM_s is a deterministic FSM. Then for every $\sigma' \in P_\sigma$, the set $out(FSM_s$ **after** $\sigma')$ contains only the output which is in σ after the subtrace σ'. This output is denoted as $out(\sigma')$ $(out(\sigma') = out(FSM_s$ **after** $\sigma'))$. Then $\sigma \in traces(FSM_i)$ iff $\forall \sigma' \in P_\sigma$: $out(FSM_i$ **after** $\sigma') = out(\sigma') = out(FSM_s$ **after** $\sigma')$.

Let us consider $F_{UIO} = \cup_{\sigma \in T_{UIO}} P_{\{\sigma\}} = P_{T_{UIO}}$. Then $T_{UIO} \subseteq traces(FSM_i)$ iff $\forall \sigma \in F_{UIO} : out(FSM_i$ **after** $\sigma') = out(\sigma) = out(FSM_s$ **after** $\sigma)$. And this means that FSM_i UIO $FSM_s \Leftrightarrow FSM_i \; ioco_{F_{UIO}} \; FSM_s$; In a similar way, the point (2) of the lemma is proved if we take $F_{UIOv} = P_{T_{UIOv}}$.

We will compare conformance relations such as UIO, UIOv and *ioconf*. We want to write for example UIOv $\geq$ UIO with the intuition that UIOv is more powerful than UIO.

We want to make such comparison statements under certain conditions, for example considering only the situation where the specification is connected, minimal and where the size of the state space of the implementation does not exceed that of the specification. So if FSM_i and FSM_s are given, we write UIOv $\geq$ UIO to mean that whenever FSM_i UIOv FSM_s we find that FSM_i UIO FSM_s. The same meaning has the notation: *ioconf* $\geq$ UIO. By UIOv $\sim$ *ioconf* we mean: *ioconf* $\geq$ UIOv $\wedge$ UIOv $\geq$ *ioconf*.

Theorem 2. *Let* FSM_i *be an implementation and let* FSM_s *be a specification. Let us assume that:*

A) FSM_s *is connected;*
B) FSM_s *is minimal;*
C) *the number of the states of* FSM_i *is less than or equal to the number of the states of* FSM_s.

If **A)** $\wedge$ **B)** $\wedge$ **C)** *then*

1. UIOv $\geq$ UIO;
2. *ioconf* $\geq$ UIO;
3. UIOv $\sim$ *ioconf*.

Proof. UIO= $ioco_{F_{UIO}}$ and UIOv= $ioco_{F_{UIOv}}$. For UIO relation, $ioco_{F_{UIO}}$ has less discriminating power than *ioconf* = $ioco_{traces(FSM_s)}$ relation (because F_{UIO} $\subseteq$ $traces(FSM_s)$) and UIOv relation (because $F_{UIO} \subseteq F_{UIOv}$). Then UIOv $\geq$ UIO $\wedge$ *ioconf* $\geq$ UIO and so the first two points of the theorem are proved.

For proving point (3), we should remember that UIOv $\sim$ *ioconf* means *ioconf* $\geq$ UIOv $\wedge$ UIOv $\geq$ *ioconf*. The first part *ioconf* $\geq$ UIOv is true because UIOv has less discriminating power than *ioconf* (because $F_{UIOv} \subseteq traces(FSM_s)$). The second part of the $\wedge$ is also true because the UIOv is complete (see Subsection 4.1) under assumptions *A)* $\wedge$ *B)* $\wedge$ *C)*, UIOv can detect all the erroneous IUTs.

An example in which $\neg(FSM_i \ ioco \ FSM_s)$ and $\neg(FSM_i \ UIOv \ FSM_s)$, but FSM_i UIO FSM_s is given in Figure 1. The following theorem relates the UIOv and *ioconf* conformance relations in the general case in which the assumption *C)* does not hold. At the end of this subsection a summary of the classification is given.

In the theorem by *ioconf* $> UIOv$ we mean:
1) $\forall FSM_i, \forall FSM_s : FSM_i \ ioconf \ FSM_s \Rightarrow FSM_i \ UIOv \ FSM_s$, and moreover
2) $\exists FSM_i, \exists FSM_s : \neg(FSM_i \ ioconf \ FSM_s) \wedge FSM_i \ UIOv \ FSM_s$.

Theorem 3. *If* $\neg$C) *then ioconf* $> $ UIOv.

Proof. For proving that *ioconf* $> $ UIOv when *C)* does not hold, we should show that both (1) and (2) are true. The first part (1) is true because UIOv has less discriminating power than *ioconf* (because UIOv= $ioco_{F_{UIOv}}$ and $F_{UIOv} \subseteq$ $traces(FSM_s)$).

The second part (2) is proved by a counter example. Let us consider the IUT i from Figure 4 and the specification *Spec* from Figure 1. Then: i UIOv *Spec* (i passes all the UIOv test cases from Figure 2) but $\neg(i \ ioconf \ Spec)$ (i can be detected with the trace $a/b.a/c.a/b.a/c \in traces(Spec)$).

The summary of the classification is given in the following table

conditions	1.	2.	3.
A) $\wedge$ *B)* $\wedge$ *C)*	UIOv $\geq$ UIO	*ioconf* $\geq$ UIO	*ioconf* $\sim$ UIOv
arbitrary			*ioconf* $> $ UIOv

The conclusion of this section is that due to powerful constraints (determinism, limit for the implementation states number) UIO and UIOv are well–known to not find all errors, so from the beginning *ioco* algorithms are more powerful. In

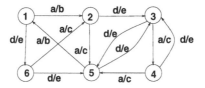

Fig. 4. An faulty IUT

addition, under these constraints, UIOv (which is complete) has a similar detection power as *ioco*, but UIO is less powerful than the *ioco* and UIOv conformance relation. So Phact has in any situation less detection power than TorX, TGV and Autolink.

5   Autolink

Autolink is the name of the algorithm which is integrated in the TAU/Telelogic tool–set [15] for generating test cases. Autolink supports the automatic generation of TTCN test suites based on an SDL–92 specification and a test purpose described in MSC–92. It is widely used because it is integrated in a commercial tool (TAU). It can be used in a batch–oriented way.

Since Autolink does not have an explicit conformance relation on which it is based, it was necessary to try to reconstruct a conformance relation for it and in this way to consolidate its conformance foundation. Based on this conformance relation we will compare Autolink with the other algorithms: TorX, TGV and UIO (UIOv) algorithms. We will express the conformance relation in terms of *ioco* theory.

This section is organized in the following way: Subsection 5.1 describes the Autolink algorithm; Subsection 5.2 gives the translation of the Autolink algorithm in *ioco* terms; Subsection 5.3 gives the classification of Autolink in comparison to other algorithms.

5.1   The Autolink Algorithm

Autolink is an algorithm for test derivation which has as inputs: 1) the specification which is built in SDL–92; 2) the test purpose which is expressed in MSC–92. It produces as an output the test case represented in TTCN.

Before explaining the algorithm, we should make some remarks about the MSC–92 which represents the test purpose. One thing to remember is that semantically an MSC–92 describes a partial ordering of the events (this means that the events of an instance are partially ordered; between two instances there is not a temporal order, except for the ordering induced by the messages). Autolink looks at the system under test as a black box to which signals are sent and received via different PCOs (Points of Control and Observation). The system under test (which is described by the SDL–92 specification) is one instance of the test purpose MSC–92. The PCOs are the other instances of the MSC–92. For generating tests, Autolink looks at the events which happen at the PCO level. So it does not look at the partial order of the events of the instance of the specification; it looks only to the partial order of the events of the PCO instances. So from an MSC–92, a set of traces can be formed (to be checked that these are traces of the specification) by taking all the possible linearizations of the signals which appear in the MSC–92 and which respect the partial order of the PCO instances and without considering the partial order of the specification instance.

Example Let us consider the MSC–92 from Figure 5. In this MSC–92 the specification is called *Example* and it forms one instance of the MSC–92. The other instances are the PCOs: A and B. It seems that the first trace to be checked in the specification is the trace *acbd*. This trace respects the partial order of the specification instance. But initially, Autolink does not consider this partial order. It can also use the traces: *abcd*, *abdc* which do not respect the partial order of the *Example* instance (in the generated test, Autolink will keep only the traces which are part of the traces of the specification). So from this MSC–92, a set of traces can be generated and it is {*acbd, abcd, abdc, bdac, badc, bacd*}.

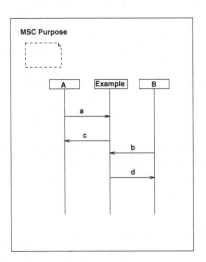

Fig. 5. A test purpose MSC–92

For generating a TTCN test, Autolink considers only a subset of traces from the set of traces generated as above. This is because Autolink applies some restrictions. For example, Autolink uses priorities between a send event and a receive event. The authors of Autolink [15] assume that the environment always sends signals to the system as soon as possible, whereas receive events occur with an undefined delay. Therefore the send event has a higher priority than the receive event in the default configuration priority of Autolink. So all the traces from the set of traces of the MSC–92 which do not respect this rule are removed from it. This will work as follows in our example: the set of inputs is $\{a, b\}$ and the set of outputs is $\{c, d\}$; the traces $acbd$ and $bdac$ are removed because the outputs (c and d) happen before the inputs (b and a) and this contradicts the rule that the input event has a higher priority than the output event; so the subset obtain is $\{abcd, abdc, badc, bacd\}$. And some other restrictions exist and are parameterized in the options configuration of Autolink.

In the explanation of Autolink, we will not enter into technical details such as how someone should generate an MSC–92 test purpose or how to configure the options of Autolink algorithm. Also we will only consider MSC–92 without references, because this is general enough to cover all the important aspects of Autolink.

Autolink can be described briefly in the following way:

1. *Validation:* the MSC–92 representing the test purpose is validated against the SDL–92 specification (this means that at least one of the traces from the set of traces of the MSC–92 is a valid trace of the SDL–92 specification);
2. *Generation:* the traces formed from the validated MSC–92 which respect the options configuration of Autolink are checked against the SDL–92 specification with a *state space exploration* algorithm and the TTCN test case is formed by all the MSC–92 traces which are also traces of the SDL–92 specification.

In the TTCN test case generated by Autolink a Timeout event gets an Inconclusive verdict. When an unexpected output is produced by the IUT (OtherwiseFail event) a Fail verdict is given by the TTCN test case. The traces of the specification which are not contained in the set of traces of the MSC–92 are cut and an Inconclusive verdict is assigned to them. We will exemplify the building of the TTCN test case in the following example

Example From the SDL–92 specification *Example*, only the process which describes the behaviour of the *Example* specification is shown in Figure 6. The process can receive from the environment the input a via channel A or input b

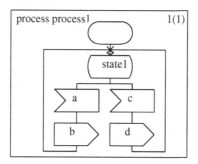

Fig. 6. The SDL–92 specification *Example*

test3 ITEX 3.4.0	Oct 12.2000	ITEX 3.4.0 sdt@wsfm04

Test Case Dynamic Behaviour

Test Case Name: Purpose
Group   :
Purpose   :
Configuration :
Default   : OtherwiseFail
Comments   :

Nr.	Label	Dynamic Description	Constraints Ref.	Verdict	Comments
1		A!a	c1		
2		B!b	c2		
3		A?c	c3		
4		B?d	c4	Pass	
5		B?d	c5		
6		A?c	c6	Pass	

Dynamic Part

Fig. 7. The test generated with Autolink

via channel B. After receiving input a, it can send to the environment output c (via channel A) and it returns in the initial state $State1$. The same things happens if input b is received: output d is sent to the environment (via channel B) and it returns to $State1$.

The test purpose is described in the MSC–92 of Figure 5. This MSC–92 is a valid test purpose because there is at least one trace of the $Example$ specification which is also part of the MSC–92 set of traces (for example the trace $acbd$).

Using the default options configuration, Autolink generated the TTCN test case that has its the dynamic behaviour presented in Figure 7. In this TTCN test, there are two traces which are present in the set of traces of MSC–92 (reduced in accordance with the restrictions imposed by the default options configuration) and which are also part of the set of traces of the $Example$ specification; these are $abcd$ and $abdc$. After performing one of these traces, an IUT will get a Pass verdict.

5.2   The Translation of Autolink in *Ioco* Terms

The first concern regarding the translation of Autolink in terms of *ioco* theory is how to transform the SDL–92 system into a labelled transition system. This can be done by simply considering that the labelled transition system of an SDL–92 system is the unfolding tree of the SDL–92 system on which the transitions in the tree are labelled with the name of the signals (inputs or outputs) and its states contain the triple $\langle state\ of\ the\ SDL\text{--}92\ system,\ last\ signal\ received,\ depth\rangle$; an exception for the last rule is the $Root$ state of the tree (initial state of the label transition system) and the states from the first level which contain the initial states names of the SDL–92 system.

Definition 15. *The labelled transition system $M' = \langle S, L, \longrightarrow, s_0 \rangle$ of an SDL–92 specification is the unfolding tree of the SDL–92 specification where: S is the set of states of the unfolding tree; the set of labels is formed by the union of the*

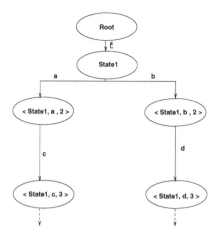

Fig. 8. The label transition system of the SDL–92 specification

set of inputs the set of outputs of the SDL–92 system and $\{\epsilon\}$; the transitions are the transition of the unfolding tree; the initial state is the Root state of the unfolding tree.

Example For the SDL–92 system of Figure 6, its labelled transition system is represented in Figure 8. The initial state s_0 is the *Root* state. Via the empty sequence ϵ the label transition system is moving from the initial state to the next state *State1*. From this state it is able, after performing the input a, to arrive in the state $\langle State1, a, 2 \rangle$ and after performing the input b in the state $\langle State1, b, 2 \rangle$. The construction of labelled transition system of this SDL–92 system proceeds in a similar style as we explained above.

The elements from the name of the states signify, taking for example the state $\langle State1, b, 2 \rangle$: *State1* is the name of the state from the SDL–92 system; b is the name of the input which is performed in this state and 2 is the depth.

As we explained in Section 5.1, the MSC–92 test purpose can be replaced by a set of traces. For an MSC–92 with name M, we will denote this set of traces as $S(M)$ (in M we assume that the first instance is the system under test and the rest of the instances are the PCOs; with this assumption, without loss of generality, we do not put the list of PCOs as a parameter of $S(M)$). The options configuration of Autolink which selects only a subset from the set of traces of the test purposes MSC–92, can be seen as a predicate P on traces. The TTCN test case will be represented as a set of traces. We will not enter into technical details regarding how Autolink implements the cutting of the uninteresting traces (traces which are traces of the specification but are not traces of the test purpose). We will consider that the finite set of the uninteresting traces is produced by a procedure which is part of Autolink and which we will call *cut*.

The signature of *cut* is : $cut(traces(Q), S(M), P)$. Every parameter of the procedure $(traces(Q), S(M), P)$ has the same meaning as the one described above. The set of traces $S_{cut} \subset traces(Q)$ which is returned by the procedure has the property that for every trace which is not part of the test purpose $(S(M))$ but is a valid trace of the specification $(traces(Q))$, there exists a prefix trace of it in the set S_{cut}.

Now the Autolink algorithm can be expressed in *ioco* terms in the following way

Autolink
Inputs:

1. the labelled transition system Q of the SDL–92 specification S;
2. the set of traces $S(M)$ of the test purpose MSC–92 with name M;
3. the predicate P defined by the option configuration of Autolink.

Output:
the set of traces T which is the test case generated.
Body of Autolink

1. *Validation:* $traces(Q) \cap S(M) \neq \emptyset$;
2. *Generation:* $T' = \{t \in traces(Q) \cap S(M) \mid P(t)\} \cup cut(traces(Q), S(M), P)$;
   (the union of the set of traces which are common in the test purpose and the specification and the set of traces from the specification which are cut because they are not present in the test purpose);
   $T = PC(T')$ (the test is the set of the prefixes of T').

For the rest of this paper we will refer to the algorithm which is described above as the Autolink algorithm.

Now we will define **passes** for the Autolink algorithm. We will not keep Definition 10 for **passes**, because it is only works for labelled transitions systems derived from finite state machines (an FSM which can only produce an output in a state). For a label transition system, in general, we will need more requirements. These requirements are added in Definition 16. However, one can easily prove that Definition 10 and Definition 16 are equivalent for labelled transitions systems derived from finite state machines.

Before defining **passes**, we will do some transformations on the verdicts contained in a test produced with the Autolink algorithm because in *ioco* theory there are only Pass and Fail verdicts. As we did in [4] we will give extra power to Autolink by assigning a Fail verdict (in place of Inconclusive) for the Timeout event. So the *null* output will be mapped onto a Timeout event. And we will also replace the Inconclusive verdict assigned to the uninteresting traces which are not present in the test purpose and are cut by the Autolink algorithm, with a Pass verdict. We should remember also that Autolink checks the outputs of the implementation only in positions of the traces where outputs exists. With these considerations the new definition of **passes** is:

Definition 16. *Let T be a test suite and w an implementation. Let $P_T = \{t \in L^* \mid \exists o \in L_O : to \in PC(T)\}$ be the set of the prefixes from the traces of T which are followed by an output. Let $P_{T,w} = \{t \mid \exists t' \in P_T, \exists o \in out(w \text{ after } t) : t = $*

$t'o\}$ *be the set of traces built from every trace from* P_T *concatenated with every output produced by the implementation* w *after performing the trace. Then:* w **passes** $T \Leftrightarrow P_{T,w} \subseteq T$.

Now we will define the test suite which is produced by Autolink

Definition 17. *Let* Q *be the labelled transition system of the SDL–92 specification* S. *Let* κ *be the set of all predicates* P *defined by the options configuration of Autolink. Let* ω *be the set of all MSC–92 which can be valid test purposes for* S. *Then:* $T_{\text{Auto}} = \cup_{P \in \kappa, M \in \omega}$ **Autolink**$(Q,\ S(M),\ P)$ *is the test suite formed by the set of test cases produced by the Autolink algorithm.*

The key observation is that for every trace of the specification one can build an MSC–92 test purpose such that its set of traces contains the specification trace. By allowing full freedom in exploiting the variations of the options configuration, there will be at least one test case which contains this trace. So we may conclude that $traces(Q) \subseteq T_{Auto}$. Since T_{Auto} is produced from the generation step in the **Body of Autolink** defined as the result of an intersection with $traces(Q)$, we know already that $T_{Auto} \subseteq traces(Q)$. So $T_{Auto} = traces(Q)$.

Now we are ready to define a conformance relation. In order to deal with predicates P, each of which represents a specific setting of the options configuration, we take the following point of view: the tester is allowed to take full freedom in exploiting the variations of the options configuration. If there exists at least one setting by which a certain error can be detected, then this is added to the algorithm error detection capability as represented by the conformance relation.

Definition 18. *Let* Spec *be the label transition system of the SDL–92 specification* S *and* Imp *be the label transition system of an implementation. Let* T_{Auto} *be the set of the test cases produced by the the Autolink algorithm on* Spec. *Then:* Spec *Auto* Imp $\Leftrightarrow$ *Imp* **passes** T_{Auto}.

It is quite easy to prove that Autolink generates sound tests and it is exhaustive with respect to Auto conformance relation. The exhaustiveness is a result of the definition itself and the soundness is easily proved because any test case is part of the set of all the test cases T_{Auto}.

5.3   Comparing Autolink to TorX, TGV, UIO,UIOv

Theorem 4 gives the translation of Auto conformance relation in *ioco* terms.

Theorem 4. *Let* Imp *be an implementation and let* Spec *be a specification.*

1. $\exists F_{Auto} \subseteq traces(\text{Spec})$, $F_{Auto} = \{t \in traces(\text{Spec}) \mid out(\text{Spec } \textbf{after } t) \neq \{null\}\}$:
   Spec *Auto* Imp $\Leftrightarrow$ Spec *ioco*$_{F_{Auto}}$ *Imp; (*F_{Auto} *is the subset of traces from* $traces(\text{Spec})$ *with the property that every trace from this subset makes the specification to produce an output);*

400   N. Goga

2. *Spec ioconf Imp* ⇒ *Spec* Auto *Imp;*
3. ∃ *Spec, Imp:* ¬(*Spec* ioconf *Imp*) ∧ *Spec* Auto *Imp.*

Proof.
1. In conformance with Definition 18 and Definition 16
    (a) *Spec* Auto *Imp* ⇔ *Imp* **passes** T_{Auto} ⇔ (b);
    (b) because $T_{Auto} = traces(Spec)$, we will replace T_{Auto} with $traces(Spec)$ in
        the relation of point (b); the relation which should hold is
        $P_{traces(Spec),Imp} = \{t \in L^* \mid \exists t' \in traces(Spec), \exists o \in out(Imp \textbf{ after } t) :$
        $out(Spec \textbf{ after } t) \neq \{null\} \wedge t = t'o\} \subseteq traces(Spec).$

    Let $F_{Auto} = \{t \in traces(Spec) \mid out(Spec \textbf{ after } t) \neq \{null\}\}$
    (all the traces which make the specification to produce an output). Now, the
    point (b) from above can be rewritten as $\forall t \in F_{AUTO} : out(Imp \textbf{ after } t) \subseteq$
    $out(Spec \textbf{ after } t)$. Then Auto=$ioco_{F_{Auto}}$. So point (1) of the theorem is
    proved.
2. The point (2) of the theorem is now easy proved because *ioconf* has more
    discriminating power than $ioco_{F_{Auto}}$ ($F_{Auto} \subseteq traces(\text{Spec})$).
3. For the point (3), let us consider the specification s and the implementation
    i from the Figure 9. The conformance relation *ioconf* is able to detect that
    after the trace a, i produces the output e which is not specified by s. Autolink
    cannot check what output is produced after the trace a, because it looks at
    the outputs only in the moment when the specification produces an output
    (different from the *null* output); this is not the case for the trace a. So
    Autolink will let i go undetected.

    The above shows that Auto has less detection power then *ioconf*. And in
    practice, in order to reach its theoretical error detection capacity an infinite
    number of MSC–92s must be created, which is impractical (typical method-
    ologies are based on interactive use of the simulator or validator to create
    MSC–92s; a similar remark applies to TGV). Because [16] shows that *ioco*
    conformance relation has more discriminating power then *ioconf*, the Au-
    tolink algorithm has also less detection power then TorX and TGV. Based
    on the observation that $F_{UIO} \subseteq F_{UIOv} \subseteq F_{Auto}$, we can conclude that in
    general Autolink is more powerful than the UIO and UIOv algorithms.

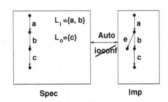

Fig. 9. The relation between *ioconf* and Auto

6   Concluding Remarks

In this paper we classified four known algorithms: TorX, TGV, Autolink (Tele-logic/TAU) and UIO (UIOv) algorithms (Phact –Conformance Kit). The clas-sification was made as a function of the conformance relation that they use, each conformance relation being expressed in terms of *ioco* theory. Also we con-solidated the conformance foundation of Autolink by reconstructing an explicit conformance relation for it. This paper treats only this criteria of classification (it looks at the error detection power of the algorithms); other criteria such as complexity or timing is out of the scope of this paper.

From the theoretical analysis it resulted that TorX and TGV have the same detection power. Autolink has less detection power because it implements a less subtle relation than the first two (some situations exists in which the first two can detect an erroneous implementation and Autolink cannot). UIO algorithms (Phact) in any situation have less detection power than Autolink, TGV and TorX. When the assumptions on which UIOv is based holds, UIOv algorithms have the same detection power as the three algorithms; but, because this happens rarely in practice, we can conclude that the three algorithms are in general more powerful than UIOv algorithms. The same classification is given by the practical experiments, as explained below.

With these algorithms, experiments were made using specification of a Con-ference Protocol in SDL–92, Lotos, Promela and FSM. These specifications were used by the algorithms to generate test cases which were run against 26 mutants as described in detail in [4,5,11]. The experimental results show the same clas-sification as the theoretical one: Phact detected 21 mutants, Autolink detected 22 mutants, TGV and TorX all *ioco* mutants.

References

1. International Standard ISO/IEC 9646-1/2/3. OSI – open system interconnec-tion, information technology – open systems interconnection conformance testing methodology and framework – part 1: General concept – part 2: Abstract test suites specification – part 3 : The tree and tabular combined notation (TTCN), June 1992.
2. A.V. Aho, A.T. Dahbura, D. Lee, and M.U. Uyari. An optimization technique for protocol conformance test generation based on UIO sequences and rural chinese postman tours. In K. Sabnani and S. Aggarwal, editors, *Protocol Specification, Testing and Verification VIII*, volume 8. Elsevier Science, 1998.
3. B. Algayres, Y. Lejeune, F. Hugonnet, and F. Hants. The AVALON project: A VALidation environment for SDL/MSC description. In O. Færgemand and A. Sarma, editors, *SDL'93 Using Objects*, Proceedings of the Sixth SDL Forum, pages 221–235, Darmstadt, 1993. Elsevier Science Publishers B.V.
4. A. Belinfante, J. Feenstra, R.G. Vries, J. Tretmans, N. Goga, L. Feijs, S. Mauw, and L. Heerink. Formal test automation: A simple experiment. In G. Csopaki, S. Dibuz, and K. Tarnay, editors, *Intenational Workshop on Testing of Comunica-tion Systems*, pages 179–196. Kluwer Academic, 1999.
5. L. Bousquet, S. Ramangalahy, S. Simon, C. Viho, A. Belinfante, and R.G. Vries. Formal test automation: the conference protocol with TGV/TorX. Proceedings of TestCom2000, Canada, 2000.

6. L.M.G. Feijs, F.A.C Meijs, J.R. Moonen, and J.J. van Wamel. Conformance testing of a multimedia system using Phact. In Alexandre Petrenko and Nina Yevtushenko, editors, *Testing of Communicating Systems*, volume 11, pages 143–210. Kluwer Academic, 1998.

7. G. Fernandez, J.C. Jard, C. Jéron, and T.Viho. Using on-the-fly verification techniques for the generation of test suites. In T. Alur and A. Henzinger, editors, *Computer-Aided Verification (CAV'96), New Brunswick, New Jersey, USA*, volume 1102 of *LNCS*. Springer–Verlag, 1996.

8. J.C. Fernandez, H. Garavel, A. Kerbrat, R. Mateescu, L. Mournier, and M. Sighireanu. CADP: A protocol validation and verification toolbox. In T. Alur and A. Henzinger, editors, *Computer-Aided Verification (CAV'96), New Brunswick, New Jersey, USA*, volume 1102 of *LNCS*. Springer–Verlag, 1996.

9. J.C. Fernandez, C. Jard, T. Jeron, and C. Viho. Using on–the–fly verification techniques for the generation of a test suites. In T. Alur and A. Henzinger, editors, *Computer-Aided Verification (CAV'96), New Brunswick, New Jersey, USA*, volume 1102 of *LNCS*. Springer–Verlag, 1996.

10. R. Groz and N. Risser. Eight years of experience in test generation from FDTs using TVEDA. In T. Mizuno, N. Shiratori, T. Higashino, and A. Togashi, editors, *FORTE/PSTV'97*. Chapman and Hall, 1997.

11. L. Heerink, J. Feenstra, and J. Tretmans. Formal test automation: the conference protocol with Phact. Proceedings of TestCom2000, Canada, 2000.

12. T. Jéron and P. Morel. Test generation derived from model-checking. In Nicolas Halbwachs and Doron Peled, editors, *Computer-Aided Verification (CAV'99)*, volume 1633 of *LNCS*, pages 108–122. Springer, 1999.

13. Zvi Kohavi. *Switching and Finite Automata Theory*. McGraw-Hill Book Company, New York, 1978.

14. J.R. Moonen, J.M.T. Romijn, O. Sies, J.G. Springintveld, L.M.G. Feijs, and R.L.C. Koymans. A two–level approach to automated conformace testing of VHDL designs. In M. Kim, S. Kang, and K. Hong, editors, *Testing of Communicating Systems*, volume 10, pages 432–447. Chapman and Hall, 1997.

15. M. Schimitt, A. Ek, J. Grabowski, D. Hogrefe, and B. Koch. Autolink – puting SDL–based test generation into practice. In A. Petrenko, editor, *Proceedings of the 11th International Workshop on Testing Comunicating Systems (IWTCS'98)*, pages 227–243. Kluwer Academic, 1998.

16. J. Tretmans. Test Generation with Inputs, Outputs and Repetitive Quiescence. Software—Concepts and Tools, 17(3), 1996. Also: Technical Report No. 96-26, Centre for Telematics and Information Technology, University of Twente, The Netherlands.

17. S.P. van de Burgt. Test sequence algorithms and formal language. In *Proceedings of Second Asian Test Symposium*, volume 8, pages 136–146. IEEE Computer Society Press, 1993.

18. S.P. van de Burgt, J. Kroon, E. Kwast, and H.J. Wilts. The RNL Conformance Kit. In J. de Meer, L.Mackert, and W. Effelsberg, editors, *Proceeding of the 2nd International Workshop on Protocol Test Systems*, volume 2, pages 279–294. North–Holland, October 1989.

19. S.T. Vuong, W.Y.L. Chan, and M.R. Ito. The UIOv–method for protocol test sequence generation. In Jan de Meer, Lothar Machert, and Wolfgang Effelsberg, editors, *International Workshop on Protocol Test Systems*, volume 2, pages 161–176. North–Holland, 1990.

Verifying Large SDL-Specifications
Using Model Checking

Natalia Sidorova[1] and Martin Steffen[2]

[1] Department of Mathematics and Computer Science
Eindhoven University of Technology
Den Dolech 2, P.O.Box 513,
5612 MB Eindhoven, The Netherlands
`n.sidorova@tue.nl`
[2] Institut für angewandte Mathematik und Informatik
Christian-Albrechts-Universität
Preußerstraße 1–9, D-24105 Kiel, Deutschland
`ms@informatik.uni-kiel.de`

Abstract. In this paper we propose a methodology for model-checking based verification of large SDL specifications. The methodology is illustrated by a case study of an industrial medium-access protocol for wireless ATM. To cope with the state space explosion, the verification exploits the layered and modular structure of the protocol's SDL specification and proceeds in a bottom-up compositional way. To make a compositional approach feasible in practice, we develop a technique for closing SDL components with a chaotic environment without incurring the state-space penalty of considering all possible combinations of values in the input queues. The compositional arguments are used in combination with abstraction techniques to further reduce the state space of the system. With debugging the system as the prime goal of the verification, we corrected the specification step by step and validated various untimed and time-dependent properties until we built and verified a model of the whole control component of the medium-access protocol. The significance of the case study is in demonstrating that verification tools can handle complex properties of a model as large as shown.

Keywords: SDL model checking; abstraction; compositional, bottom-up verification; verification case study.

1 Introduction

Formal methods, most notably model checking, are increasingly accepted as an important part of the software design process [6]. There is a clear tendency to provide validation facilities in the commercial SDL-design tools such as ObjectGEODE [18] and SDT [22]. Currently, these tools allow validation of SDL specifications by means of exhaustive testing. Due to the high cost of errors in the telecommunication system design, complementary ways of debugging and verification are needed. In this paper, we describe the verification methodology

R. Reed and J. Reed (Eds.): SDL 2001, LNCS 2078, pp. 403–421, 2001.

we applied to a large industrial software product, namely the control layer of the wireless ATM communication protocol *Mascara* [24].

Formal verification of SDL-specifications via model checking [5] is an area of active investigation [3,10,11,8,23] (notably, the last two mentioned citations are developments of the telecommunication industry itself). The "push-button" appeal is responsible for the increasing acceptance of model checking by industry with its promise to allow for fully automatic checking of a program or a system — the model — against a logical specification: typically a formula of some temporal logic. As model checking is based on state-space exploration, the size of a system that can be checked is limited, and it is often held that only relatively small systems can be verified with a model checker.

The limitations of model checking by the system size implies that verification is possible only using abstractions and/or compositional techniques. These techniques allow construction of a verification model whose state space is smaller than the one of the original system. However, providing a formal proof of correctness for each abstraction or composition step is prohibitively costly. If the aim is primarily debugging, performing these steps at a semi-formal level does not cause difficulties, as spotted errors can easily be validated afterwards. The errors can be checked against the concrete model by the designers so that spurious errors can be detected. But if a property holds for the verification model, one cannot claim that the property necessarily holds for the system under consideration as well, although the obtained result argues in favour of correctness of the system design. Therefore, we see that the primary goal of verification is advanced debugging and finding potential errors in its design and thus increasing its reliability, rather than proving the overall correctness of the product.

For the verification of Mascara, we use the Vires tool-set on the SDL specification, automatically translating the SDL-code into the input language of a discrete-time extension of the well-known *Spin* model-checker. As Mascara is too large to be verified by any existing verifier as a whole, we exploit the protocol's layered structure and perform a bottom-up, compositional verification. In a number of cases, the the basis of abstraction of a component are its proved correctness requirements. This abstraction replaces the real component at the next step when a slice at an upper hierarchical level of the protocol is considered for verification. Doing so we were able to reach the point where the whole control entity of Mascara together with a simple abstraction of the rest of the protocol was taken into account.

The rest of the paper is organised as follows: In Sects. 2 and 3 we briefly survey the protocol and the set of design and model check tools we used in the case study. In Sect. 4 we present the methodology and the techniques applied in the verification, and in Sect. 5 we highlight results of the investigation. We conclude in Sect. 6 by evaluating the results and discussing related work.

2   Mascara: A Wireless ATM Medium-ACcess Protocol

Located between the ATM-layer and the physical medium, Mascara is a medium-access layer or, in the context of the ISDN reference model, a transmission convergence sub-layer for wireless ATM communication [1,14] in local area networks. It has been developed within the *WAND*[1] project [24], a joint European initiative by various telecommunication companies to specify and implement a wireless access system for ATM-LANs.

Besides the standard transmission convergence sub-layer tasks (such as cell delineation, transmission frame adaptation, header error control, cell-rate decoupling), operating over radio-links (a necessarily shared physical medium) adds to the complexity of the protocol. Mascara has to arbitrate *medium access* to the radio environment of a variable number of mobile ATM-stations.[2] It has to provide enhanced error detection and correction mechanisms at various levels to counter the comparatively high bit-error rate of air-borne data-transmission. Last but not least, it has to cater for *mobility* features, allowing a mobile terminal to switch its association with an *access point* in a *handover*.

2.1   Overall Structure

From the perspective of verification, Mascara is a large protocol.[3] It is itself composed of various protocol layers and sub-entities (cf. Fig. 1).

The *layer control protocol* together with the *message encapsulation unit* assists in various ways the information exchange between the Mascara layer and entities located within the upper layers. The *segmentation and reassembly* unit does exactly what its name implies: cutting peer-to-peer control messages (also called MPDUs) into ATM-cell size and putting them together upon reception. All three mentioned top-level entities are comparatively unsophisticated and straightforward, as they mainly perform data transformations. The *WDLC*-layer, operating already on cell-level, is reminiscent to conventional (non-ATM) data-link protocols and responsible, per virtual channel, for error- and flow-controlled cell-transmission. The lowest level of Mascara is the *data-pump* including a real-time scheduler, which forms a large portion of the protocol's code-size. Despite its raw size, the functionality offered to the Mascara-layers above is rather simple: the data-pumps of two communicating stations act as a duplex, lossy FIFO-buffer. The other large part of Mascara, making up almost half of the SDL-code, is its *control entity,* on which we concentrate here. For a more thorough coverage of Mascara's structure and internals, consult the specification material provided by the Wand consortium [24].

---

[1] Wireless ATM Network Demonstrator.

[2] Hence the acronym "*M*obile *A*ccess *S*cheme based on *C*ontention *a*nd *R*eservation for *A*TM".

[3] Over 300 pages of (graphical) SDL.

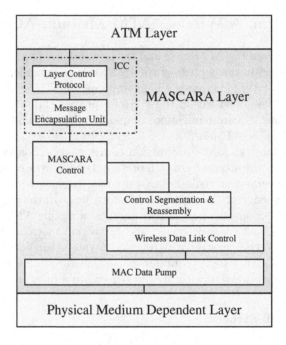

Fig. 1. Top-level functional entities

2.2   Mascara Control

As the name suggests, the *Mascara control* entity (MCL) is responsible for the
protocol's control and signalling tasks. It offers its services to the ATM-layer
above while using the services of the underlying segmentation and reassembly
entity, the sliding-window entities (WDLC's), and in general the low-layer data-
pump.

Being responsible for signalling, MCL maintains and manages *associations*
linking access points with mobile terminals, and *connections* (that is, the basic
data and signalling transfer channels, corresponding to ATM virtual channels).
Mascara control falls into four sub-entities, each divided in various sub-processes
themselves. The two important and complex ones are the *dynamic control* (DC)
and the *steady-state control* (SSC). The division of work between the dynamic
and the steady-state control is roughly as follows: SSC monitors in various ways
current associations and the quality of the radio environment in order to ensure an optimal transmission quality, to keep informed about alternative access
points, and to initiate timely changes of associations (so-called *handovers*). The
dynamic control's task, on the other hand, is to set-up and tear down the associations and connections while managing the related administrative work like
address management, resource allocation, etc. Of minor complexity are the *radio
control* entity (RCL, with the *radio control manager* RCM as its most important
process) and the *generic Mascara control* (GMC).

3   Model Checking Environment

Dealing with a protocol of Mascara's size, formal validation results with acceptable effort are possible only with appropriate tool support including editing and specification, validation, and of course model checking support.

The tool-set we use for the verification experiments on Mascara is a combination of well-established tools and a number of tools developed within Vires (see Fig. 2). Since developing a state-of-the-art model checker from scratch is a daunting task, it was decided to use a powerful model checker as the starting point rather than to design a new one. The model checker was enhanced with the ability to deal with time, for Mascara relies heavily on timers.

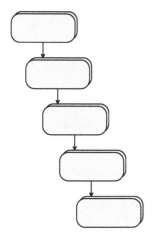

Fig. 2.   Tool-set for Mascara verification

The tool-set we used especially features:

– ObjectGEODE [18], a Telelogic tool-set for analysis, design, verification, and validation through *simulation*, as well as C/C++ code generation and testing of real-time and distributed applications. Targeted especially for telecommunication software and safety-critical systems, ObjectGEODE integrates complementary object-oriented and real-time approaches based on SDL [19] and MSCs [17], and recently UML.
– *sdl2if* and *if2pml*, which are the chain of translators rendering SDL into the Intermediate Format IF [4], a language for timed asynchronous systems, and IF into DT Promela [2] — a discrete-time extension of Promela (the input language of the model checker *Spin*), respectively. Both tools were developed within Vires.
– LIVE [16], used to optimise IF specifications by static-analysis techniques. It transforms an IF specification into a semantically equivalent one by adding

systematic resets of non-live variables. The transformation preserves the behaviour while reducing dramatically the global state space (and further, the exploration time). In our experiments, LIVE reduced the state space of the models by a factor of 8 on the average.

- *Spin*, a software package for the specification and verification of concurrent systems [12]. The core of *Spin* is a state-of-the-art enumerative on-the-fly model checker, which can be used to report unreachable code, deadlocks, unspecified receptions, race conditions, and the like. Correctness properties can be specified as system or process invariants (using assertions) or as general *linear-time temporal logic* requirements, either directly in LTL-syntax or indirectly as Büchi automata (called never claims).
- *DTSpin* [2], a discrete-time extension of *Spin*, intended for model checking concurrent systems that depend on timing parameters. It is completely compatible with the standard, untimed version of *Spin*.

4   Methodology

This section describes the methodological aspects of the verification process. The size of the protocol renders any direct, brute-force attempt of model checking out of question. To achieve the main goal, namely debugging the given real-life Mascara protocol, we faced a number of problems, where the most important have been: How to *break-up* the complex program into smaller entities and how to proceed in verifying them? How to *close* the smaller components in order to feed them into the model checker environment? And how to simplify and *abstract* them further in case the components are too large to be accepted by the model checker. We address these questions in turn.

4.1   Bottom-Up Compositional Verification

Our prime goal was to *apply* formal methods, in particular model checking, to industrial protocols: Mascara in this case. With the given overall protocol specification in SDL-92 (Specification and Description Language) [19] as a starting point, we chose to proceed bottom-up to be able to debug and clean up the single smaller entities with relative ease before proceeding to composed and larger ones. The layered and structured design of Mascara with blocks of processes greatly facilitated this compositional, bottom-up approach to verification.

We started with relatively small blocks of processes from the global specification. First, a model has to be closed by adding an *environment* specification. This environment should be an abstraction of the rest of the protocol. Constructing this abstraction is discussed later. After debugging and verifying a number of properties for simple components, we proceed with considering blocks composed from the verified ones (or their abstractions). Conceptually, the approach corresponds to the rely/guarantee or assumption/commitment paradigm of compositional verification, where the abstractions take the role of the assumptions about the environment.

Using a bottom-up approach in the verification, one gains a lot. Even some magical model checker that allows the whole protocol to be fed into it, and get the result by just pressing the proverbial button would be of limited use, for it is very well possible, for instance, that some components of the system under consideration are deadlocked, but not the whole system. The model checker tells then that the system is deadlock-free and one should remember to check that no component of the system is deadlocked. The formulation of such a property is not straightforward and involves fairness restrictions and other non-trivial conditions. Going bottom-up, one detects such deadlocks at the very first steps without much effort — the model checker just finds them automatically.

4.2   Closing the Model

Sub-models cut out of a global model cannot be verified as stand-alone processes, since they are not self-contained: the specification of a sub-model relies on the cooperation of the rest of the protocol. It should be noted that Mascara itself, like many other protocols, is an open model in the sense that it relies on the existence of an environment whose behaviour is not specified in the protocol. To model-check an open model the user must first transform it into a closed one. Closing models is often performed for exhaustive testing open systems, where processes are introduced within the model to feed it with signal inputs. The way inputs are sent to the model is controlled by these processes and then superfluous or non-significant inputs sequences can be avoided [15,9].

Adding Chaos   For the purpose of model checking, the way the model is closed should be well-considered to alleviate the state-space explosion problem: adding even a simple process increases unavoidably the state vector and, worse still, in general the state space. Basically, there are two extreme options how to implement an "outside" environment. One is to construct a simple process behaving chaotically: sending and receiving arbitrary signals in arbitrary order. In the context of verification of SDL with its asynchronous message-passing communication model, this immediately leads to a combinatorial explosion caused by considering all combinations of messages in the input queues, even if most of them cannot be dealt with by the processes and they are discarded. Another option is to tailor the environment process in such a way that it sends the "relevant" signals only: that is, the ones to which the model under investigation can possibly react. While easing the state-space explosion in the input-queues, it can make the environmental process itself rather complicated and large, multiplying thus the overall state space. At least as detrimental from a practical point of view is that a tailor-made environment requires insight into the model, analysing when it can and when it cannot handle messages. This takes time and is error-prone for large systems such at the components of Mascara.

To avoid both problems, we chose an alternative way: we model the environment as simple chaos but not a separate process external to the model. Instead, the chaotic environment is *embedded* into the model itself by a simple SDL source code transformation. The main idea is quite simple. Since we assume the environment to be chaotic, we must assure that whenever a process is in a state where it can take an input from the environment, it must have a possibility to take this branch. That can be done by replacing this input by the unconditionally enabled None input (thereby abstracting from the sent data at the same time). Outputs to the environment are just removed. For input, the replacement with None effectively removes the (chaotic) data reception from the input action, in this way influencing variable instances in the process appearing as input parameters in those inputs. Therefore, the actions potentially influenced by reception from outside and variable instances whose values consequently cannot be relied on must be eliminated, too. This is done by *data-flow* analysis of the model (see [21] for the semantical underpinning of the approach).

Chaotic Timers When closing a component, not only must all non-deterministic behaviour with respect to signal exchange be captured, but also all timed behaviour, which plays a crucial role in telecommunication protocols. The time semantics chosen for Mascara uses discrete-valued timer variables [3]. Ordinary transitions are instantaneous: they take zero time, and time can progress by incrementing all active timers only when all input queues are empty and there is no None-input enabled.[4]

Now closing the component by adding all possible signal-exchanges renders input from the outside continuously enabled. Especially by incorporating the chaos as sketched above, the branches input-guarded by the newly introduced None-inputs are unconditionally enabled, which means that time may not pass in this situation any longer, for the enabled input actions take priority over the time progress. So due to adding just chaotic sending and receiving of messages, time-outs possible in the original system may not occur after the transformation, in which case time can never pass, a so-called *zero-time loop* occurs. In other words, the simple approach of replacing environment-inputs by None-inputs fails to respect the discrete time semantics of SDL.

In order to preserve the timed behaviour, we must take into account that in any state time the new None-inputs do not forestall potential time-progress. For this purpose, one additional timer is introduced for every process receiving messages from the environment and at every process state, an input from this timer may be taken. This timer takes values Now or Now + 1, where Now means that the timer transition is enabled and messages from the environment may arrive, and Now + 1 means that no messages from the environment will come until the next time slice starts. The decision to set the timer to Now + 1 is taken non-deterministically – the time-out may occur after an arbitrarily many "inputs from the environment". Hence all the behaviour of the original specification is preserved. The pattern of the transformation is shown in Fig. 3.

Another issue concerns the influence of chaotic data received from the outside to the *values* of timers. Like ordinary variables, timer variables can be influenced

---

[4] More precisely, to allow timer increments, all queues must be empty except saved messages.

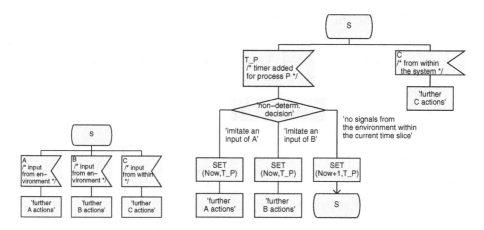

Fig. 3. Transformation of inputs: before (left) and after (right) the transformation

by the reception of chaotic data from outside, but unlike ordinary data variables, we cannot just remove timers whose exact values cannot be relied on. Timers instantiated to an undefined "chaotic" value can expire at an arbitrary moment in time. Therefore, they are treated similar to the ones for inputs from the environment. The operation of setting a timer to an undefined value is transformed into setting it to the Now + 1 value, and correspondent inputs of timer messages are transformed into timer expiration after which a choice is made either to set this timer to Now + 1 and return to the same process state, delaying the timer expiration, or to take the sequence of actions following the actual timer expiration according to the specification. The transformation is shown schematically in Fig. 4.

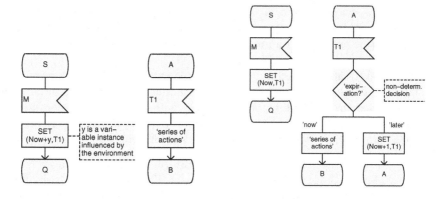

Fig. 4. Transformation of timers: before (left) and after (right) the transformation

4.3   Abstraction

One of the main tools of our methodological arsenal was *abstraction*. Abstraction is a rather general technique; intuitively it means replacing one semantic model by an abstract, in general simpler, one. To allow transfer of verification results from the abstract model to the concrete one, both must be related by a safe abstraction relation. The concept of *safe abstraction* is well-developed and has applications in many areas of semantics, program analysis, and verification (see [7] for the seminal, original contribution). For *safety properties* in linear-time temporal logic, often paraphrased as "never something bad will happen", the abstract system must at least show all the traces of the concrete one to be used as a safe abstraction. To find safe abstractions of a reactive, parallel system such as a protocol, it is helpful to distinguish between the *data* of a program (the values stored and transmitted) and its *control* (the control flow within the processes and their communication behaviour) and also respectively between *data* and *control abstractions*. A third abstraction we routinely used is timer abstraction.

Data Abstraction   Often, the behaviour of a program does not depend on the *specific* values of its data. In this case, many properties of the program stated over the full, often infinite, data domain can be equivalently expressed over finite domains of enough elements. For instance, being interested in a proof that an entity of Mascara handles addresses of mobile terminals correctly and does not give away the same address twice, a two-valued domain of addresses would suffice. This approach is known as *data independence* technique [26].

Control Abstraction   Given the amount of various entities and processes of the protocol, using data abstraction alone will not yield. The processes of the specification are given in great detail, to serve as the basis for an implementation, and they often possess internally non-obvious behaviour (for instance loops, jumps, conditions depending on data-values, and the like). To deal with this complexity we used a specific type of control abstraction. After a wholesale entity has being verified against a set of its requirements in the chaotic environment, we replace this entity with an abstraction which was the simplest entity for which this requirements holds.

We illustrate this technique on a simple entity of Mascara, the *radio control* (RCL). Seen from the outside, RCL builds Mascara-control's interface with the lower-layer physical radio modem. Its task is to operate the modem to tune into the terminal with a known frequency upon request, if possible. A property the RCL should guarantee can be phrased as the following simple *response property:*

> "Whenever, after initialisation, the radio control manager receives a request `Acquire_New_AP(newchannel)`, the RCM-process responds either positively or negatively (`Acquire_New_AP_ok` or`Acquire_New_AP_ko`). Moreover, the answer is sent in a given amount of time after getting the request."

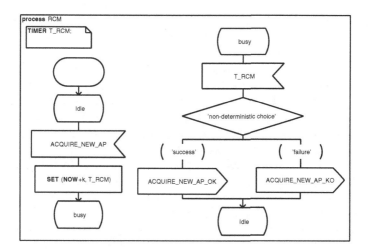

Fig. 5. Abstract radio control manager

The entity must be ready to react upon requests at any time, so it was closed in a *chaotic* environment. To reduce the state space of the verification model, we used data independence limiting the data domain of the parameter **newchannel** with 2 values. We checked the model for absence of zero-time cycles first, afterwards the proper initialisation of the component was checked. Coding the above property in LTL, we could finally verify that the concrete RCL satisfied the property.

Since initialisation of RCL is a confirmed service, and the other entities are initialised only after the initialisation confirmation has been received from radio control, we can abstract away from the initialisation phase in radio control.

After having verified the above LTL-property, one can exploit in the following experiments an abstract variant of RCL which is just one process, radio control manager (Fig. 5). The more sophisticated decisions of the concrete radio control [5] are captured in the manually given, abstract version simply by a non-deterministic choice between a positive or negative decision and the abstraction contains all the information the other components need to in order be verified.

Timer Abstraction Another abstraction we apply to cope with the state-space explosion is timer abstraction. A timer whose value is expected to have no influence on the truth of the property can be abstracted by assuming that it can take any value: it becomes a timer of "chaotic nature" (as compared with Sect. 4.2). Operations on this timer are replaced according to the patterns described for the chaotic timer.

It would seem obvious to verify all non-timed properties with an abstracted-time model and the timed ones with a concrete model, but our experiments show

---

[5] RCL, a small part of Mascara control, takes 9 SDL-pages of the specification.

that abstracting the timers may be ambivalent, both with respect to the state space and concerning the ability to transfer results from the abstract model to the concrete protocol.

First, the experiments shows that often the state space of the abstracted model is larger than the one of the concrete model when small values for timer delays are taken. In case the behaviour of the protocol strongly depends on timers, abstracting the values of timers may add much behaviour and thus potentially results in a larger state space. But of course, investigating the protocol for various timer settings will require the checking of infinite many combinations of timer settings, and moreover, even when restricting to "representative cases", choosing larger timer setting may in many cases increase the state space beyond the tractable limits. Using just abstract timers, the state space often happened to be manageable for the model checker.

Second, if some functional property is proved with the abstracted-time model, it is shown for all possible values of timers. On the other hand, if the property is disproved or a deadlock in the model is found, the next step is to check whether the erroneous trace given by *Spin* is a real error in the system or a spurious error caused by adding erroneous behaviour either by abstracting from time or by a too abstract environment specification. It can happen that the property fails to hold for the concrete model, however the erroneous trace given by *Spin* is one of the added behaviour. This behaviour cannot be reproduced for the SDL model with SDL simulation tools and we cannot conclude whether the property holds or not. In such a situation one should just redo the experiment using *DTSpin*: one cannot force *Spin* to give the trace from the non-added behaviour, but *DTSpin* guarantees that timers are expiring in the correct order. In our experiments we encountered several cases when using *DTSpin* instead of *Spin*, gave a chance to get a real erroneous trace and disprove the property.

5   Verification Results

In this section we shortly survey the verification results. Following the bottom-up, compositional approach sketched above, we obtained a number of results about Mascara control. Starting from MT target cell (MTC, an important part of the steady-state control), we proceeded investigating the steady-state control and the dynamic control, the two largest sub-blocks of Mascara-control (as compared with Sect. 2.2), in isolation, and finally, we verified properties of a model of the whole Mascara control.

Dealing with the various set-ups, we basically follow a bottom-up approach not only proceeding from smaller entities to larger, combined ones, but also advancing from simpler to more complex properties. After a number of reachability checks, we use the built-in Spin features for finding deadlocks and livelocks. The Message Sequence Charts, which are given by Spin and which corresponds to erroneous traces, are analysed on the original model with the help of the ObjectGEODE simulator. After correcting discovered structural errors, we proceed to more advanced properties, like *safety*, *liveness*, and *response* properties.

5.1   Reachability Checks

Enumerating the whole state space, the *Spin* model checker reports on unreachable code and we use this report as a guideline for formulating reachability properties to check. The report of *Spin* tells which code is unreachable, but it gives no hint why this code is unreachable. Analysing the unreachable code allows to find a reachable point in the specification suspected as the predecessor of an unreachable state. The reachability checks are easily done by just checking *assertion violations* where assertions are inserted at the reachable predecessors of unreachable states. Running *Spin* with an assertion-violation check gives the trace which can be used to look at this reachable state, scrutinising the values of different parameters, states of other processes, etc., to get a clue of what is wrong with the specification. In this way, we found a number of "obviously reachable" states being unreachable and thus a couple of unexpected errors of various kinds.

The reachability checks ensure that the more complicated LTL-properties investigated later are not trivially satisfied.[6] Depending on the entity, typical properties checked were:

- successful/unsuccessful association is possible;
- termination of association is possible;
- successful connection set-up is possible;
- incommunicado cycle is successfully completed.

Used in this way, reachability checking is employed as a sophisticated debugging facility with the assertions used to steer the checker to the critical points of the system. Besides weeding-out errors, we found it likewise very helpful, to use assertion checking (or, a little more complicated, checking LTL-formulas) in a dual way: marking the property of interest as "undesirable" while hoping for its satisfaction — the corresponding "error trace" is useful illustrating characteristic *desired* scenarios. They can be compared with the scenarios provided during the specification phase, thus giving a better understanding of the behaviour of the protocol, and thus enhancing the confidence in the specification.

5.2   Errors Found

Quite a number of errors discovered in Mascara were "just" *programming errors,* including such classics as uninitialised variables (even uninitialised variables due to a typo), forgotten branches in case distinctions, mal-considered limit cases in loops, and the like. Concerning the communication behaviour, we encountered most commonly

- race conditions;
- ambiguous receiver;

---

[6] Indeed, we started to perform reachability checks regularly after "proving" a sophisticated property only to learn later, that the premise of the implication of this property was unexpectedly false, since unreachable.

- unspecified reception;
- variables out of range.

as general errors at each stage of the verification process. Some of the found error turned out to be false errors caused by the too abstract environment. In this case, the experiment was redone with a more refined version of environment. Reproducing the erroneous trace on the original version of the protocol in the ObjectGEODE simulator, those errors confirmed to be real errors in the protocol design, were reported to the developers of Mascara.

Race conditions denote a situation where two signals are sent to an entity "at the same time" such that, due to SDL's asynchronous communication model, the order of reception is undetermined; here we mean more specifically that an unexpected reception order results in an error. Especially prone for this type of error turned out to be the initialisation phases of processes: often, the initialisation signals are given as *unconfirmed* messages. When a number of processes is asynchronously spawned, initialised, and starts communicating under the assumption that the rest of the processes is ready as well, messages may get lost.

Unspecified reception means that a process receives a message in a state where no such message is foreseen; the default reaction in SDL-92 then is to discard the message. The discarding feature is often used on purpose in Mascara's specification, since it saves code, but in some cases the discard is caused by unforeseen behaviour. Given the amount of asynchronous communication activities in the protocol, resulting errors are very hard to detect by code inspection. Signals in the specification with more than one potential receiving process ("ambiguous receiver") also had been a significant source of errors in MCL.

After constructing a small verification model (small compared to the overall specification), we witnessed in several cases state-space explosion without obvious reasons. It turned out that the specification contained some variable that could infinitely decrease or grow. For instance, being informed about deassociation of the same mobile terminal twice — from two different sources — an access point may (under some circumstances) decrease the counter of associated mobile terminals by two instead of one. Thus, the number of associated terminals may become negative. We found it helpful to regularly check that all variables in the model are bounded (their bounds are usually known or can be easily determined).

Besides quite a number of instances of these general errors at each level and besides spurious property violations due to abstraction, errors more specific to Mascara-control model were found and corrected. To exclude "false negatives", each erroneous behaviour was checked against the full SDL-specification by simulation or at least by code inspection and reported to the developers. In the following section, we show one of the more complex properties we verified.

5.3 Time-Dependent Safety Property: Unique MAC-Addresses

To illustrate up to which extent we could go with the verification, we describe one of the most involved properties verified. It concerns the cooperation of the complete control entity (MT- and AP-side), the interaction of various independently working protocols — notably association handover, the incommunicado protocol, and the "I'm-alive" protocol — and it takes into account the settings of several timers. To maintain an established association between a mobile terminal and an access point, it is important to determine when the association *breaks down* (as opposed to terminating an association properly by deassociating). Driven by various timers, both sides continuously check whether their current association is still functioning.

To determine that an association has gone for good, a mobile terminal and an access point must act independently and rely on their *local* timers, since if the connection is lost, no further communication is possible in the worst case. An important *safety requirement* here is that *"never the access point relinquishes an association before the mobile terminal does"*. This requirement is important for the correct working of Mascara control, especially the correct management of addresses by the dynamic control entity, for if the AP gives up the association, its dynamic control is free to reuse the various addresses allocated to that association for new ones. MT still clambers to reactivate the temporarily broken connection and if it succeeds in doing so, the same addresses will be in use for two different MT's, leading to errors. The property as LTL-formula reads
$\Box(\varphi_{mt-lost} \implies \varphi_{ap-lost})$,
where proposition $\varphi_{mt-lost}$ describes sending the signal MT_Lost, whereby AP's I'm-alive-agent entity gives-up the association. Similarly, $\varphi_{ap-lost}$ captures all situations where the mobile terminal gives up the association by signalling AP_Lost or HO_ind, both from the MHI-entity.

We established this safety property, if the inequation $\min(\tau_{AP}) > \max(\tau_{MT})$ is satisfied, where τ_{AP} and τ_{MT} are the respective times for the two sides of the association. The two times are bounded according to the following two inequations.

$$\tau_{AP} \geq (Max\_Time\_Periods + 1) * T_{iaa\_poll} + (IAA\_Max - 1) * T_{frame\_start}$$
$$\tau_{MT} \leq (Max\_Cellerrors) * T_{GDP\_period} + (Max\_AP\_Index + 1) * T_{rcm}$$

In the inequations, T_{iaa_poll}, T_{frame_start}, T_{GDP_period}, and T_{rcm} are the values of 4 timers determining the behaviour of the above-mentioned protocols, the remaining parameters are program constants of the responsible processes (especially loop bounds). It should be noted that the inequations are not immediate from the SDL-code of MCL: while it is comparatively easy to *identify* the timers that can influence satisfaction of the property by looking at the processes involved, what makes it complicated is the *interference* of the timed reactions: the activities of the various protocols can especially *suspend* other processes temporarily and thus postpone expiration of other timers. With *Spin/DTSpin* it is not possible to automatically derive the equations. Therefore, we verified satisfaction of the safety requirement, resp. checked its violation, for various combinations of values according to the inequations, especially for a number of border-cases, to validate our intuition about the correct interplay of the timers involved.

6   Conclusion

With SDL as the language of choice for the design of telecommunication applications, there is a growing need for formal verification techniques targeted towards SDL and of course corresponding integrated tool support. Currently, most of the work in the field relies on testing and/or validating the design via simulation. For instance in [20], an ATM user-to-network interface is validated using the SDT tool set [22]. The state space is explored by so called bit-state hashing and by random walk traversal. In our work, on the contrary, we use the full state space exploration of the *Spin* model checker, but abstraction techniques instead to deal with the state-space explosion problem. Similarly, [9] explores a number of heuristics or state-saving techniques, especially partial-order techniques, to counter the complexity of state exploration of SDL-specifications, but in the context of simulation. With similar goals and facing similar problems, [13] uses the SDL reachability analyser *Emma* for model-checking telecommunication software. Their tool is based on Petri-nets and it uses (as *Spin* does) partial-order techniques. Unlike our approach, where we rely on the discrete-time semantics as implemented in *DTSpin*, in the work of Husberg and Manner time is modeled by complete non-determinism; so time-dependent properties as the one shown in Sect. 5.3, cannot be treated. Similarly, the works in [11,23], also using the *Spin* model-checker, doesn't deal with timing aspects.

A major part of the verification effort expended can be seen as *debugging* the specification. A rightful question then is why to use model checking instead of simulation if model checking is not directly applicable to a large-size model while simulation is. We believe that both methods have their place and well complement each other. Indeed, at the first stage of debugging it is easier and better to use simulation, not model checking. The simple error situations like getting deadlocked already at the initial phase of functioning can be quickly detected by simulation. However, after a number of errors that can be found by simulation are corrected, the model checker shows its strength. For instance, model checker reports about unreachable code which immediately indicates the area of potential problems. Next, the erroneous trace given by a simulator can be very long, and one can not force a simulator to give a shortest one; with a model checker, one can (as most model checkers include a "shortest trail" option). These options significantly simplify the analysis of the cause of an error. Another argument is that only quite a restricted set of temporal properties can be verified via simulation. Model checking enlarge the facilities of debugging in this sense.

One conclusion to draw from our experience of working on the Mascara protocol is that by using state-of-the-art model checking support together with quite a simple methodological approach, one can already achieve a lot. The straightforward approach of using a chaotic closing together with rather simple abstractions has a number of methodological and practical advantages. First, allowing all possible traces by the non-deterministic environment, the safety of the abstraction is immediate. Secondly, closing the model by an environment process takes time; closing it with a more or less chaotic environment can be done fast and routinely. Thirdly, leaving the structure of the entity under investigation untouched allows fast spotting of potential errors, in case the model checker finds a property violation on the abstract level. Moreover, only when retaining the internal process structure it is possible to detect errors concerning the internal loops, conditions, etc., at all. Used in this way, model checking can provide valuable support in increasing the software reliability. As for future work, we expect that the process of verification will greatly benefit from automating some of the routine, but tedious tasks.

References

1. The ATM forum. http://www.atmforum.com/, 2000.
2. D. Bošnački and D. Dams. Integrating real time into Spin: A prototype implementation. In *Proceedings of Formal Description Techniques and Protocol Specification, Testing, and Verification (FORTE/PSTV'98)*. Kluwer Academic Publishers, 1998.
3. D. Bošnački, D. Dams, L. Holenderski, and N. Sidorova. Verifying SDL in Spin. In *TACAS 2000, LNCS 1785*. Springer-Verlag, 2000.
4. M. Bozga, J.-C. Fernandez, L. Ghirvu, S. Graf, J.-P. Krimm, and L. Mounier. IF: An intermediate representation and validation environment for timed asynchronous systems. In Wing et al. [25].
5. E. M. Clarke, O. Grumberg, and D. Peled. *Model Checking*. MIT Press, 1999.
6. E. M. Clarke and J. M. Wing. Formal methods: State of the art and future directions. *ACM Computing Surveys*, December 1996.
7. P. Cousot and R. Cousot. Abstract interpretation: A unified lattice model for static analysis of programs by construction or approximation of fixpoints. In *4th POPL, Los Angeles, CA*. ACM, January 1977.
8. A. Barnard F. Regensburger. Formal verification of SDL systems at the Siemens mobile phone department. In *Proceedings of TACAS '98, LNCS 1384*, pages 439–455. Springer-Verlag, 1998.
9. J. Grabowski, R. Scheurer, D. Toggweiler, and D. Hogrefe. Dealing with the complexity of state space exploration algorithms. In *Proceedings of the 6th GI/ITG technical meeting on 'Formal Description Techniques for Distributed Systems'*. Universität Erlangen-Nürnberg, 1996.
10. U. Hinkel. Verification of SDL specifications on the basis of stream semantics. In Y. Lahav, A. Wolisz, J. Fischer, and E. Holz, editors, *Proc. of the 1st Workshop of the SDL Forum Society on SDL and MSC*. Humboldt-Universität zu Berlin, 1998.
11. G. Holzmann and J. Patti. Validating SDL specifications: an experiment. In Ed Brinksma, editor, *International Workshop on Protocol Specification, Testing and Verification IX (Twente, The Netherlands)*, pages 317–326. North-Holland, 1989. IFIP TC-6 International Workshop.
12. G. Holzmann. *Design and Validation of Computer Protocols*. Prentice Hall, 1991.
13. N. Husberg and T. Manner. Emma: Developing an industrial reachability analyser for SDL. In Wing et al. [25], pages 642–662.
14. International Telecommunications Union (ITU) , series I recommendations – Integrated services digital networks (ISDN). http://www.itu.int/itudoc/itu-t/rec/i/index.html, 2000.
15. Ph. Leblanc. Simulation, verification and validation of models. white paper. http://www.telelogic.com/download/ObjectGeode/wp_simuv.zip, February 1998.

16. L. Ghirvu. M. Bozga, J. Cl. Fernandez. State space reduction based on Live. In Agostino Cortesi and Gilberto Filé, editors, *Proceedings of SAS '99, LNCS 1694*. Springer-Verlag, 1999.
17. Message sequence charts (MSC). ITU-TS Recommendation Z.120, 1996.
18. ObjectGeode 4. `http://www.csverilog.com/products/geode.htm`, 2000.
19. Specification and Description Language SDL, blue book. CCITT Recommendation Z.100, 1992.
20. S. M. Shahrier and R. M. Jenevein. SDL specification and verification of a distributed access generic optical network interface for SMDS networks. Technical Report TR-97-18, University of Texas at Austin, Department of Computer Sciences, July 1997.
21. N. Sidova and M. Steffen. Embedding chaos. Technical Report TR-ST-01-2, Lehrstuhl für Software-Technologie, Institut für Informatik und praktische Mathematik, Christian-Albrechts-Universität Kiel, March 2000, submitted for publication.
22. Telelogic Malmö AB. *SDT 3.1 User Guide and SDT 3.1 Reference Manual*, 1997.
23. H. Tuominen. Embedding a dialect of SDL in Promela. In Dennis Dams, Rob Gerth, Stefan Leue, and Mieke Massink, editors, *Theoretical and Practical Aspects of SPIN Model Checking, Proceedings of 5th and 6th Internaional SPIN Workshops, Trento/Toulouse, LNCS 1680*, pages 245–260. Springer-Verlag, 1999.
24. A wireless ATM network demonstrator (WAND), ACTS project AC085. `http://www.tik.ee.ethz.ch/~wand/`, 1998.
25. J. Wing, J. Woodcock, and J. Davies, editors. *Proceedings of Symposium on Formal Methods (FM 99), LNCS 1708*. Springer-Verlag, September 1999.
26. P. Wolper. Expressing interesting properties of programs in propositional temporal logic. In *Proceedings of 13th POPL*, pages 184–193. ACM, January 1986.

Applying SDL Specifications and Tools
to the Verification of Procedures*

Wenhui Zhang

Institute for Energy Technology, P.O.Box 173, N-1751 Halden, Norway
`Wenhui.Zhang@hrp.no`

Abstract. Verification of operating procedures (that is, specifications of manual control actions) by model checking has been discussed in [16]. The modelling language Promela and the model checker Spin were used in that report. In order to be able to apply model checking in a wider scope, modelling languages with graphical interface and verification tools used in industrial context are preferable (for example, to facilitate collaboration with process experts). In this paper, we discuss how to use SDL to model systems consisting of operating procedures and the controlled processes. Verification of procedures against correctness specifications is done by using the tool SDT. We conclude the paper with a short discussion of the integration of formal verification with the procedure design process.

1 Introduction

Operating procedures are documents telling operators what to do in various situations. They may be used in normal operation situations or for bringing a system from an unsafe state to a safe state. The main contents of a procedure (an operating procedure is sometimes referred to as a *procedure* for short, and for clarity, a procedure used as a subroutine in SDL is referred to as an *SDL-procedure*) are sequences of instructions. An operator could for instance be instructed to read some relevant information, to perform a sequence of actions, to check a system state and then go to a suitable instruction according to the system state, and to wait for a condition. Instructions may be grouped into steps and sub-steps.

Operating procedures are widely used in process industries including the nuclear power industry. The correctness of such procedures is of great importance to the safe operation of nuclear power plants. Traditional methods for the verification and validation of procedures are reviews and simulation. By means of reviews, the designer or members of a review team walk through the document to check errors, violation of standards or other problems. As the number of execution paths of a procedure could be large, this technique may not be practical

---

* The research described in this paper was carried out at the OECD Halden Reactor Project at the Institute for Energy Technology, Norway. The author thanks G. Dahll, T. Sivertsen and K. Stølen for their helpful suggestions and comments. The author also thanks anonymous referees for their constructive criticisms that helped in improving this paper.

R. Reed and J. Reed (Eds.): SDL 2001, LNCS 2078, pp. 421–437, 2001.

for large procedures. Simulation can be compared with software testing with a limited number of executions (in the environment provided by a simulator). As simulations are time-consuming and the number of simulations is limited, this approach may not be able to provide the desirable level of confidence on the correctness of procedures.

In the recent years, procedures have been computerised in order to enhance safety and reduce operator stress [13,7,10]. Computerised procedures provide a better opportunity for formal verification as they are written in formal or semi-formal languages with better structure and formality. Earlier work on using formal methods for verification of operating procedures includes investigating techniques based on algebraic specifications and related theorem proving tools [14,15] and on techniques based on model checking [16]. In the work reported in [16], the modelling language Promela and the model checking tool Spin [8,9] were used. In this paper, we present an approach to modelling and verification of operating procedures by using SDL.

Motivation: The main motivation of investigating how to use SDL for the verification of operating procedures is that SDL is a modelling language with graphical interface and verification tools used in industrial context. For the purpose of verification of procedures in a wider scope, a collaboration between formal methods experts and process experts (with knowledge on how procedures and the controlled systems work) is necessary, and an intuitive graphical syntax instead of a textual mathematical syntax seems to be crucial.

SDL specifications: SDL [2,4] is a specification and description language standardised as ITU (International Telecommunication Union) recommendation Z.100. The basis for description of behaviour is communicating extended state machines that are represented by processes. Communication is represented by signals and can take place between processes, or between processes and the environments of the system model. Some aspects of communication between processes are closely related to the description of system structure. An extended state machine consists of a number of states and a number of transitions connecting the states. One of the states is designated the initial state.

In SDL, each feature has a graphical syntax and a textual syntax. The sequential components are called processes. Processes can be composed into blocks, blocks can be composed into larger blocks and form systems. Systems may communicate with external environments. SDL supports object-oriented notions such as classes, objects encapsulation and inheritance.

A tool that can be used to create SDL descriptions and to verify properties of such descriptions is SDT [2,3]. There are different ways for validating a system description. Properties of a description can be specified by using user-defined rules, assertions, message sequence charts and observes processes, and verified accordingly. An example of user-defined rules is

$$\text{Exists P:Proc} \mid (\text{P->var=1})$$

which is true for all system states where there exists a process P of type "Proc" with a variable "var" that is equal to 1 (comparing with CTL [5], one may consider a user-defined rule as a formula quantified by the combination of the operators E and F). An assertion is a test performed at run-time (for example checking that the value of a specific variable is within the expected range). Assertions are described by introducing #CODE directives with calls to the C function xAssertError in a TASK. An example of an assertion in C Code is as follows (the box represents the keyword TASK):

```
"/* #CODE
#ifdef XASSERT
xAssertError("Violation of Conditions");
#endif */
```

Although the C-code in the assertion may not be attractive to people (such as a process expert) not familiar with details of SDL specifications, it is possible to create SDL-procedures that hide the details of such assertions (as done for representing actions, see Sect. 3). In the verification, assertions are checked during state space exploration. Whenever $xAssertError()$ is called during the execution of a transition, a report is generated. The advantage of using assertions, as opposed to using user-defined rules, is that in-line assertions are computed much more efficiently by the validator than the user-defined rules (in situations where assertions are applicable). Message sequence charts and observer processes are not used, as they are not necessary for formulating the types of correctness requirements considered in this paper (see Sect. 4).

Operating procedures: The basic elements of specifications of operating procedures are waiting for a condition, performing an action, and checking a condition and then according to the condition jumping to appropriate instructions. Potential problems of specifications consisting of these basic elements could be: dead code, wrong conditions, wrong actions or sequences of actions, wrong references (indicating places to jump to), and missing cases in case analysis (when checking a condition). Consequences of these problems may be violation of conditions, violation of invariants, and unreached goals. Errors that are difficult to detect by the traditional means are for instance incomplete exception handling and unforeseen side-effects.

Designing operating procedures (or generally, control systems) requires that many different situations be considered. There could be difficulties related to identifying different alternative behaviours of the plant (the controlled system), determining all the necessary preconditions of a successful control action and understanding all the possible consequences of a control action. In order to present a systematic approach for verification, we consider the language PROLA [6,11] designed for the specification of sequences of control actions. The basic elements of this language are as follows:

message	Read a message
auto-action	Perform an action
man-action	Perform a manual action
auto-check	Check the state of (a unit or a variable) of the system
man-check	Perform a manual check
wait	Wait for a specified state or for a condition to become true
goto	Go to a specified instruction
gosub	Go to a subroutine
return	Return from a subroutine to the previous gosub-instruction
finish	End of the procedure
monitor	Initiate a monitoring process (as a background process)
initiate	Initiate a procedure

The basic constructs of the language PROLA resemble those of guarded command languages. However PROLA is not a formal language. There are no strict rules for writing instructions. The interpretation of an instruction is sometimes situation dependent and is done by an operator. For the verification of a specification, the basic approach is to formalise the specification, formulate logic formulas (or assertions) for correctness requirements (such as invariants and goals), create an abstract model of relevant plant processes, and specify a set of the possible initial states of the plant. After we have these specifications, we use SDT to verify the specification against the logical formulas, the plant processes and the initial states.

2   Structure of SDL Models

For modelling and verification, the operating procedure is modelled as an SDL process. The actions in the specification are modelled by sending a message to an intermediate process which is modelled separately. The global state of the plant is modelled as a process with methods for accessing the state. The plant processes are modelled as SDL-processes. The initial state is modelled by an SDL-procedure which is to be called at the beginning of the process containing the global state.

The first level SDL specification is illustrated in Fig. 1. The specification consists of two parts: one for the operating procedure (called **Proc** in the figure) and the other (called **Sys**) for the rest of the necessary processes. In addition there are specifications of channels and constants.

Within the block **Sys**, there are several processes including the process containing the global state (called **SystemStates**), the intermediate process (called **SysInteraction**) and the plant processes (for instance, processes represent the change of states of pump speeds, cycling valves and steam levels). **SysInteraction** is the process that connects **Proc** and the rest of **Sys**. An illustration of this specification is shown in Fig. 2.

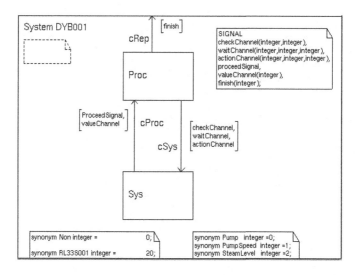

Fig. 1. Illustration of a first level specification

As shown in Fig. 1, numerical constants are defined to represent physical objects (such as pumps, valves). States of objects are also represented as numbers. Within the process **SystemStates**, the array

$$\text{varSystemState:} \; objects \rightarrow states$$

is defined to represent the states of objects. Methods for accessing and updating this array are "GetValue" and "PutValue" shown in Fig. 2. In the next section, we outline our approach for the specification of **Proc** which is the most important process in the specification, namely the procedure (i.e. the controlling process). For the brevity in this paper, details of other processes are not presented.

3   Modelling Procedures with SDL

This section presents an approach to model specifications written in PROLA. For modelling, we do not distinguish between "man-action" and "auto-action". These two types of instructions are covered in Sect. 3.1. We do not distinguish between "man-check" and "auto-check" either. These two types of instructions are covered in Sect. 3.2. The type "wait" is covered in Sect. 3.3, "initiate" and "monitor" in Sect. 3.4. The contents of instructions of type "message" are ignored (or only modelled as comments). In such cases, the instructions are to be treated as actions that change the state of the operator (which needs to be represented explicitly by a state variable). The types "goto", "gosub", "return" and "finish" are not covered separately. They are either very simple or are involved as parts in techniques for modelling the other instruction types.

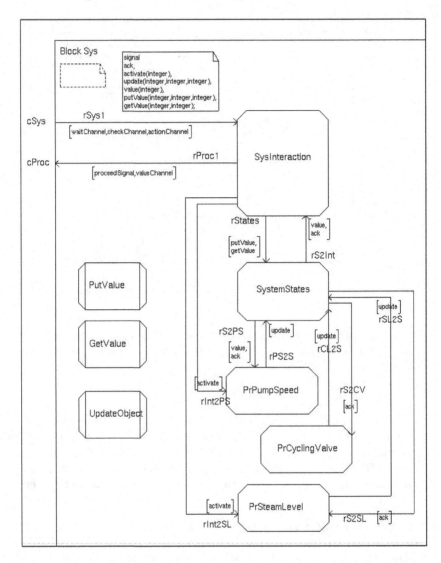

Fig. 2. Illustration of the block **Sys**

3.1　Actions and Sequences of Actions

The purpose of an action is to change the state of an object. An action can be modelled by sending a signal to **SysInteraction**. The process will then handle the signal accordingly (either according to the abstract process model or an assumption on the effect of the action). In order to be sure that the signal is processed before the execution of the procedure proceeds, the procedure will wait for a signal from **SysInteraction**. We use the channel actionChannel for the purpose of sending action-signals and use the channel proceedSignal for receiving the proceed-signal. For modelling actions, we define an SDL-procedure *Action(objtype, objid, objstate)* where *objid* represents the object to which the action is performed, *objstate* is the state of the object the action want to achieve and *objtype* is the type of the object. The definition of the SDL-procedure is as follows:

> *procedure Action; fpar objtype, objid, objstate integer;*
> *start; output actionChannel(objtype, objid, objstate); nextstate S1;*
> *state S1; input proceedSignal; return;*
> *endstate;*
> *endprocedure Action;*

We also define a macro or an SDL-procedure for each type of actions. For instance, we may define an SDL-procedure for repairing valves (of the type ControlValve) and an SDL-procedure for opening valves (of the type MainValve) as follows:

> *procedure RepairValve; fpar objid integer;*
> *start; call Action(ControlValve, objid, Normal); return;*
> *endprocedure RepairValve;*
> *procedure OpenMainValve; fpar objid integer;*
> *start; call Action(MainValve, objid, Open); return;*
> *endprocedure OpenMainValve;*

The graphical representations of the SDL-procedures *Action*, *RepairValve* and *OpenMainValve* are shown in Fig. 3. Graphical representations of other SDL specifications are omitted in the rest of this section, except the example of a part of an operating procedure represented by graphical SDL notations in Sect. 3.5.

Sequential Composition: As the description language is based on processes, it is easy to model sequences of actions which are concatenation of actions.

3.2   Checking System States

For modelling instructions for checking system states, an SDL-procedure is defined: *Check(objtype, objid, objstate)*, which sends a request with an object-identifier (i.e. *objtype* and *objid*) to **SysInteraction** and gets the state of the object. Modelling checking a system state and then going to a given instruction according to the state is illustrated by the following example.

Example: The following is an instruction for checking a system state and the corresponding SDL specification.

The instruction:
INSTRUCTION 2: "Check steam generator control valves"   MANCHECK: "Which steam generator control valve has ... ?"   IF "RL33S002" THEN GOTO 3-1-1   IF "RL35S002" THEN GOTO 4-1-1   IF "NONE" THEN GOTO 2 3

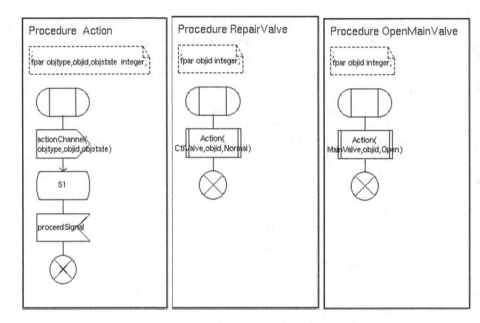

Fig. 3. Action, RepairValve and OpenMainValve

The corresponding SDL specification:
call Check(ControlValve,RL33S002,vRL33S002);
call Check(ControlValve,RL35S002,vRL35S002);
decision vRL33S002 = Defected;
(false) : decision vRL35S002 = Defected;
(false) : ...; (true) : ...; enddecision; (true) : ...; enddecision;

The destination instructions of the goto-instructions are not shown in the above SDL specification. They can either be specified directly (i.e. replace "..." in the above model with the instructions) or be specified as SDL-procedures (i.e. replace "..." with calls to SDL-procedures).

3.3 Waiting for Conditions

A wait-instruction can be modelled by sending a signal to **SysInteraction** and waiting for a signal to proceed. We use the channel waitChannel for the purpose of sending a wait-signal and use the channel proceedSignal (the same one used for action-instructions) for receiving the proceed-signal. For this purpose, we define an SDL-procedure $Wait(objtype, objid, objstate)$ where $objid$ represents an object or a property associated with an object such as a pump or pump-speed, $objstate$ is the state of the object for which the process is waiting and $objtype$ is the type of the object. For instance, waiting for the pump YD13Y013 to reach its maximum speed can simply be specified as:

   call Wait(PumpSpeed,YD13Y013,MaxSpeed).

3.4   Subroutines and Monitoring Processes

A procedure may initiate another procedure as a part of its execution. This is like calling a subroutine in programs. There are two possibilities, one is to model the procedure as an SDL-procedure and the other possibility is to model it as a process to be initiated by the main process. In the latter case, some mechanism must be implemented so that the main process waits until the subroutine process is finished. One such mechanism could be that the latter process sends a signal to the former right before it finishes.

A procedure may need to invoke another procedure as a background monitoring process (i.e. to execute it in parallel with the current process). This case can be dealt with in a similar way as in the previous case. The difference is that the calling process continues its execution independent of the called process.

3.5   Example of Procedure Specifications

The following is a part of a specification of control actions of a procedure:

```
STEP 3-1-1 "Start to isolate the steam generator"
INSTRUCTION 1 "Observe the fast-closing stop valve RL33S003"
INSTRUCTION 2 WAIT FOR RL33S003 IS CLOSING
INSTRUCTION 3 ACTION CLOSE RL33S001
INSTRUCTION 4 "Progress notification"
STEP 3-1-2 "Manipulate the primary loop"
INSTRUCTION 1 WAIT FOR YB13L001 < 3.38
INSTRUCTION 2 ACTION STOP YD13D001
INSTRUCTION 3 ACTION CLOSE YA13S002; CLOSE YA13S001
STEP 3-1-3 "Continue isolation of the steam generator"
INSTRUCTION 1 ACTION CLOSE RA13S004; CLOSE RY13S001
INSTRUCTION 2 GOSUB 7
```

A translation of the specification to an SDL specification is shown in Fig. 4. The steps STEP 3-1-1, 3-1-2 and 3-1-3 are implemented in one SDL-procedure. At the end of this SDL-procedure, it calls another SDL-procedure which implements the subroutine (in accordance with the instruction GOSUB 7).

In the translation, the emphasis is on the intuitiveness of the SDL specification. The semantics of different instructions is hidden behind special SDL-procedures, so that the specification is easily readable by those (with no formal methods background) who design control action sequences, in order to ease communication between procedure designers and formal methods experts.

3.6   Discussion

The SysInteraction process: It is an important process that receives messages from procedure process and acts and responds accordingly. For action-instructions, in most cases (as default), **SysInteraction** changes the state of

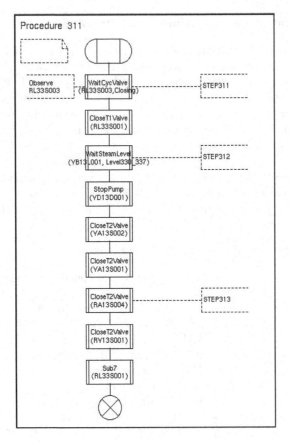

Fig. 4. Illustration of SDL specification of control actions

the object in accordance with the purpose of the action. However, other responses (one may check a condition in the system in order to decide whether the state is to be changed or remain unchanged, or one may start a process as the response to an action) can also be specified. For wait-instructions, **SysInteraction** may test whether the condition for which the procedure process is waiting is satisfied and then send a signal back to the procedure process. It may as well force the system (by changing the system state) to satisfy the condition and then send a signal back to the procedure process. The latter possibility is used when the system is not sufficiently modelled (there are always aspects that are too complicated to be included in a model and for practical reasons, we may also make assumptions to simplify models). For check-instructions, **SysInteraction** may return the checked system state or may assign a random value (among a set of potential values) to the system state. The latter possibility is used when the checked system state does not have a meaningful value (e.g. variables referred to in manual checks are often too complicated to be modelled in such a way that they have accurate values).

Manual and automatic actions: In the result of the modelling of procedures (i.e. translation of PROLA specifications to SDL specifications), we do not distinguish between manual actions and automatic actions. Both are modelled by sending action requests to **SysInteraction** which then decides the effect of the action requests. In the process of modelling, there are some differences between how the modelling could be done. For the manual actions, the actions are normally specified in a natural language (with possibility for situation dependent interpretation), one has to analyse the specification and determine the action and the potential effect. This requires a combination of formal methods knowledge and process knowledge. Cooperation between these two types of experts is necessary. For the automatic actions, the actions are specified in a language which has a fixed syntax, the translation can be done with help from a computer program. Such an attempt is discussed in [11] (however the target language is not SDL). The same is true for manual checks and automatic checks.

4 Representing Correctness Requirements

The main purpose of modelling a procedure using SDL is to analyse properties of the procedure. We outline our approach to express properties of procedures by specification methods and error reporting methods supported by SDT. An analysis of different types of errors can be found in [16]. We consider the following types of properties: unreachable instructions, fail-stops, infinite loops, conditions for performing actions, invariants and goals.

Unreachable Instructions: All instructions are implemented by sending messages to **SysInteraction** and receiving a response from the process. We may add an extra parameter identifying the instruction when sending a message to **SysInteraction**. With this piece of information, we can define a user rule to check whether a certain instruction is reachable. For instance, we can determine whether the instruction identified by the constant 10 is reachable by checking:

Exists P:SysInteraction | (P->CurrentInstruction=10)

Fail-Stops and Infinite Loops: Presence of fail-stops or infinite loops can be checked by using the SDT command "exhaustive-exploration". Fail-stops may be associated with deadlocks, however, the concept of deadlock is at system level. The fact that the process representing the procedure is not executable (a fail-stop) does not constitute a deadlock when there are executable processes in the system. Originally, there are two types of situations that can lead to a fail-stop. One is waiting for conditions and the other is incomplete case-analysis. The former is modelled as a loop (in **SysInteraction**), so when such conditions are not satisfiable, there will be a loop instead of a deadlock. For the second case, there are several solutions for turning such a fail-stop into other errors. One is to turn it into a loop by attaching a loop to it. Another solution is to create a special instruction to be attached to it and check whether this instruction is reachable. The third solution is to create an error report when such a situation occurs. The following task can be used to create such a report:

TASK " /* #CODE xAssertError("No Instructions"); */;

Conditions for Performing Actions: Conditions for performing actions can be specified as local safety properties related to specific instructions. For specifying conditions, we may create an SDL-procedure for checking and reporting violations of such conditions as follows:

procedure Assert; fpar objtype, objid, objstate integer, b boolean;
dcl currentState integer;
start; call Check(objtype, objid, currentState);
decision b; (true) : decision currentState = objstate;
(false) :
grst0 : task " /*#CODE xAssertError("Violation of Conditions"); */;
grst1 : return;
(true) : join grst1; enddecision;
(false) :
decision currentState = objstate; (true) : join grst0; (false) : join grst1;
enddecision; enddecision;
endprocedure Assert;

This SDL-procedure takes object-type, object-identifier, object-state and a boo-lean variable b as parameters. The boolean variable indicates whether the object should or should not be in the specified state. The SDL-procedure checks the state of the object, compares it with the specified state and gives report if there is an error. For instance, if we want to specify that the condition "RL33S001 is not open" holds before performing an "open-valve" action, we add a call to *Assert*(MainValve,RL33S001,Open,false) before calling *OpenMainValve*(RL33S001). To specify that the condition should hold for any "OpenMainValve" action, we re-define the SDL-procedure *OpenMainValve(v)* by adding a call to *Assert*(MainValve,v,Open,false) inside the SDL-procedure body of *OpenMainValve(v)*.

Invariants: Invariants are global safety properties. To verify invariant, we use the command "define-rule" to define a user rule. For instance, consider that we want to check whether the invariant "either the valve YA13S002 is Open or the pump YD13D001 is Stopped" holds. The interesting situations are those that satisfy:

(P->InitializationCompleted=1 and
(P->vYA13S002=3 or P->vYD13D001=0))

where P is an instance (and the only instance) of **SystemStates**, the number 3 represents the state "Open" and 0 represents the state "Stopped". Previously, we have explained that the array varSystemState is used to represent the state of the system and object-identifiers are numerical constants. However this variable cannot be used in such a user-defined rule and certain changes have to be made in order to let the states of relevant objects be available to user-defined rules. Here we have two additional variables vYA13S002 and vYD13D001 which have respectively the same value as varSystemState(YA13S002) and varSystemState(YD13D001). The condition "InitializationCompleted=1" is added to make sure that the state of the system before "InitializationCompleted=1" is not relevant. By default, SDT checks whether there are some states satisfying a user-defined rule. What we want is whether all relevant states satisfy the invariant. Hence we modify the formula and check whether the following user-rule is not satisfiable.

Exists P:SystemStates |
(P->InitializationCompleted=1 and
not (P->vYA13S002=3 or P->vYD13D001=0))

Goals: Goals may be decomposed into two parts: one is that the procedure terminates and the other is that the desirable states are achieved given that the procedure has terminated. The first part has been taken care of by checking that there are no fail-stops and infinite loops. In the following, we discuss the second part. One technique to verify whether *"the desirable states are achieved given that the procedure has terminated"* is to have a special variable indicating that procedure has terminated and then prove this part of the goals (say G) by checking that the following is not satisfiable:

Exists P:SystemStates | (P->ProcedureCompleted=1 and not G).

For instance, the unsatisfiability of the following formula represents that the goal "the 4 control valves are normal or supervisor is notified" is reached at the end (after the value of "ProcedureCompleted" has been set to 1) of all executions of the procedure:

Exists P:SystemStates |
(P->ProcedureCompleted=1 and
not ((P->vRL33S002=0 and P->vRL35S002=0
and P->vRL72S002=0 and P->vRL74S002=0) or P->vSupervisor=1))

where 0 is the constant representing the state Normal and 1 is the constant representing both the state Yes and the state Notified.

5   A Case Study

The case study is verification of an operating procedure with seeded errors. The purpose of this case study is to illustrate the use of proposed modelling and verification approach. The operating procedure considered here is "PROCEDURE D-YB-001 — The steam generator control valve opens and remains open" [16]. It is a disturbance procedure to be applied when one of the steam generator control valves opens and remains open.

The Procedure: The primary goal of the procedure is to check the control valves (there are 4 such valves), identify and repair the valve that has a problem (i.e. opens and remains open). After the defective valve is identified, the basic action sequence of the operator is as follows:

Start isolating the steam generator.
Manipulate the primary loop.
Continue isolation of the steam generator.
Repair the steam generator control valve and prepare to increase power.
Increase the power.

There are 93 instructions (each of the instructions consists of one or several actions) involving around 40 plant units including 4 control valves, 4 cycling valves, 4 steam level meters, 4 pumps, 4 pump speed meters and 4 protection signals.

In order to demonstrate error detection, we seeded 3 errors into the procedure: an incomplete case analysis, a wrong action and a wrong sequence of actions. The model used in this case study consists of the following processes: 1 procedure process for modelling the procedure (with the 3 seeded errors); 3 processes for representing the changes of respectively cycling valves, pump speed meters, and steam level meters; 1 process for representing the system state; and 1 process for representing the interaction between the procedure process and the rest of the processes.

Verification: The error detection is as follows:

- Error 1 was detected by exhaustive exploration with reports on a satisfied assertion.
  Explanation: The erroneous specification is illustrated in Fig. 5 where 4 objects are checked and the values are stored in respectively $v1$, $v2$, $v3$ and $v4$. A decision on what to do is made according to the values of the objects. The error is that there are no instructions for coping with the case where none of the objects has the value "Defected".
- Error 2 was detected by exhaustive exploration with reports on loops.
  Explanation: The erroneous specification is a misplace of *OpenT1Valve* instead of *CloseT1Valve* (the correct specification is shown in Fig. 4, which is a subroutine called in the second column of Fig. 5). The error causes the condition in one of the subsequent wait-instructions never to be satisfied.
- Error 3 was detected by checking the user-defined *Invariants* rule of Sect. 4 on page 432 with reports that the rule was satisfied.
  Explanation: The erroneous specification is a swapping of the order of the two actions *StopPump*(YD13D001) and *CloseT2Valve*(YA13S002) shown in Fig. 4.

The seeded errors and the detection of these show that the verification approach is effective for detecting potential errors of the discussed types in operating procedures.

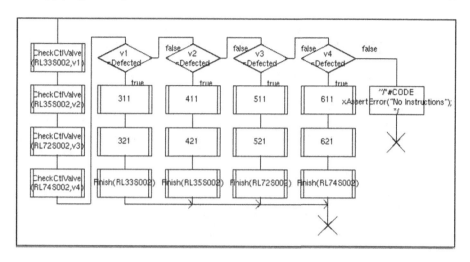

Fig. 5. Illustration of an incomplete case-analysis

6   Discussion and Concluding Remarks

We have demonstrated a verification approach for operating procedure specifications by formalising specifications and assumptions in SDL. One of the
advantages of SDL is that it has graphical notations. Processes and sequences
of actions can be represented intuitively. It is therefore easily comprehensible
and is good for communication with other members of the verification team.
In the proposed approach for modelling of operating procedures, the semantics
of different instructions is hidden behind special SDL-procedures, so that the
specifications are easily readable by those who design control action sequences.

For verification of procedures, traditionally, action models are used. An action
model includes two parts: a pre-condition describing the state conditions of that
must hold for the action to be applicable and an effect describing the state
changes which are initiated by the execution of the action. If we have such a
model of an action, the pre-condition is in our analysis the same as a condition
for performing the action that can be checked by using an assertion, and the
effect can be used as a post-condition (which can be implemented in the same
way as a pre-condition) in order to check whether the effect is achieved in the
SDL model.

In addition to assertions which are used to specify correctness requirements
for individual actions and instructions, user-defined rules are used to specify
global correctness requirements. One limitation of user-defined rules is that array
variables cannot be used in such rules.

For the location of detected errors, error reports and message sequence charts
are provided by the validator. They help understanding of the problem and
correcting the errors.

The approach of formal verification is effective for detecting potential errors of the discussed types in operating procedures, if we have appropriate models (with respect to correctness and abstractness) of relevant processes, as demonstrated by the examples in the previous section. The potential benefit and importance of formal verification is evidenced by the limitations and costs of the traditional verification and validation (verification by reviews and validation by tests on plant simulators). An integration of formal verification into the procedure design process is important to achieve reasonable allocation of resources for the verification and validation activities. A procedure design process with formal verification could be as follows:

Phase	Task
1.	specify relevant plant processes.
2.	specify possible initial plant states.
3.	write the intended procedure.
4.	specify additional correctness requirements.
5.	formalise and verify the specifications, and interpret the result.
6.	verify by reviews and other traditional techniques.
7.	validate by tests on plant simulators.

We may consider phases 3, 5, 6 and 7 as the main phases, while phases 1, 2 and 4 as helpful phases where the tasks are originally parts of the main phases. By separating these tasks and making them their own phases, phases 1 and 2 can benefit both phases 3 and 5, and phase 4 can benefit phases 5, 6 and 7.

In addition to a reasonable process, it is also important to develop tools with functionalities (e.g. as being considered in the new version of the computerised operation manuals system COPMA, developed at the OECD Halden Reactor Project [6,1,12]) that can benefit both formal verification and traditional verification and validation activities. Therefore, in future research and for the purpose of integration, it is necessary to put together different works, such as work on specialised editors for operating procedures (for integrating correctness requirements specifications and other relevant information with procedure specifications), work on computer-based systems and the use of such systems to keep track on operator actions and decisions in tests performed by operators on simulators (for automatically testing specifications of correctness requirements in simulation), work on translation of procedures written in semi-formal languages to formal languages (for example [11]), and work on formal verification of procedures (for verifying selected correctness requirements under given assumptions).

References

1. F. Bakkehøi. Assosiasjon av prosedyrelle aksjoner til XML baserte dokumenter. M.Sc. Thesis (in Norwegian). Institute of Informatics, University of Oslo. 2000.
2. R. Bræk and A. Sarma (editors). Proceedings of the 7th SDL Forum (SDL'95), North-Holland, 1995.

3. Anders Ek, Jens Grabowski, Dieter Hogrefe, Richard Jerome, Beat Koch and Michael Schmitt. Towards the Industrial Use of Validation Techniques and Automatic Test Generation Methods for SDL Specifications. Proceedings of the 8th SDL Forum (SDL'97):245-259, North-Holland, 1997.
4. J. Ellsberger, D. Hogrefe and A. Sarma. SDL - Formal Object-oriented Language for Communicating Systems. Prentice Hall Europe, 1997, ISBN 0-13-621384-7.
5. E. A. Emerson. Temporal and Modal Logic. Handbook of Theoretical Computer Science (B):997-1072. 1990.
6. F. Handelsby, E. Ness and J. Teigen. OECD Halden Reactor Project: COPMA II On-Line Functional Specifications. Report: HWR-319, Institute for Energy Technology, Norway. 1992.
7. D. G. Hoecker, K. M. Corker, E. M. Roth, M. H. Lipner and M. S. Bunzo. Man-Machine Design and Analysis System (MIDAS) Applied to a Computer-Based Procedure-Aiding System. Proceedings of the Human Factors and Ergonomics Society 38th Annual Meeting 1: 195-199. 1994.
8. G. J. Holzmann. Design and Validation of Computer Protocols. Prentice Hall, New Jersey, 1991.
9. G. J. Holzmann. The model checker Spin. IEEE Transactions on Software Engineering 23(5): 279-295. May 1997.
10. M. H. Lipner and S. P. Kerch. Operational Benefits of an Advanced Computerised Procedure System. 1994 IEEE Conference Record: Nuclear Science Symposium and Medical Imaging Conference:(1068-1072). 1995.
11. C. H. Nguyen. Modellering av operatør prosedyrer. M.Sc. Thesis (in Norwegian). Institute of Informatics, University of Oslo. 2000.
12. S. Nilsen and W. Zhang. XML Tagging Support in Procedure V&V. Institute for Energy Technology, Halden, Norway. Presented at the Workshop on Computerised Procedures, Halden, Norway. Nov. 30 and Dec. 1, 2000.
13. L. Reynes and G. Beltranda. A Computerised Control Room to Improve Nuclear Power Plant Operation and Safety. Nuclear Safety 31(4):504-511. 1990.
14. T. Sivertsen and H. Valisuo. Algebraic Specification and Theorem Proving used in Formal Verification of Discrete Event Control Systems. Report: HWR-260, Institute for Energy Technology, Norway. 1989.
15. K. Ylikoski and G. Dahll. Verification of Procedures. Report: HWR-318, Institute for Energy Technology, Norway. 1992.
16. W. Zhang. Model Checking Operator Procedures. Proceedings of the 6th International SPIN Workshop on Practical Aspects of Model Checking. (Springer LNCS 1680: 200-215). Toulouse, France. 1999.

Author Index

Amyot, D.   268
Arthaud, R.   52

Baeten, J.C.M.   328
Baker, P.   148
Bauer, N.   107
Beek, H.M.A. van   328
Böhme, H.   250
Bordeleau, F.   268
Born, M.   250
Bozga, M.   223
Bræk, R.   90

Cameron, D.   268
Campara, D.   19
Camus, J.-L.   1
Courzakis, V.   316

Dörfel, M.   203
Dubois, F.   250

Fischer, J.   250
Floch, J.   90

Geppert, B.   72
Goga, N.   379
Gotzhein, R.   72
Grabowski, J.   129
Graf, S.   223
Graubmann, P.   129

Hélouët, L.   348
Haugen, Ø.   38
Hofmann, R.   203
Hogrefe, D.   168
Holz, E.   250

Johansen, U.   90
Jonsson, B.   365

Kath, O.   250
Kerbrat, A.   182

Koch, B.   168

Larmouth, J.   241
Le Sergent, T.   1
Löwis, M. von   316
Luukkala, V.   288

Münzenberger, R.   203
Mansurov, N.   19
Mauw, S.   328
Miga, A.   268
Monkewich, O.   300
Mounier, L.   223

Neubauer, B.   250
Neukirchen, H.   168

Ober, I.   182, 223

Padilla, G.   365
Probert, R.   300

Roessler, F.   72
Roux, J-L.   223
Rudolph, E.   129, 148

Sales, I.   300
Sanders, R.   90
Schieferdecker, I.   148
Schröder, R.   316
Sidorova, N.   403
Sipilä, J.   288
Slomka, F.   203
Steffen, M.   403
Stoinski, F.   250

Vincent, D.   223

Wiles, A.   123
Woodside, M.   268

Zhang, W.   421

Lecture Notes in Computer Science

For information about Vols. 1–1998
please contact your bookseller or Springer-Verlag

Vol. 1999: W. Emmerich, S. Tai (Eds.), Engineering Distributed Objects. Proceedings, 2000. VIII, 271 pages. 2001.

Vol. 2000: R. Wilhelm (Ed.), Informatics: 10 Years Back, 10 Years Ahead. IX, 369 pages. 2001.

Vol. 2001: G.A. Agha, F. De Cindio, G. Rozenberg (Eds.), Concurrent Object-Oriented Programming and Petri Nets. VIII, 539 pages. 2001.

Vol. 2002: H. Comon, C. Marché, R. Treinen (Eds.), Constraints in Computational Logics. Proceedings, 1999. XII, 309 pages. 2001.

Vol. 2003: F. Dignum, U. Cortés (Eds.), Agent Mediated Electronic Commerce III. XII, 193 pages. 2001. (Subseries LNAI).

Vol. 2004: A. Gelbukh (Ed.), Computational Linguistics and Intelligent Text Processing. Proceedings, 2001. XII, 528 pages. 2001.

Vol. 2006: R. Dunke, A. Abran (Eds.), New Approaches in Software Measurement. Proceedings, 2000. VIII, 245 pages. 2001.

Vol. 2007: J.F. Roddick, K. Hornsby (Eds.), Temporal, Spatial, and Spatio-Temporal Data Mining. Proceedings, 2000. VII, 165 pages. 2001. (Subseries LNAI).

Vol. 2009: H. Federrath (Ed.), Designing Privacy Enhancing Technologies. Proceedings, 2000. X, 231 pages. 2001.

Vol. 2010: A. Ferreira, H. Reichel (Eds.), STACS 2001. Proceedings, 2001. XV, 576 pages. 2001.

Vol. 2011: M. Mohnen, P. Koopman (Eds.), Implementation of Functional Languages. Proceedings, 2000. VIII, 267 pages. 2001.

Vol. 2012: D.R. Stinson, S. Tavares (Eds.), Selected Areas in Cryptography. Proceedings, 2000. IX, 339 pages. 2001.

Vol. 2013: S. Singh, N. Murshed, W. Kropatsch (Eds.), Advances in Pattern Recognition – ICAPR 2001. Proceedings, 2001. XIV, 476 pages. 2001.

Vol. 2014: M. Moortgat (Ed.), Logical Aspects of Computational Linguistics. Proceedings, 1998. X, 287 pages. 2001. (Subseries LNAI).

Vol. 2015: D. Won (Ed.), Information Security and Cryptology – ICISC 2000. Proceedings, 2000. X, 261 pages. 2001.

Vol. 2016: S. Murugesan, Y. Deshpande (Eds.), Web Engineering. IX, 357 pages. 2001.

Vol. 2018: M. Pollefeys, L. Van Gool, A. Zisserman, A. Fitzgibbon (Eds.), 3D Structure from Images – SMILE 2000. Proceedings, 2000. X, 243 pages. 2001.

Vol. 2019: P. Stone, T. Balch, G. Kraetzschmar (Eds.), RoboCup 2000: Robot Soccer World Cup IV. XVII, 658 pages. 2001. (Subseries LNAI).

Vol. 2020: D. Naccache (Ed.), Topics in Cryptology – CT-RSA 2001. Proceedings, 2001. XII, 473 pages. 2001

Vol. 2021: J. N. Oliveira, P. Zave (Eds.), FME 2001: Formal Methods for Increasing Software Productivity. Proceedings, 2001. XIII, 629 pages. 2001.

Vol. 2022: A. Romanovsky, C. Dony, J. Lindskov Knudsen, A. Tripathi (Eds.), Advances in Exception Handling Techniques. XII, 289 pages. 2001

Vol. 2024: H. Kuchen, K. Ueda (Eds.), Functional and Logic Programming. Proceedings, 2001. X, 391 pages. 2001.

Vol. 2025: M. Kaufmann, D. Wagner (Eds.), Drawing Graphs. XIV, 312 pages. 2001.

Vol. 2026: F. Müller (Ed.), High-Level Parallel Programming Models and Supportive Environments. Proceedings, 2001. IX, 137 pages. 2001.

Vol. 2027: R. Wilhelm (Ed.), Compiler Construction. Proceedings, 2001. XI, 371 pages. 2001.

Vol. 2028: D. Sands (Ed.), Programming Languages and Systems. Proceedings, 2001. XIII, 433 pages. 2001.

Vol. 2029: H. Hussmann (Ed.), Fundamental Approaches to Software Engineering. Proceedings, 2001. XIII, 349 pages. 2001.

Vol. 2030: F. Honsell, M. Miculan (Eds.), Foundations of Software Science and Computation Structures. Proceedings, 2001. XII, 413 pages. 2001.

Vol. 2031: T. Margaria, W. Yi (Eds.), Tools and Algorithms for the Construction and Analysis of Systems. Proceedings, 2001. XIV, 588 pages. 2001.

Vol. 2032: R. Klette, T. Huang, G. Gimel'farb (Eds.), Multi-Image Analysis. Proceedings, 2000. VIII, 289 pages. 2001.

Vol. 2033: J. Liu, Y. Ye (Eds.), E-Commerce Agents. VI, 347 pages. 2001. (Subseries LNAI).

Vol. 2034: M.D. Di Benedetto, A. Sangiovanni-Vincentelli (Eds.), Hybrid Systems: Computation and Control. Proceedings, 2001. XIV, 516 pages. 2001.

Vol. 2035: D. Cheung, G.J. Williams, Q. Li (Eds.), Advances in Knowledge Discovery and Data Mining – PAKDD 2001. Proceedings, 2001. XVIII, 596 pages. 2001. (Subseries LNAI).

Vol. 2037: E.J.W. Boers et al. (Eds.), Applications of Evolutionary Computing. Proceedings, 2001. XIII, 516 pages. 2001.

Vol. 2038: J. Miller, M. Tomassini, P.L. Lanzi, C. Ryan, A.G.B. Tettamanzi, W.B. Langdon (Eds.), Genetic Programming. Proceedings, 2001. XI, 384 pages. 2001.

Vol. 2039: M. Schumacher, Objective Coordination in Multi-Agent System Engineering. XIV, 149 pages. 2001. (Subseries LNAI).

Vol. 2040: W. Kou, Y. Yesha, C.J. Tan (Eds.), Electronic Commerce Technologies. Proceedings, 2001. X, 187 pages. 2001.

Vol. 2041: I. Attali, T. Jensen (Eds.), Java on Smart Cards: Programming and Security. Proceedings, 2000. X, 163 pages. 2001.

Vol. 2042: K.-K. Lau (Ed.), Logic Based Program Synthesis and Transformation. Proceedings, 2000. VIII, 183 pages. 2001.

Vol. 2043: D. Craeynest, A. Strohmeier (Eds.), Reliable Software Technologies – Ada-Europe 2001. Proceedings, 2001. XV, 405 pages. 2001.

Vol. 2044: S. Abramsky (Ed.), Typed Lambda Calculi and Applications. Proceedings, 2001. XI, 431 pages. 2001.

Vol. 2045: B. Pfitzmann (Ed.), Advances in Cryptology – EUROCRYPT 2001. Proceedings, 2001. XII, 545 pages. 2001.

Vol. 2047: R. Dumke, C. Rautenstrauch, A. Schmietendorf, A. Scholz (Eds.), Performance Engineering. XIV, 349 pages. 2001.

Vol. 2048: J. Pauli, Learning Based Robot Vision. IX, 288 pages. 2001.

Vol. 2051: A. Middeldorp (Ed.), Rewriting Techniques and Applications. Proceedings, 2001. XII, 363 pages. 2001.

Vol. 2052: V.I. Gorodetski, V.A. Skormin, L.J. Popyack (Eds.), Information Assurance in Computer Networks. Proceedings, 2001. XIII, 313 pages. 2001.

Vol. 2053: O. Danvy, A. Filinski (Eds.), Programs as Data Objects. Proceedings, 2001. VIII, 279 pages. 2001.

Vol. 2054: A. Condon, G. Rozenberg (Eds.), DNA Computing. Proceedings, 2000. X, 271 pages. 2001.

Vol. 2055: M. Margenstern, Y. Rogozhin (Eds.), Machines, Computations, and Universality. Proceedings, 2001. VIII, 321 pages. 2001.

Vol. 2056: E. Stroulia, S. Matwin (Eds.), Advances in Artificial Intelligence. Proceedings, 2001. XII, 366 pages. 2001. (Subseries LNAI).

Vol. 2057: M. Dwyer (Ed.), Model Checking Software. Proceedings, 2001. X, 313 pages. 2001.

Vol. 2059: C. Arcelli, L.P. Cordella, G. Sanniti di Baja (Eds.), Visual Form 2001. Proceedings, 2001. XIV, 799 pages. 2001.

Vol. 2060: T. Böhme, H. Unger (Eds.), Innovative Internet Computing Systems. Proceedings, 2001. VIII, 183 pages. 2001.

Vol. 2062: A. Nareyek, Constraint-Based Agents. XIV, 178 pages. 2001. (Subseries LNAI).

Vol. 2064: J. Blanck, V. Brattka, P. Hertling (Eds.), Computability and Complexity in Analysis. Proceedings, 2000. VIII, 395 pages. 2001.

Vol. 2065: H. Balster, B. de Brock, S. Conrad (Eds.), Database Schema Evolution and Meta-Modeling. Proceedings, 2000. X, 245 pages. 2001.

Vol. 2066: O. Gascuel, M.-F. Sagot (Eds.), Computational Biology. Proceedings, 2000. X, 165 pages. 2001.

Vol. 2068: K.R. Dittrich, A. Geppert, M.C. Norrie (Eds.), Advanced Information Systems Engineering. Proceedings, 2001. XII, 484 pages. 2001.

Vol. 2070: L. Monostori, J. Váncza, M. Ali (Eds.), Engineering of Intelligent Systems. Proceedings, 2001. XVIII, 951 pages. 2001. (Subseries LNAI).

Vol. 2071: R. Harper (Ed.), Types in Compilation. Proceedings, 2000. IX, 207 pages. 2001.

Vol. 2072: J. Lindskov Knudsen (Ed.), ECOOP 2001 – Object-Oriented Programming. Proceedings, 2001. XIII, 429 pages. 2001.

Vol. 2073: V.N. Alexandrov, J.J. Dongarra, B.A. Juliano, R.S. Renner, C.J.K. Tan (Eds.), Computational Science – ICCS 2001. Part I. Proceedings, 2001. XXVIII, 1306 pages. 2001.

Vol. 2074: V.N. Alexandrov, J.J. Dongarra, B.A. Juliano, R.S. Renner, C.J.K. Tan (Eds.), Computational Science – ICCS 2001. Part II. Proceedings, 2001. XXVIII, 1076 pages. 2001.

Vol. 2075: J.M. Colom, M. Koutny (Eds.), Applications and Theory of Petri Nets 2001. Proceedings, 2001. XII, 403 pages. 2001.

Vol. 2077: V. Ambriola (Ed.), Software Process Technology. Proceedings, 2001. VIII, 247 pages. 2001.

Vol. 2078: R. Reed, J. Reed (Eds.), SDL 2001: Meeting UML. Proceedings, 2001. XI, 439 pages. 2001.

Vol. 2081: K. Aardal, B. Gerards (Eds.), Integer Programming and Combinatorial Optimization. Proceedings, 2001. XI, 423 pages. 2001.

Vol. 2082: M.F. Insana, R.M. Leahy (Eds.), Information Processing in Medical Imaging. Proceedings, 2001. XVI, 537 pages. 2001.

Vol. 2083: R. Goré, A. Leitsch, T. Nipkow (Eds.), Automated Reasoning. Proceedings, 2001. XV, 708 pages. 2001. (Subseries LNAI).

Vol. 2084: J. Mira, A. Prieto (Eds.), Connectionist Models of Neurons, Learning Processes, and Artificial Intelligence. Proceedings, 2001. Part I. XXVII, 836 pages. 2001.

Vol. 2085: J. Mira, A. Prieto (Eds.), Bio-Inspired Applications of Connectionism. Proceedings, 2001. Part II. XXVII, 848 pages. 2001.

Vol. 2089: A. Amir, G.M. Landau (Eds.), Combinatorial Pattern Matching. Proceedings, 2001. VIII, 273 pages. 2001.

Vol. 2091: J. Bigun, F. Smeraldi (Eds.), Audio- and Video-Based Biometric Person Authentication. Proceedings, 2001. XIII, 374 pages. 2001.

Vol. 2092: L. Wolf, D. Hutchison, R. Steinmetz (Eds.), Quality of Service – IWQoS 2001. Proceedings, 2001. XII, 435 pages. 2001.

Vol. 2096: J. Kittler, F. Roli (Eds.), Multiple Classifier Systems. Proceedings, 2001. XII, 456 pages. 2001.

Vol. 2097: B. Read (Ed.), Advances in Databases. Proceedings, 2001. X, 219 pages. 2001.

Vol. 2099: P. de Groote, G. Morrill, C. Retoré (Eds.), Logical Aspects of Computational Linguistics. Proceedings, 2001. VIII, 311 pages. 2001. (Subseries LNAI).

Vol. 2110: B. Hertzberger, A. Hoekstra, R. Williams (Eds.), High-Performance Computing and Networking. Proceedings, 2001. XVII, 733 pages. 2001.